Annals of Mathematics Studies
Number 164

Discrete Orthogonal Polynomials

Asymptotics and Applications

J. Baik
T. Kriecherbauer
K. T.-R. McLaughlin
P. D. Miller

PRINCETON UNIVERSITY PRESS
PRINCETON AND OXFORD
2007

Copyright © 2007 by Princeton University Press

Published by Princeton University Press, 41 William Street, Princeton, New Jersey 08540

In the United Kingdom: Princeton University Press, 3 Market Place, Woodstock, Oxfordshire OX20 1SY

All Rights Reserved

Library of Congress Cataloging-in-Publication Data

Baik, J., 1973–
    Discrete orthogonal polynomials : asymptotics and applications / J. Baik ... [et al.].
      p. cm.
    Includes bibliographical references and index.
    ISBN-13: 978-0-691-12733-0 (alk. paper)
    ISBN-10: 0-691-12733-6 (alk. paper)
    ISBN-13: 978-0691-12734-7 (pbk. : alk. paper)
    ISBN-10: 0-691-12734-4 (pbk. : alk. paper)
    1. Orthogonal polynomials–Asymptotic theory. I. Title.

QA404.5 .D57 2007
515′.55—dc22
                                                                                                                                                                   2006050587

British Library Cataloging-in-Publication Data is available

This book has been composed in LaTeX

The publisher would like to acknowledge the author of this volume for providing the camera-ready copy from which this book was printed.

Printed on acid-free paper. ∞

pup.princeton.edu

Printed in the United States of America

10 9 8 7 6 5 4 3 2

Contents

Preface   vii

Chapter 1. Introduction   1
- 1.1 Motivating applications   1
- 1.2 Discrete orthogonal polynomials   8
- 1.3 Assumptions   10
- 1.4 Goals and methodology   11
- 1.5 Outline of the rest of the book   22
- 1.6 Research background   23

Chapter 2. Asymptotics of General Discrete Orthogonal Polynomials in the Complex Plane   25
- 2.1 The equilibrium energy problem   25
- 2.2 Elements of hyperelliptic function theory   31
- 2.3 Results on asymptotics of discrete orthogonal polynomials   33
- 2.4 Equilibrium measures for some classical discrete orthogonal polynomials   41

Chapter 3. Applications   49
- 3.1 Discrete orthogonal polynomial ensembles and their particle statistics   49
- 3.2 Dual ensembles and hole statistics   51
- 3.3 Results on asymptotic universality for general weights   52
- 3.4 Random rhombus tilings of a hexagon   57
- 3.5 The continuum limit of the Toda lattice   60

Chapter 4. An Equivalent Riemann-Hilbert Problem   67
- 4.1 Choice of Δ: the transformation from $\mathbf{P}(z; N, k)$ to $\mathbf{Q}(z; N, k)$ 67
- 4.2 Removal of poles in favor of discontinuities along contours: the transformation from $\mathbf{Q}(z; N, k)$ to $\mathbf{R}(z)$ 69
- 4.3 Use of the equilibrium measure: the transformation from $\mathbf{R}(z)$ to $\mathbf{S}(z)$ 70
- 4.4 Steepest descent: the transformation from $\mathbf{S}(z)$ to $\mathbf{X}(z)$ 78
- 4.5 Properties of $\mathbf{X}(z)$ 79

Chapter 5. Asymptotic Analysis   87
- 5.1 Construction of a global parametrix for $\mathbf{X}(z)$ 87
- 5.2 Error estimation   99

Chapter 6. Discrete Orthogonal Polynomials: Proofs of Theorems Stated in §2.3   105
- 6.1 Asymptotic analysis of $\mathbf{P}(z; N, k)$ for $z \in \mathbb{C} \setminus [a, b]$ 105
- 6.2 Asymptotic behavior of $\pi_{N,k}(z)$ for z near a void of $[a, b]$: the proof of Theorem 2.9   107
- 6.3 Asymptotic behavior of $\pi_{N,k}(z)$ for z near a saturated region of $[a, b]$ 108
- 6.4 Asymptotic behavior of $\pi_{N,k}(z)$ for z near a band   110
- 6.5 Asymptotic behavior of $\pi_{N,k}(z)$ for z near a band edge   112

Chapter 7. Universality: Proofs of Theorems Stated in §3.3   115
- 7.1 Relation between correlation functions of dual ensembles   115

7.2	Exact formulae for $K_{N,k}(x,y)$	118
7.3	Asymptotic formulae for $K_{N,k}(x,y)$ and universality	124

Appendix A. The Explicit Solution of Riemann-Hilbert Problem 5.1 — 135

A.1 Steps for making the jump matrix piecewise-constant: the transformation from $\dot{\mathbf{X}}(z)$ to $\mathbf{Y}^\sharp(z)$ — 135
A.2 Construction of $\mathbf{Y}^\sharp(z)$ using hyperelliptic function theory — 137
A.3 The matrix $\dot{\mathbf{X}}(z)$ and its properties — 141

Appendix B. Construction of the Hahn Equilibrium Measure: the Proof of Theorem 2.17 — 145

B.1 General strategy: the one-band ansatz — 145
B.2 The void-band-void configuration — 146
B.3 The saturated-band-void configuration — 149
B.4 The void-band-saturated configuration — 150
B.5 The saturated-band-saturated configuration — 151

Appendix C. List of Important Symbols — 153

Bibliography — 163

Index — 167

Preface

This work develops a general framework for the asymptotic analysis of systems of polynomials orthogonal with respect to measures supported on finite sets of nodes. Starting from a purely discrete interpolation problem for rational matrices whose solution encodes the polynomials, we show how the poles can be removed in favor of discontinuities along certain contours, turning the problem into an equivalent Riemann-Hilbert problem that we analyze with the help of an appropriate equilibrium measure related to weighted logarithmic potential theory. For a large class of general weights and general distributions of nodes (not necessarily uniform), we calculate leading-order asymptotic formulae for the polynomials, with error bound inversely proportional to the number of nodes. We obtain a number of asymptotic formulae that are valid in different overlapping regions whose union is the entire complex plane. We prove exponential convergence of zeros to the nodes of orthogonalization in saturated regions where the equilibrium measure achieves a certain upper constraint. Two of the asymptotic formulae for the polynomials display features distinctive of discrete weights: one formula (uniformly valid near the endpoints of the interval of accumulation of nodes where the upper constraint is active) is written in terms of the Euler gamma function and another formula (uniformly valid near generic band edges where the upper constraint becomes active) is written in terms of both Airy functions $\text{Ai}(z)$ and $\text{Bi}(z)$ (by contrast $\text{Ai}(z)$ appears alone at band edges where the lower constraint becomes active, as with continuous weights). We illustrate our methods with the Krawtchouk polynomials and two families of polynomials belonging to the Hahn class. We calculate the equilibrium measure for the Hahn weight.

We then use our results to examine several relevant application problems. We investigate universality of a number of statistics derived from discrete orthogonal polynomial ensembles (discrete analogues of random matrix ensembles) using asymptotics for the discrete orthogonal polynomials. In particular, we establish the universal nature of the discrete sine and Airy kernels as models for the correlation functions in certain regimes, and we prove convergence of distributions governing extreme particles near band edges to the Tracy-Widom law. We apply these results to the problem of computing asymptotics of statistics for random rhombus tilings of a large hexagon. This problem is described in terms of discrete orthogonal polynomial ensembles corresponding to Hahn-type polynomials. Therefore, combining the universality theory with our specific calculations of the equilibrium measure for the Hahn weights yields new error estimates and edge fluctuation phenomena for this statistical model. We also apply our theory to obtain new results for the continuum limit of the Toda lattice, including strong asymptotics after shock formation.

J. Baik was supported in part by the National Science Foundation under grant DMS-0208577. T. Kriecherbauer was supported in part by the Deutsche Forschungsgemeinschaft under grant SFB/TR 12. K. T.-R. McLaughlin was supported in part by the National Science Foundation under grants DMS-9970328 and DMS-0200749. K. T.-R. McLaughlin wishes to thank T. Paul, F. Golse, and the staff of the École Normal Superieur, Paris, for their kind hospitality. P. D. Miller was supported in part by the National Science Foundation under grant DMS-0103909 and by a grant from the Alfred P. Sloan Foundation.

We wish to thank several individuals for their comments: A. Borodin, P. Deift, M. Ismail, K. Johansson, and A. Kuijlaars. We also thank J. Propp for providing us with Figure 1.3.

Ann Arbor, Bochum, and Tucson, May 2005.

Discrete Orthogonal Polynomials

Chapter One

Introduction

1.1 MOTIVATING APPLICATIONS

The main aim of this monograph is to deduce asymptotic properties of polynomials that are orthogonal with respect to pure point measures supported on finite sets and use them to establish various statistical properties of discrete orthogonal polynomial ensembles, a special case of which yields new results for a random rhombus tiling of a large hexagon. Throughout this monograph, the polynomials that are orthogonal with respect to pure point measures will be referred to simply as *discrete orthogonal polynomials*. We begin by introducing several applications in which asymptotics of discrete orthogonal polynomials play an important role.

1.1.1 Discrete orthogonal polynomial ensembles

In order to illustrate some concrete applications of discrete orthogonal polynomials and also to provide some motivation for the scalings we study in this book, we give here a brief introduction to discrete orthogonal polynomial ensembles. More details can be found in Chapter 3.

General theory

In the theory of random matrices [Meh91, Dei99], the main object of study is the joint probability distribution of the eigenvalues. In unitary-invariant matrix ensembles, the eigenvalues are distributed as a Coulomb gas in the plane confined on the real line at the inverse temperature $\beta = 2$ subject to an external field. In recent years, various problems in probability theory have turned out to be representable in terms of the same Coulomb gas system with the condition that the particles are further confined to a discrete set. Such a system is called a *discrete orthogonal polynomial ensemble*. More precisely, consider the joint probability distribution of finding k particles at positions $x_1, \ldots, x_k$ in a discrete set X to be given by the following expression (we are using the symbol $\mathbb{P}(\text{event})$ to denote the probability of an event):

$$p^{(k)}(x_1, \ldots, x_k) := \mathbb{P}(\text{there are particles at each of the nodes } x_1, \ldots, x_k)$$
$$= \frac{1}{Z_k} \prod_{1 \leq i < j \leq k} (x_i - x_j)^2 \cdot \prod_{j=1}^{k} w(x_j), \tag{1.1}$$

where Z_k is a normalization constant (or *partition function*) chosen so that

$$\sum_{\substack{x_1 < \cdots < x_k \\ x_j \in X}} p^{(k)}(x_1, \ldots, x_k) = 1.$$

Note that the particles are all indistinguishable from each other.

Discrete orthogonal polynomial ensembles arise in a number of specific contexts (see, for example, [Bor001, Joh00, Joh01, Joh02]), with particular choices of the weight function $w(\cdot)$ related (in cases we are aware of) to classical discrete orthogonal polynomials. For instance:

- The Meixner weight

$$w(x) = \binom{x + M - N}{x} q^x, \qquad \text{for } x = 0, 1, 2, \ldots,$$

arises in the directed last-passage site percolation model in the two-dimensional finite lattice $\mathbb{Z}_M \times \mathbb{Z}_N$ with independent geometric random variables as passage times for each site [Joh00]. The rightmost node occupied by a particle in the ensemble, $x_{\max} := \max_j x_j$, is a random variable having the same distribution as the last passage time to travel from the site $(0,0)$ to the site $(M-1, N-1)$.

- The Charlier weight

$$w(x) = \frac{t^x}{x!}, \qquad \text{for } x = 0, 1, 2, \ldots,$$

arises in the longest random word problem [Joh01].

- The Krawtchouk weight

$$w(x) = \binom{N-1}{x} p^x q^{N-1-x}, \qquad \text{for } x = 0, 1, \ldots, N-1,$$

arises in the random domino tiling of the Aztec diamond [Joh01, Joh02].

- The Hahn weight

$$w(x) = \binom{x+\alpha}{x}\binom{N-1+\beta-x}{N-1-x}, \qquad \text{for } x = 0, 1, 2, \ldots, N-1,$$

arises in the random rhombus tiling of a hexagon [Joh01, Joh02]. See also §3.4 for more details.

The first two cases (Meixner and Charlier) are examples of the *Schur measure* [BorO01, Oko01] on the set of partitions. On the other hand, in special limiting cases the Meixner and Charlier ensembles both become the *Plancherel measure*, which describes the longest increasing subsequence of a random permutation[1] [BaiDJ99, BorOO00, Joh01]. Clearly it would be of some theoretical interest to determine properties of the ensembles that are more or less independent of the particular choice of weight function, at least within some class. Such properties are said to support the conjecture of *universality* within the class of weight functions under consideration.

As is well known in random matrix theory [Meh91, TraW98], many quantities such as correlation functions and gap probabilities are expressible in terms of the polynomials that are orthogonal with respect to the weight w. In a similar way, the asymptotic study of discrete orthogonal polynomial ensembles is translated to the asymptotic study of discrete orthogonal polynomials. In many applications, the parameters of the weight are varying, and at the same time the location of the particle of interest also varies. For example, in the Hahn case, the interesting case turns out to be when $N \to \infty$ and the location of a particle is scaled as $x = aN + \xi$ or $x = aN + \xi N^{1/3}$ for some constant $a > 0$. Equivalently, upon scaling by $1/N$, the discrete set X/N asymptotically fills out an interval, while x/N scales as $x/N = a + \xi/N$ or $x/N = a + \xi N^{-2/3}$. Under this scaling limit (Plancherel-Rotach asymptotics), the asymptotics of Meixner, Charlier, and Krawtchouk polynomials were obtained using integral representations [IsmS98, Joh00, Joh02]. However, even though the Hahn polynomials are also classical polynomials, their integral representation does not seem to be so straightforward to analyze asymptotically using the classical steepest-descent method, and the asymptotics have not yet been obtained. The subject of this book is a general class of weights that contains Krawtchouk and Hahn weights (and a lot more) as special cases. For these general weights, we will compute the asymptotics of the associated discrete orthogonal polynomials for all values of the variable z in the complex plane. By specializing to suitable scaling regimes, the desired asymptotics of the associated discrete orthogonal polynomial ensembles are obtained.

Random tilings

The discrete orthogonal polynomial ensemble with the Hahn weight represents a probability distribution for certain events in a random rhombus tiling of a hexagon, and new results on the asymptotics of random

INTRODUCTION

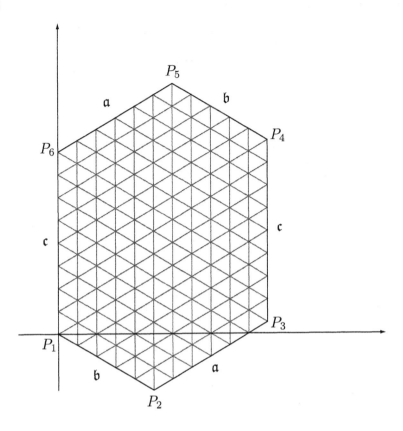

Figure 1.1 *The* $\mathfrak{abc}$*-hexagon with vertices* $P_1, \ldots, P_6$, *and the lattice* $\mathcal{L}$.

tilings will be presented in Chapter 3 using this connection. Here, we briefly introduce the subject of random rhombus tilings of a hexagon.

Let $\mathfrak{a}$, $\mathfrak{b}$, and $\mathfrak{c}$ be positive integers and consider the hexagon illustrated in Figure 1.1 having the following vertices (written as points in the complex plane):

$$\begin{aligned} P_1 &= 0, & P_2 &= \mathfrak{b}e^{-i\pi/6}, & P_3 &= P_2 + \mathfrak{a}e^{i\pi/6}, \\ P_4 &= P_3 + i\mathfrak{c}, & P_5 &= P_4 + \mathfrak{b}e^{5\pi i/6}, & P_6 &= i\mathfrak{c}. \end{aligned}$$

All interior angles of this hexagon are equal and measure $2\pi/3$ radian, and the lengths of the sides are, starting with the side (P_1, P_2) and proceeding in counterclockwise order, $\mathfrak{b}, \mathfrak{a}, \mathfrak{c}, \mathfrak{b}, \mathfrak{a}, \mathfrak{c}$. We call this the $\mathfrak{abc}$-*hexagon*. Denote by $\mathcal{L}$ the part of the set of lattice points

$$\left\{ ke^{i\pi/6} + je^{-i\pi/6} \right\}_{k,j \in \mathbb{Z}} = \left\{ \frac{\sqrt{3}}{2}n + \frac{i}{2}n' \right\}_{n, n' \in \mathbb{Z}}$$

that lies within the hexagon, including the sides (P_6, P_1), (P_1, P_2), (P_2, P_3), and (P_3, P_4) but excluding the sides (P_4, P_5) and (P_5, P_6). The lattice $\mathcal{L}$ can also be seen in Figure 1.1.

Consider covering the $\mathfrak{abc}$-hexagon with rhombus tiles having sides of unit length. The rhombus tiles come in three different types (orientations), which we refer to as type I, type II, and type III as shown in Figure 1.2. Tiles of types I and II are sometimes collectively called *horizontal rhombi*, while tiles of type III

---

[1] Strictly speaking, this is not a discrete orthogonal polynomial ensemble in the sense we have described because as a consequence of the limiting process involved in the definition, the number of particles k is not fixed in advance but is itself a random variable.

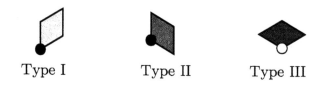

Figure 1.2 *The three types of rhombus tiles; the position of each tile is indicated with a dot.*

are sometimes called *vertical rhombi*. The position of each rhombus tile in the hexagon is a specific lattice point in $\mathcal{L}$ defined as indicated in Figure 1.2.

MacMahon's formula [Mac60] gives the total number of all possible rhombus tilings of the $\mathfrak{abc}$-hexagon as the expression

$$\prod_{i=1}^{\mathfrak{a}} \prod_{j=1}^{\mathfrak{b}} \prod_{k=1}^{\mathfrak{c}} \frac{i+j+k-1}{i+j+k-2}.$$

Consider the set of all rhombus tilings equipped with uniform probability. Hence we choose a tiling of the $\mathfrak{abc}$-hexagon at random. It is of some current interest to determine the behavior of various corresponding statistics of this ensemble in the limit as $\mathfrak{a}, \mathfrak{b}, \mathfrak{c} \to \infty$.

In the scaling limit of $n \to \infty$, where

$$\mathfrak{a} = \mathfrak{A}n, \qquad \mathfrak{b} = \mathfrak{B}n, \qquad \mathfrak{c} = \mathfrak{C}n, \tag{1.2}$$

with fixed $\mathfrak{A}, \mathfrak{B}, \mathfrak{C} > 0$, the regions near the six corners are *frozen* or *polar zones* (*i.e.*, regions in which only one type of tile is present), while toward the center of the hexagon is a *temperate zone* (*i.e.*, a region containing all three types of tiles). The random tiling shown in Figure 1.3 dramatically illustrates the two types of regions, and the asymptotically sharp nature of the boundary between them, the *Arctic circle*. Indeed, it was shown by Cohn, Larsen, and Propp [CohLP98] that in such a limit, upon scaling by $1/n$, the expected shape of the boundary separating the polar zones from the temperate zone is given by the inscribed ellipse. The next interesting problem would be to compute the limiting fluctuation of the boundary. While this problem had remained unsolved until our work, an analogous problem had been solved in the context of rectangle tilings of Aztec diamonds; it is proved in [Joh01] that the fluctuation of the boundary between the polar zones and the temperate zone in the Aztec diamond tiling model is governed (in a proper scaling limit) by the Tracy-Widom law in random matrix theory [TraW94]. Indeed, Johansson [Joh01, Joh02] expressed the induced probability for certain configurations of rectangles in an Aztec diamond and of rhombi in the $\mathfrak{abc}$-hexagon in terms of particular discrete orthogonal polynomial ensembles. The weights corresponding to the Aztec diamond are Krawtchouk weights, and those corresponding to the $\mathfrak{abc}$-hexagon are Hahn or associated Hahn weights. By applying the classical steepest-descent method to the integral representation of the Krawtchouk polynomials, Johansson obtained the Tracy-Widom distribution for the Aztec diamond. One of the results implied by our analysis of general discrete orthogonal polynomial ensembles (see Theorem 3.14) is that the same Tracy-Widom law holds for rhombus tilings of the $\mathfrak{abc}$-hexagon. More details of the connection between the statistics of rhombus tilings and the Hahn discrete orthogonal polynomial ensembles are discussed along with our asymptotic results in §3.4 and, specifically, §3.4.2.

1.1.2 The continuum limit of the Toda lattice

In Flaschka's variables, the Toda lattice equations are the following coupled nonlinear ordinary differential equations:

$$\frac{da_k}{dt} = 2b_k^2 - 2b_{k-1}^2 \tag{1.3}$$

and

$$\frac{db_k}{dt} = (a_{k+1} - a_k)b_k. \tag{1.4}$$

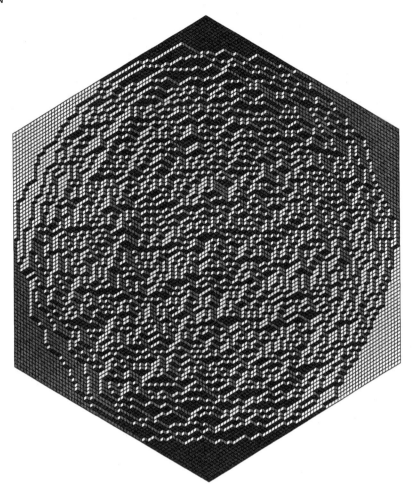

Figure 1.3 *A rhombus tiling of a large* abc-*hexagon with* a = b = c = 64. *The tiles are shaded as in Figure 1.2. Image provided by J. Propp.*

Here k is an integer index. The finite Toda lattices arise by "cutting the chain", with the imposition of boundary conditions at, say, $k = 0$ and $k = N - 1$. For one type of boundary condition, we may assume that with $a_k = a_{N,k}$ and $b_k = b_{N,k}$, (1.4) holds for $k = 0, \ldots, N - 2$ and (1.3) holds for $k = 1, \ldots, N - 2$, while

$$\frac{da_{N,0}}{dt} = 2b_{N,0}^2 \quad \text{and} \quad \frac{da_{N,N-1}}{dt} = -2b_{N,N-2}^2. \tag{1.5}$$

Thus we obtain a first-order system of nonlinear differential equations for unknowns $a_{N,0}(t), \ldots, a_{N,N-1}(t)$ and $b_{N,0}(t), \ldots, b_{N,N-2}(t)$.

One way to view the Toda lattice equations (1.3) and (1.4) is as a numerical scheme for integrating the hyperbolic system

$$\frac{\partial A}{\partial T} = 4B \frac{\partial B}{\partial c} \quad \text{and} \quad \frac{\partial B}{\partial T} = B \frac{\partial A}{\partial c}. \tag{1.6}$$

Here $T = t/N$ is a rescaled time, $1/N$ is a small lattice spacing, and $c = k/N$. Of course, as (1.6) is a nonlinear hyperbolic system of partial differential equations, initially smooth solutions of (1.6) can develop *shocks* (derivative singularities) in finite time. When this occurs, the Toda lattice equations should no longer be viewed as a viable numerical method for studying weak solutions of (1.6), as rapid oscillations develop in the numerical solution that are (it turns out) inconsistent even in an averaged sense with viscosity solutions of the hyperbolic system. See Figures 1.4 and 1.5.

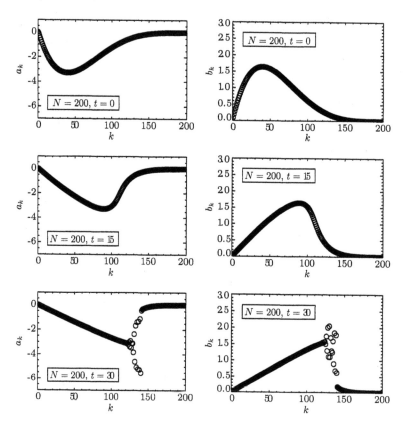

Figure 1.4 *A solution of the Toda lattice equations with smooth initial data on the interval $c = k/N \in [0, 1]$ sampled over $N = 200$ points.*

The proof of convergence of the Toda numerical scheme for (1.6), for times T less than the shock time, is one of the results of the analysis carried out by Deift and McLaughlin [DeiM98]. Their analysis of the Toda lattice equations with smooth initial data goes beyond the shock time and characterizes precisely the average properties of the rapid oscillations that subsequently occur. Indeed, it turns out that, while rapidly varying (on the fast time scale t and on the grid scale k), these oscillations can nonetheless be characterized by slowly varying quantities that satisfy an enlarged set of hyperbolic nonlinear partial differential equations (the Whitham equations) generalizing the hyperbolic system (1.6). The continuum limit of the Toda lattice has also been considered from a geometric point of view [BloGPU03] and from the point of view of orthogonal polynomials [AptV01].

The method used in [DeiM98] exploits the fact that the Toda lattice equations (1.3) and (1.4) comprise a completely integrable system. Specifically, this fact implies closed-form formulae for $a_{N,k}(t)$ and $b_{N,k}(t)$ in terms of initial data via ratios of Hankel-type determinants. Deift and McLaughlin analyzed these determinantal formulae in the continuum limit $N \to \infty$ for initial data sampling fixed smooth functions $A(c)$ and $B(c) > 0$ given on the interval $c \in [0, 1]$. Using the Lax-Levermore method, they showed that the large-N asymptotics are characterized by the solution of a constrained variational problem for a certain extremal measure on an interval, which they solved by converting it into a scalar Riemann-Hilbert problem.

INTRODUCTION

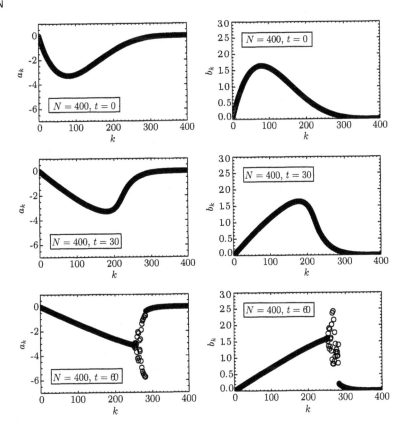

Figure 1.5 *A solution of the Toda lattice equations with smooth initial data on the interval* $c = k/N \in [0, 1]$ *sampled over* $N = 400$ *points.*

The number of subintervals where the extremal measure is unconstrained (this number depends generally on c and T) turns out to be related to the size of the system of Whitham equations needed to describe the slowly varying features of the microscopic oscillations. In particular, when there is only one such subinterval, the limit $N \to \infty$ is strong, and the hyperbolic system (1.6) governs the limit. Shock formation corresponds to the splitting of one unconstrained subinterval into two. Once this occurs, the analysis in [DeiM98] establishes the continuum limit only in a weak (averaged) sense. One of the applications that we will describe in this monograph is the strengthening of the asymptotics obtained in [DeiM98] after the shock time; we will obtain strong (locally uniform) asymptotics for the $a_{N,k}(t)$ and $b_{N,k}(t)$ even in the oscillatory region appearing after a shock has occurred, for example, in the irregular regions of the final plots in Figures 1.4 and 1.5.

Dynamical stability of solutions of the Toda lattice equations may be studied by means of the linearized Toda equations. Substituting $a_{N,k}(t) \to a_{N,k}(t) + \hat{a}_{N,k}(t)$ and $b_{N,k}(t) \to b_{N,k}(t)(1 + \hat{b}_{N,k}(t))$ into (1.3) and (1.4) and keeping only the linear terms gives

$$\frac{d\hat{a}_{N,k}}{dt} = 4b_{N,k}^2 \hat{b}_{N,k} - 4b_{N,k-1}^2 \hat{b}_{N,k-1} \qquad (1.7)$$

and

$$\frac{d\hat{b}_{N,k}}{dt} = \hat{a}_{N,k+1} - \hat{a}_{N,k}. \qquad (1.8)$$

Linear stability analysis is interesting in the large-N limit both before and after shock formation. Unfortunately, the detailed analysis in [DeiM98] does not directly provide information about a complete basis for solutions of the linearized equations (1.7) and (1.8) corresponding to given smooth functions $A(c)$ and $B(c)$. One of the consequences of the analysis described in this book is an asymptotic description of the solutions of the linearized Toda equations in the continuum limit both before and after shock time. In some applications the linearized problem can have meaning as a linear system in its own right with prescribed time-dependent potentials $a_{N,k}(t)$ and $b_{N,k}(t)$; it is mathematically convenient then to imagine that the potentials originate from a Toda solution, in which case there is a complete basis for time-dependent modes (*i.e.*, a time-dependent spectral transform) for the linear problem. This approach is described for other completely integrable systems in [MilA98] and [MilC01], with corresponding physical applications described in references therein. See §3.5 for details of the results implied by application of the analysis in this monograph to the continuum limit of the Toda lattice.

1.2 DISCRETE ORTHOGONAL POLYNOMIALS

The common thread in all of these applications is the role played by systems of discrete orthogonal polynomials. Let $N \in \mathbb{N}$ and consider N distinct real *nodes* $x_{N,0} < x_{N,1} < \cdots < x_{N,N-1}$ to be given; together the nodes make up the support of the pure point measures we consider. We use the notation

$$X_N := \{x_{N,n}\}_{n=0}^{N-1}, \qquad \text{where } x_{N,j} < x_{N,k} \text{ whenever } j < k,$$

for the support set. We focus in this book on finite discrete sets of nodes; the analysis for infinite discrete node sets (as in the case of Meixner and Charlier weights) is not totally different, but details will be described elsewhere. Along with nodes we are given positive *weights* $w_{N,0}, w_{N,1}, \ldots, w_{N,N-1}$, which are the magnitudes of the point masses located at the corresponding nodes. We will occasionally use the alternate notation $w(x)$, $x \in X_N$, for a weight on the set of nodes X_N; thus

$$w(x_{N,n}) = w_{N,n}, \qquad n = 0, 1, 2, \ldots, N-1. \tag{1.9}$$

One should not infer from this notation that $w(x)$ has any meaning for any x, complex or real, other than for $x \in X_N$; even if w has a convenient functional form, we will evaluate $w(x)$ only when $x \in X_N$. The discrete orthogonal polynomials associated with this data are polynomials $\{p_{N,k}(z)\}_{k=0}^{N-1}$, where $p_{N,k}(z)$ is of degree exactly k with a positive leading coefficient and where

$$\sum_{n=0}^{N-1} p_{N,k}(x_{N,n}) p_{N,l}(x_{N,n}) w_{N,n} = \delta_{kl}. \tag{1.10}$$

Writing $p_{N,k}(z) = c_{N,k}^{(k)} z^k + \ldots + c_{N,k}^{(0)}$, we introduce distinguishing notation for the positive leading coefficient,

$$\gamma_{N,k} := c_{N,k}^{(k)},$$

and we denote by $\pi_{N,k}(z)$ the associated monic polynomial,

$$\pi_{N,k}(z) := \frac{1}{\gamma_{N,k}} p_{N,k}(z).$$

The discrete orthogonal polynomials exist and are uniquely determined by the orthogonality conditions because the inner product associated with (1.10) is positive-definite on span $(1, z, z^2, \ldots, z^{N-1})$ but is degenerate on larger spaces of polynomials. The polynomials $p_{N,k}(z)$ may be built from the monomials by a Gram-Schmidt process. A general reference for properties of orthogonal polynomials specific to the discrete case is the book by Nikiforov, Suslov, and Uvarov [NikSU91].

One well-known elementary property of the discrete orthogonal polynomials is an exclusion principle for the zeros that forbids more than one zero from lying between adjacent nodes.

Proposition 1.1. *Each discrete orthogonal polynomial $p_{N,k}(z)$ has k simple real zeros. All these zeros lie in the range $x_{N,0} < z < x_{N,N-1}$, and no more than one zero lies in the closed interval $[x_{N,n}, x_{N,n+1}]$ between any two consecutive nodes.*

Proof. From the Gram-Schmidt process it follows that the coefficients of $p_{N,k}(z)$ are all real. Suppose that $p_{N,k}(z)$ vanishes to the nth order for some nonreal z_0. Then it follows that $p_{N,k}(z)$ also vanishes to the same order at z_0^* and thus that $p_{N,k}(z)/[(z-z_0)^n(z-z_0^*)^n]$ is a polynomial of lower degree, $k - 2n \geq 0$. By orthogonality, we must have on the one hand,

$$\sum_{n=0}^{N-1} p_{N,k}(x_{N,n}) \cdot \frac{p_{N,k}(x_{N,n})}{|x_{N,n} - z_0|^{2n}} \cdot w_{N,n} = 0.$$

On the other hand, the left-hand side is strictly positive because $k < N$, so $p_{N,k}(z)$ cannot vanish at all of the nodes. So we have a contradiction and the roots must be real.

The necessarily real roots are simple for a similar reason. If z_0 is a real root of $p_{N,k}(z)$ of order greater than 1, the quotient $p_{N,k}(z)/(z-z_0)^2$ is a polynomial of degree $k - 2 \geq 0$, which must be orthogonal to $p_{N,k}(z)$ itself,

$$\sum_{n=0}^{N-1} p_{N,k}(x_{N,n}) \cdot \frac{p_{N,k}(x_{N,n})}{(x_{N,n} - z_0)^2} \cdot w_{N,n} = 0,$$

but the left-hand side is manifestly positive, which gives the desired contradiction.

If a simple real zero z_0 of $p_{N,k}(z)$ satisfies either $z_0 \leq x_{N,0}$ or $z_0 \geq x_{N,N-1}$, then we repeat the above argument considering the polynomial $p_{N,k}(z)/(z-z_0)$ of degree $k - 1 \geq 0$, to which $p_{N,k}(z)$ must be orthogonal but for which the inner product is strictly of one sign.

Finally, if more than one zero of $p_{N,k}(z)$ lies between the consecutive nodes $x_{N,n}$ and $x_{N,n+1}$, then we can certainly select two of them, say z_0 and z_1, and construct the polynomial $p_{N,k}(z)/[(z-z_0)(z-z_1)]$ of degree $k - 2 \geq 0$. Again, this polynomial must be orthogonal to $p_{N,k}(z)$, but the corresponding inner product is of one definite sign, leading to a contradiction. $\square$

Another well-known and important feature of all systems of orthogonal polynomials, which is present whether the weights are discrete or continuous, is the existence of a three-term recurrence relation. See [Sze91] for details. There are constants $a_{N,0}, a_{N,1}, \ldots, a_{N,N-2}$ and positive constants $b_{N,0}, b_{N,1}, \ldots, b_{N,N-2}$ such that

$$zp_{N,k}(z) = b_{N,k} p_{N,k+1}(z) + a_{N,k} p_{N,k}(z) + b_{N,k-1} p_{N,k-1}(z) \tag{1.11}$$

holds for $k = 1, \ldots, N - 2$, while for $k = 0$ one has

$$zp_{N,0}(z) = b_{N,0} p_{N,1}(z) + a_{N,0} p_{N,0}(z), \tag{1.12}$$

and for $k = N - 1$,

$$zp_{N,N-1}(z) = \gamma_{N,N-1} \prod_{n=0}^{N-1} (z - x_{N,n}) + a_{N,N-1} p_{N,N-1}(z) + b_{N,N-2} p_{N,N-2}(z). \tag{1.13}$$

Necessarily, one has $b_{N,k} = \gamma_{N,k}/\gamma_{N,k+1}$. Therefore the vectors $\mathbf{v}_j := (p_{N,0}(x_{N,j}), \ldots, p_{N,N-1}(x_{N,j}))^T$ form an orthonormal basis of eigenvectors (with corresponding eigenvalues $z = x_{N,j}$) of the symmetric tridiagonal $N \times N$ Jacobi matrix constructed from the sequences $\{a_{N,0}, \ldots, a_{N,N-1}\}$ and $\{b_{N,0}, \ldots, b_{N,N-2}\}$.

Our goal is to establish the asymptotic behavior of the polynomials $p_{N,k}(z)$ or their monic counterparts $\pi_{N,k}(z)$ in the limit of large degree, assuming certain asymptotic properties of the nodes and the weights. In particular, the number of nodes must necessarily increase to admit polynomials of arbitrarily large degree, and the weights we consider involve an exponential factor with the exponent proportional to the number of nodes (such weights are sometimes called *varying weights*). We will obtain pointwise asymptotics with a precise error bound uniformly valid in the whole complex plane. Our assumptions about the nodes and weights include as special cases all relevant classical discrete orthogonal polynomials but are significantly more general; in particular, we will consider nodes that are not necessarily equally spaced.

1.3 ASSUMPTIONS

1.3.1 Basic assumptions

We will establish rigorous asymptotics for the discrete orthogonal polynomials subject to the following fundamental assumptions.

The nodes

We suppose the existence of a *node density function* $\rho^0(x)$ that is real-analytic in a complex neighborhood of a closed interval $[a, b]$ and satisfies

$$\int_a^b \rho^0(x)\,dx = 1$$

and

$$\rho^0(x) > 0 \text{ strictly}, \quad \text{for all } x \in [a, b]. \tag{1.14}$$

The nodes are then defined precisely in terms of the density function $\rho^0(x)$ by the quantization rule

$$\int_a^{x_{N,n}} \rho^0(x)\,dx = \frac{2n+1}{2N}, \tag{1.15}$$

for $N \in \mathbb{N}$ and $n = 0, 1, 2, \ldots, N-1$. Thus the nodes lie in a bounded open interval (a, b) and are distributed with density $\rho^0(x)$.

The weights

Without loss of generality, we write the weights in the form

$$w_{N,n} = (-1)^{N-1-n} e^{-NV_N(x_{N,n})} \prod_{\substack{m=0 \\ m \neq n}}^{N-1} (x_{N,n} - x_{N,m})^{-1} = e^{-NV_N(x_{N,n})} \prod_{\substack{m=0 \\ m \neq n}}^{N-1} |x_{N,n} - x_{N,m}|^{-1}. \tag{1.16}$$

No generality has been sacrificed with this representation because the family of functions $\{V_N(x)\}$ is *a priori* specified only at the nodes; in other words, given positive weights $\{w_{N,n}\}$, one may solve (1.16) uniquely for the N quantities $\{V_N(x_{N,n})\}$. However, we now assume that for each sufficiently large N, $V_N(x)$ may be taken to be a real-analytic function defined in a complex neighborhood G of the closed interval $[a, b]$ and that

$$V_N(x) = V(x) + \frac{\eta(x)}{N}, \tag{1.17}$$

where $V(x)$ is a fixed real-analytic *potential function* defined in G and

$$\limsup_{N \to \infty} \sup_{z \in G} |\eta(z)| < \infty. \tag{1.18}$$

Note that in general the correction $\eta(z)$ may depend on N, although $V(x)$ may not. In some cases (*e.g.*, Krawtchouk polynomials; see §2.4.1) it is possible to take $V_N(x) \equiv V(x)$ for all N, in which case $\eta(x) \equiv 0$. However, the freedom of assuming $\eta(x) \not\equiv 0$ is useful in handling other cases (*e.g.*, the Hahn and associated Hahn polynomials; see §2.4.2). While (1.16) may be written for any system of positive weights, the condition that (1.17) should hold restricts attention to systems of weights that have analytic continuum limits in a certain precise sense.

◁ **Remark:** The familiar examples of classical discrete orthogonal polynomials correspond to nodes that are equally spaced, say, on $(a, b) = (0, 1)$ (in which case we have $\rho^0(x) \equiv 1$). In this special case, the product factor on the right-hand side of (1.16) becomes simply

$$\prod_{\substack{m=0 \\ m \neq n}}^{N-1} |x_{N,n} - x_{N,m}|^{-1} = \frac{N^{N-1}}{n!(N-n-1)!}.$$

INTRODUCTION

Using Stirling's formula to take the continuum limit of this factor (*i.e.*, considering $N \to \infty$ with $n/N \to x$) shows that in these cases the leading term in formula (1.16) is a continuous weight on $(0, 1)$:

$$w_{N,n} \sim w(x) := C \left(\frac{e^{-V(x)}}{x^x (1-x)^{1-x}} \right)^N \tag{1.19}$$

as $N \to \infty$ and $n/N \to x \in (0, 1)$, where C is independent of x. However, the process of taking the continuum limit of the weight first to arrive at a formula like (1.19) and then obtaining asymptotics of the polynomials of degree proportional to N as $N \to \infty$ is not equivalent to the scaling limit process we will consider here. Our results will display new phenomena because we simultaneously take the continuum limit as the degree of the polynomial grows. ▷

Our choice of the form (1.16) for the weights is motivated by several specific examples of classical discrete orthogonal polynomials. The form (1.16) is sufficiently general for us to carry out useful calculations related to proofs of universality conjectures arising in certain types of random tiling problems, random growth models, and last-passage percolation problems. Also, the form (1.16) is appropriate for study of the Toda lattice in the continuum limit by inverse spectral theory.

The degree

We assume that the degree k of the polynomial of interest is tied to the number N of nodes by a relation of the form

$$k = cN + \kappa,$$

where $c \in (0, 1)$ is a fixed parameter and κ remains bounded as $N \to \infty$.

1.3.2 Simplifying assumptions of genericity

In order to keep our exposition as simple as possible, we make further assumptions that exclude certain nongeneric triples $(\rho^0(x), V(x), c)$. These assumptions depend on the functions $\rho^0(x)$ and $V(x)$, and on the parameter c, in an implicit manner that is easier to describe once some auxiliary quantities have been introduced. They will be given in §2.1.2.

In regard to these particular assumptions, we want to stress two points. First, the excluded triples are nongeneric in the sense that any perturbation of, say, the parameter c will immediately return us to the class of triples for which all of our results are valid. The discussion at the beginning of §5.1.2 provides some insight into the generic nature of our assumptions. Second, the discrete orthogonal polynomials corresponding to nongeneric triples can be analyzed by the same basic method that we use here, with many of the same results. To do this, the proofs we present will require modifications to include additional local analysis near certain isolated points in the complex z-plane. Some such modifications have already been described in detail in the context of asymptotics for polynomials orthogonal with respect to continuous weights in §5 of [DeiKMVZ99b]. The remaining modifications have to do with nongeneric behavior near the endpoints of the interval $[a, b]$, and while the corresponding local analysis has not been done before, it can be expected to be of a similar character.

1.4 GOALS AND METHODOLOGY

Given an interval $[a, b]$, appropriate fixed functions $\rho^0(x)$ and $V(x)$, appropriate sequences $\eta(x) = \eta_N(x)$ and $\kappa = \kappa_N$, and a constant $c \in (0, 1)$, we wish to find accurate asymptotic formulae, valid in the limit $N \to \infty$ with rigorous error bounds, for the polynomial $\pi_{N,k}(z)$. These formulae should be uniformly valid in overlapping regions of the complex z-plane. We will also require asymptotic formulae for related quantities, like the zeros of $\pi_{N,k}(z)$, the three-term recurrence coefficients, and the reproducing kernels $K_{N,k}(x, y)$.

1.4.1 The basic interpolation problem

Given a natural number N, a set X_N of nodes, and a set of corresponding weights $\{w_{N,n}\}$, consider the possibility of finding the matrix $\mathbf{P}(z; N, k)$ solving the following problem, where k is an integer.

Interpolation Problem 1.2. *Find a 2×2 matrix $\mathbf{P}(z; N, k)$ with the following properties:*

1. **Analyticity**: $\mathbf{P}(z; N, k)$ *is an analytic function of z for $z \in \mathbb{C} \setminus X_N$.*

2. **Normalization**: *As $z \to \infty$,*
$$\mathbf{P}(z; N, k) \begin{pmatrix} z^{-k} & 0 \\ 0 & z^k \end{pmatrix} = \mathbb{I} + O\left(\frac{1}{z}\right). \tag{1.20}$$

3. **Singularities**: *At each node $x_{N,n} \in X_N$, the first column of $\mathbf{P}(z; N, k)$ is analytic and the second column of $\mathbf{P}(z; N, k)$ has a simple pole, where the residue satisfies the condition*
$$\operatorname*{Res}_{z=x_{N,n}} \mathbf{P}(z; N, k) = \lim_{z \to x_{N,n}} \mathbf{P}(z; N, k) \begin{pmatrix} 0 & w_{N,n} \\ 0 & 0 \end{pmatrix} = \begin{pmatrix} 0 & w_{N,n} P_{11}(x_{N,n}; N, k) \\ 0 & w_{N,n} P_{21}(x_{N,n}, N, k) \end{pmatrix}, \tag{1.21}$$
for $n = 0, \ldots, N - 1$.

This problem is a discrete version of the Riemann-Hilbert problem appropriate for orthogonal polynomials with continuous weights that was first used by Fokas, Its, and Kitaev in [FokIK91] (see also [DeiKMVZ99a, DeiKMVZ99b]). The discrete version was first studied by Borodin [Bor00] (see also [Bor03] and [BorB03]). The solution of this problem encodes all quantities of relevance to a study of the discrete orthogonal polynomials, as we will now see.

Proposition 1.3. *Interpolation Problem 1.2 has a unique solution when $0 \leq k \leq N - 1$. The solution is*
$$\mathbf{P}(z; N, k) = \begin{pmatrix} \pi_{N,k}(z) & \sum_{n=0}^{N-1} \dfrac{w_{N,n} \pi_{N,k}(x_{N,n})}{z - x_{N,n}} \\ \gamma_{N,k-1} p_{N,k-1}(z) & \sum_{n=0}^{N-1} \dfrac{w_{N,n} \gamma_{N,k-1} p_{N,k-1}(x_{N,n})}{z - x_{N,n}} \end{pmatrix}, \tag{1.22}$$
for $0 < k \leq N - 1$, while
$$\mathbf{P}(z; N, 0) = \begin{pmatrix} 1 & \sum_{n=0}^{N-1} \dfrac{w_{N,n}}{z - x_{N,n}} \\ 0 & 1 \end{pmatrix}. \tag{1.23}$$

Proof. Consider the first row of $\mathbf{P}(z; N, k)$. According to (1.21), the function $P_{11}(z; N, k)$ is an entire function of z. Because $k \geq 0$, it follows from the normalization condition (1.20) that in fact $P_{11}(z; N, k)$ is a monic polynomial of degree exactly k. Similarly, from the characterization (1.21) of the simple poles of $P_{12}(z; N, k)$, we see that $P_{12}(z; N, k)$ is necessarily of the form
$$P_{12}(z; N, k) = e_1(z) + \sum_{n=0}^{N-1} \frac{w_{N,n} P_{11}(x_{N,n}; N, k)}{z - x_{N,n}},$$
where $e_1(z)$ is an entire function. The normalization condition (1.20) for $k \geq 0$ immediately requires, via Liouville's Theorem, that $e_1(z) \equiv 0$, and then when $|z| > \max_n |x_{N,n}|$, we have by geometric series expansion that
$$P_{12}(z; N, k) = \sum_{m=0}^{\infty} \left(\sum_{n=0}^{N-1} P_{11}(x_{N,n}; N, k) x_{N,n}^m w_{N,n} \right) \frac{1}{z^{m+1}}.$$

INTRODUCTION

According to the normalization condition (1.20), $P_{12}(z; N, k) = o(z^{-k})$ as $z \to \infty$; therefore it follows that the monic polynomial $P_{11}(z; N, k)$ of degree exactly k must satisfy

$$\sum_{n=0}^{N-1} P_{11}(x_{N,n}; N, k) x_{N,n}^m w_{N,n} = 0, \quad \text{for } m = 0, 1, 2, \ldots, k-1.$$

As long as $k \leq N-1$, these conditions uniquely identify $P_{11}(z; N, k)$ with the monic discrete orthogonal polynomial $\pi_{N,k}(z)$. The existence and uniqueness of $\pi_{N,k}(z)$ for such k is guaranteed given distinct orthogonalization nodes and positive weights (which implies that the inner product is a positive-definite quadratic form).

The second row of $\mathbf{P}(z; N, k)$ is studied similarly. The matrix element $P_{21}(z; N, k)$ is seen from (1.21) to be an entire function of z, which according to the normalization condition (1.20) must be a polynomial of degree at most $k-1$ (for the special case of $k = 0$ these conditions immediately imply that $P_{21}(z; N, 0) \equiv 0$). The characterization (1.21) implies that $P_{22}(z; N, k)$ can be expressed in the form

$$P_{22}(z; N, k) = e_2(z) + \sum_{n=0}^{N-1} \frac{w_{N,n} P_{21}(x_{N,n}; N, k)}{z - x_{N,n}},$$

where $e_2(z)$ is an entire function. If $k = 0$, then $P_{22}(z; N, 0) = e_2(z)$, and then according to the normalization condition (1.20) we must take $e_2(z) \equiv 1$. On the other hand, if $k > 0$, then (1.20) implies that $P_{22}(z; N, k)$ decays for large z, and therefore we must take $e_2(z) \equiv 0$ in this case. Expanding the denominator in a geometric series for $|z| > \max_n |x_{N,n}|$, we find

$$P_{22}(z; N, k) = \sum_{m=0}^{\infty} \left(\sum_{n=0}^{N-1} P_{21}(x_{N,n}; N, k) x_{N,n}^m w_{N,n} \right) \frac{1}{z^{m+1}}.$$

Imposing the normalization condition (1.20) amounts to insisting that $P_{22}(z; N, k) = z^{-k} + O(z^{-k-1})$ as $z \to \infty$; therefore

$$\sum_{n=0}^{N-1} P_{21}(x_{N,n}; N, k) x_{N,n}^m w_{N,n} = 0, \quad \text{for } m = 0, 1, 2, \ldots, k-2, \tag{1.24}$$

and

$$\sum_{n=0}^{N-1} P_{21}(x_{N,n}; N, k) x_{N,n}^{k-1} w_{N,n} = 1. \tag{1.25}$$

With the use of (1.24), the condition (1.25) can be replaced by

$$\sum_{n=0}^{N-1} P_{21}(x_{N,n}; N, k) \pi_{N,k-1}(x_{N,n}) w_{N,n} = 1$$

or, equivalently,

$$\sum_{n=0}^{N-1} \left[\frac{1}{\gamma_{N,k-1}} P_{21}(x_{N,n}; N, k) \right] p_{N,k-1}(x_{N,n}) w_{N,n} = 1. \tag{1.26}$$

The conditions (1.24) and (1.26) therefore uniquely identify the quotient $P_{21}(z; N, k)/\gamma_{N,k-1}$ with the orthogonal polynomial $p_{N,k-1}(z)$.

The interpolation problem is thus solved uniquely by the matrix explicitly given by (1.22) for $k > 0$ and by (1.23) for $k = 0$. □

◁ **Remark:** In fact, Interpolation Problem 1.2 can also be solved for $k = N$, with a unique solution of the form (1.22), if we define

$$\pi_{N,N}(z) := \prod_{n=0}^{N-1} (z - x_{N,n}),$$

which of course is not in the finite family of orthogonal polynomials as it is not normalizable. ▷

The constants in the three-term recurrence relations (see equations (1.11)–(1.13)) are also encoded in the matrix $\mathbf{P}(z; N, k)$ solving Interpolation Problem 1.2.

Corollary 1.4. *Let k be fixed with $1 \leq k \leq N-2$ and let $s_{N,k}$, $y_{N,k}$, $r_{N,k}$, and $u_{N,k}$ denote certain terms in the large-z expansion of the matrix elements of* $\mathbf{P}(z; N, k)$,

$$z^k P_{12}(z; N, k) = \frac{s_{N,k}}{z} + \frac{y_{N,k}}{z^2} + O\left(\frac{1}{z^3}\right),$$

$$\frac{1}{z^k} P_{11}(z; N, k) = 1 + \frac{r_{N,k}}{z} + O\left(\frac{1}{z^2}\right),$$

$$\frac{1}{z^k} P_{21}(z; N, k) = \frac{u_{N,k}}{z} + O\left(\frac{1}{z^2}\right),$$

as $z \to \infty$. Then

$$\begin{aligned} \gamma_{N,k} &= \frac{1}{\sqrt{s_{N,k}}}, & \gamma_{N,k-1} &= \sqrt{u_{N,k}}, \\ a_{N,k} &= r_{N,k} + \frac{y_{N,k}}{s_{N,k}}, & b_{N,k} &= \sqrt{s_{N,k+1} u_{N,k+1}}. \end{aligned} \quad (1.27)$$

Also, $a_{N,k} = r_{N,k} - r_{N,k+1}$.

Proof. By expansion of the explicit solution given by (1.22) in Proposition 1.3 in the limit of large z, we deduce the identity

$$s_{N,k} = \sum_{n=0}^{N-1} \pi_{N,k}(x_{N,n}) x_{N,n}^k w_{N,n} = \frac{1}{\gamma_{N,k}^2} \quad (1.28)$$

(because $z^k = \pi_{N,k}(z) + O(z^{k-1})$, and using the definition of the normalization constants $\gamma_{N,k}$), the identity

$$y_{N,k} = \sum_{n=0}^{N-1} \pi_{N,k}(x_{N,n}) x_{N,n}^{k+1} w_{N,n} = -\frac{c_{N,k+1}^{(k)}}{\gamma_{N,k}^2 \gamma_{N,k+1}} \quad (1.29)$$

(because $z^{k+1} = \pi_{N,k+1}(z) - c_{N,k+1}^{(k)} \gamma_{N,k+1}^{-1} \pi_{N,k}(z) + O(z^{k-1})$), the identity

$$r_{N,k} = \frac{c_{N,k}^{(k-1)}}{\gamma_{N,k}}, \quad (1.30)$$

and the identity

$$u_{N,k} = \gamma_{N,k-1}^2. \quad (1.31)$$

Similarly, by expansion of the three-term recurrence relation in the limit of large z, we deduce the identity

$$\gamma_{N,k} z^{k+1} + c_{N,k}^{(k-1)} z^k = b_{N,k} \gamma_{N,k+1} z^{k+1} + (b_{N,k} c_{N,k+1}^{(k)} + a_{N,k} \gamma_{N,k}) z^k + O(z^{k-1})$$

as $z \to \infty$. Therefore

$$b_{N,k} = \frac{\gamma_{N,k}}{\gamma_{N,k+1}} \quad \text{and} \quad a_{N,k} = \frac{c_{N,k}^{(k-1)}}{\gamma_{N,k}} - \frac{c_{N,k+1}^{(k)}}{\gamma_{N,k+1}}.$$

Comparing with (1.28)–(1.31) completes the proof. □

1.4.2 Exponentially deformed weights and the Toda lattice

Let the nodes X_N be fixed. Suppose now that the weights $\{w_{N,n}\}$ are allowed to depend smoothly on a parameter t so as to remain positive for all t. By the solution of Interpolation Problem 1.2, this deformation of the weights induces a corresponding deformation in the orthogonal polynomials and all related quantities (e.g., three-term recurrence coefficients $a_{N,k}$, $b_{N,k}$ and norming constants $\gamma_{N,k}$). In particular, the dynamics induced on the recurrence coefficients can be written as a system of nonlinear differential equations for the $a_{N,k}$ and $b_{N,k}$. The collection of all possible differential equations obtainable in this way is the *Toda lattice hierarchy*.

INTRODUCTION

The Toda lattice hierarchy is spanned by the flows

$$w_{N,n}(t) := w_{N,n} \exp(2x_{N,n}^p t), \qquad n = 0, \ldots, N-1, \tag{1.32}$$

for $p = 1, 2, 3, \ldots$. Here we show how to derive the differential equations satisfied by the recurrence coefficients in the simplest case of $p = 1$. From the solution $\mathbf{P}(z; N, k, t)$ of Interpolation Problem 1.2 with weights of the form (1.32), define the *Jost matrix*

$$\mathbf{M}(z; N, k, t) := \mathbf{P}(z; N, k, t) \begin{pmatrix} e^{zt} & 0 \\ 0 & e^{-zt} \end{pmatrix}.$$

A simple calculation then shows that

$$\operatorname*{Res}_{z = x_{N,n}} \mathbf{M}(z; N, k, t) = \lim_{z \to x_{N,n}} \mathbf{M}(z; N, k, t) \begin{pmatrix} 0 & w_{N,n} \\ 0 & 0 \end{pmatrix}, \tag{1.33}$$

for $n = 0, \ldots, N-1$. In other words, the relations that constrain the residues at the simple poles of $\mathbf{M}(z; N, k, t)$ are independent of t as well as k. Note that this does not imply that $\mathbf{M}(z; N, k, t)$ is independent of t and k; indeed, as $z \to \infty$, $\mathbf{M}(z; N, k, t)$ exhibits exponential behavior in both t and k according to the normalization condition satisfied by $\mathbf{P}(z; N, k, t)$. On the other hand, (1.33) does imply that the matrices

$$\mathbf{L}(z; N, k, t) := \mathbf{M}(z; N, k+1, t) \mathbf{M}(z; N, k, t)^{-1}$$

and

$$\mathbf{B}(z; N, k, t) := \frac{d\mathbf{M}}{dt}(z; N, k, t) \mathbf{M}(z; N, k, t)^{-1}$$

are entire analytic functions of z for each k and t. Moreover, both of these matrix-valued functions are, by Liouville's Theorem, polynomials in z because they have polynomial asymptotics as $z \to \infty$. Writing

$$\begin{pmatrix} r_{N,k}(t) & s_{N,k}(t) \\ u_{N,k}(t) & v_{N,k}(t) \end{pmatrix} := \lim_{z \to \infty} z \left[\mathbf{P}(z; N, k, t) \begin{pmatrix} z^{-k} & 0 \\ 0 & z^k \end{pmatrix} - \mathbb{I} \right],$$

a direct calculation using the normalization condition satisfied by the matrix $\mathbf{P}(z; N, k, t)$ shows that

$$\mathbf{L}(z; N, k, t) = \begin{pmatrix} z + r_{N,k+1}(t) - r_{N,k}(t) & -s_{N,k}(t) \\ u_{N,k+1}(t) & 0 \end{pmatrix} + O(z^{-1})$$

and

$$\mathbf{B}(z; N, k, t) = \begin{pmatrix} z & -2s_{N,k}(t) \\ 2u_{N,k}(t) & -z \end{pmatrix} + O(z^{-1})$$

as $z \to \infty$. By Liouville's Theorem, we therefore have, exactly,

$$\mathbf{L}(z; N, k, t) = \begin{pmatrix} z + r_{N,k+1}(t) - r_{N,k}(t) & -s_{N,k}(t) \\ u_{N,k+1}(t) & 0 \end{pmatrix}$$

and

$$\mathbf{B}(z; N, k, t) = \begin{pmatrix} z & -2s_{N,k}(t) \\ 2u_{N,k}(t) & -z \end{pmatrix}.$$

The simultaneous linear equations

$$\begin{aligned} \mathbf{M}(z; N, k+1, t) &= \mathbf{L}(z; N, k, t) \mathbf{M}(z; N, k, t), \\ \frac{d\mathbf{M}}{dt}(z; N, k, t) &= \mathbf{B}(z; N, k, t) \mathbf{M}(z; N, k, t), \end{aligned} \tag{1.34}$$

satisfied by $\mathbf{M}(z; N, k, t)$, are said to make up a *Lax pair* for the Toda lattice. By computing the "shifted derivative" $d\mathbf{M}(z; N, k+1, t)/dt$ two different ways using the Lax pair and equating the results, one finds that

$$\left[\frac{d\mathbf{L}}{dt}(z; N, k, t) + \mathbf{L}(z; N, k, t) \mathbf{B}(z; N, k, t) - \mathbf{B}(z; N, k+1, t) \mathbf{L}(z; N, k, t) \right] \mathbf{M}(z; N, k, t) = \mathbf{0}. \tag{1.35}$$

Now it also follows from the conditions of Interpolation Problem 1.2 that $\det(\mathbf{P}(z; N, k, t)) \equiv 1$, so the matrix $\mathbf{M}(z; N, k, t)$ is a fundamental solution matrix for the Lax pair. Therefore we deduce from (1.35) the *compatibility condition*

$$\frac{d\mathbf{L}}{dt}(z; N, k, t) + \mathbf{L}(z; N, k, t)\mathbf{B}(z; N, k, t) - \mathbf{B}(z; N, k+1, t)\mathbf{L}(z; N, k, t) = \mathbf{0}. \quad (1.36)$$

This condition is the *zero-curvature representation* of the Toda lattice equations. Although the matrix elements for $\mathbf{L}(z; N, k, t)$ and $\mathbf{B}(z; N, k, t)$ depend on z, it is easy to check that the combination on the left-hand side of (1.36) is independent of z and is a matrix with three nonzero elements. The result of these observations is that (1.36) is equivalent to the following three differential equations:

$$\begin{aligned} \frac{d}{dt}(r_{N,k} - r_{N,k+1}) &= 2s_{N,k+1}u_{N,k+1} - 2s_{N,k}u_{N,k}, \\ \frac{ds_{N,k}}{dt} &= 2(r_{N,k} - r_{N,k+1})s_{N,k}, \\ \frac{du_{N,k+1}}{dt} &= -2(r_{N,k} - r_{N,k+1})u_{N,k+1}. \end{aligned} \quad (1.37)$$

According to Corollary 1.4, the quantities appearing in these equations are related to the three-term recurrence coefficients

$$a_{N,k} = r_{N,k} - r_{N,k+1} \quad \text{and} \quad b_{N,k}^2 = s_{N,k+1}u_{N,k+1},$$

so from (1.37) we obtain a closed set of equations governing the dynamics of the recurrence coefficients:

$$\frac{da_{N,k}}{dt} = 2b_{N,k}^2 - 2b_{N,k-1}^2 \quad \text{and} \quad \frac{db_{N,k}}{dt} = (a_{N,k+1} - a_{N,k})b_{N,k}. \quad (1.38)$$

These are the *Toda lattice equations*. We have derived these equations assuming that $k = 1, \ldots, N-2$, but a similar analysis for $k = 0$ yields

$$\frac{da_{N,0}}{dt} = 2b_{N,0}^2 \quad \text{and} \quad \frac{db_{N,0}}{dt} = (a_{N,1} - a_{N,0})b_{N,0},$$

and for $k = N - 1$ yields

$$\frac{da_{N,N-1}}{dt} = -2b_{N,N-2}^2.$$

In other words, (1.38) holds for $k = 0, \ldots, N-1$ if we define $b_{N,-1} = b_{N,N-1} = 0$.

Now we consider the *squared eigenfunctions* associated with the Lax pair (1.34). Let $\mathbf{C}$ be a constant matrix with trace zero and define the matrix of squared eigenfunctions as

$$\mathbf{W}(z; N, k, t) := \mathbf{M}(z; N, k, t)\mathbf{C}\mathbf{M}(z; N, k, t)^{-1}.$$

It follows that $\mathbf{W}(z; N, k, t)$ has trace zero as well and therefore may be written as

$$\mathbf{W}(z; N, k, t) = \begin{pmatrix} f_{N,k}(z;t) & g_{N,k}(z;t) \\ h_{N,k}(z;t) & -f_{N,k}(z;t) \end{pmatrix}. \quad (1.39)$$

From the Lax pair, it follows that $\mathbf{W}(z; N, k, t)$ also obeys a system of linear simultaneous equations even though its elements consist of quadratic forms in the elements of $\mathbf{M}(z; N, k, t)$:

$$\begin{aligned} \mathbf{W}(z; N, k+1, t) &= \mathbf{L}(z; N, k, t)\mathbf{W}(z; N, k, t)\mathbf{L}(z; N, k, t)^{-1}, \\ \frac{d\mathbf{W}}{dt}(z; N, k, t) &= [\mathbf{B}(z; N, k, t), \mathbf{W}(z; N, k, t)], \end{aligned} \quad (1.40)$$

where $[\mathbf{A}, \mathbf{B}] := \mathbf{AB} - \mathbf{BA}$ denotes the matrix commutator. Substituting the form (1.39) into the difference equation in (1.40) (first multiplying on the right by $\mathbf{L}(z; N, k, t)$) yields four equations:

$$\begin{aligned} (z - a_{N,k})f_{N,k+1} + u_{N,k+1}g_{N,k+1} &= (z - a_{N,k})f_{N,k} - s_{N,k}h_{N,k}, \\ -s_{N,k}f_{N,k+1} &= (z - a_{N,k})g_{N,k} + s_{N,k}f_{N,k}, \\ (z - a_{N,k})h_{N,k+1} - u_{N,k+1}f_{N,k+1} &= u_{N,k+1}f_{N,k}, \\ -s_{N,k}h_{N,k+1} &= u_{N,k+1}g_{N,k}. \end{aligned}$$

INTRODUCTION

Here we have used $a_{N,k} = r_{N,k} - r_{N,k+1}$. The last of these equations may be solved by introducing a sequence $\phi_{N,k}$ and writing

$$g_{N,k} := s_{N,k}\phi_{N,k} \quad \text{and} \quad h_{N,k} := -u_{N,k}\phi_{N,k-1}. \tag{1.41}$$

This choice also makes two of the remaining three equations identical, so only two equations remain:

$$\begin{aligned}(z - a_{N,k})(f_{N,k+1} - f_{N,k}) + b_{N,k}^2 \phi_{N,k+1} - b_{N,k-1}^2 \phi_{N,k-1} &= 0, \\ f_{N,k+1} + f_{N,k} + (z - a_{N,k})\phi_{N,k} &= 0. \end{aligned} \tag{1.42}$$

Here we have used $b_{N,k}^2 = s_{N,k+1}u_{N,k+1}$. Similarly, using (1.39) and (1.41) in the differential equation in (1.40) yields only two distinct scalar equations (up to shifts in k), namely,

$$\frac{df_{N,k}}{dt} = -2b_{N,k-1}^2(\phi_{N,k} - \phi_{N,k-1}) \quad \text{and} \quad \frac{d\phi_{N,k}}{dt} = 2(z - a_{N,k})\phi_{N,k} + 4f_{N,k}.$$

Using the second equation in (1.42) to eliminate $z - a_{N,k}$ yields the linear system

$$\frac{df_{N,k}}{dt} = -2b_{N,k-1}^2(\phi_{N,k} - \phi_{N,k-1}) \quad \text{and} \quad \frac{d\phi_{N,k}}{dt} = 2f_{N,k} - 2f_{N,k+1}.$$

By taking finite differences of these equations with respect to k and introducing new variables

$$\hat{a}_{N,k} := f_{N,k+1} - f_{N,k} \quad \text{and} \quad \hat{b}_{N,k} := -\frac{1}{2}(\phi_{N,k+1} - \phi_{N,k}),$$

the equations become

$$\frac{d\hat{a}_{N,k}}{dt} = 4b_{N,k}^2\hat{b}_{N,k} - 4b_{N,k-1}^2\hat{b}_{N,k-1} \quad \text{and} \quad \frac{d\hat{b}_{N,k}}{dt} = \hat{a}_{N,k+1} - \hat{a}_{N,k}.$$

We recognize these as the linearized Toda lattice equations. Although the linearized Toda lattice equations are themselves independent of z, we note that the functions $\hat{a}_{N,k}(z;t)$ and $\hat{b}_{N,k}(z;t)$ that satisfy them do indeed depend nontrivially on z through the solution of Interpolation Problem 1.2, and so by variation of z we obtain an infinite family of solutions of the linearized problem. Of course, only $2N - 1$ of them can be linearly independent, but there is sufficient freedom in the choice of the parameter z to construct a complete basis of solutions of the linearized problem for each N.

1.4.3 Triangularity of residue matrices and dual polynomials

The matrices that encode the residues in Interpolation Problem 1.2 are upper-triangular. An essential aspect of our methodology will be to modify the matrix $\mathbf{P}(z; N, k)$ in order to selectively reverse the triangularity of the residue matrices near certain individual nodes $x_{N,n}$. Let $\Delta \subset \mathbb{Z}_N$, where

$$\mathbb{Z}_N := \{0, 1, 2, \ldots, N - 1\},$$

and denote the number of elements in Δ by $\#\Delta$. We will reverse the triangularity for nodes $x_{N,n}$ for which $n \in \Delta$. Consider the matrix $\mathbf{Q}(z; N, k)$ related to the solution $\mathbf{P}(z; N, k)$ of Interpolation Problem 1.2 as follows:

$$\mathbf{Q}(z; N, k) := \mathbf{P}(z; N, k)\left[\prod_{n \in \Delta}(z - x_{N,n})\right]^{-\sigma_3} = \mathbf{P}(z; N, k)\begin{pmatrix}\prod_{n \in \Delta}(z - x_{N,n})^{-1} & 0 \\ 0 & \prod_{n \in \Delta}(z - x_{N,n})\end{pmatrix}. \tag{1.43}$$

Here σ_3 is a Pauli matrix:

$$\sigma_3 := \begin{pmatrix} 1 & 0 \\ 0 & -1 \end{pmatrix}.$$

It is easy to check that the matrix $\mathbf{Q}(z; N, k)$ so defined is an analytic function of z for $z \in \mathbb{C} \setminus X_N$ that satisfies the normalization condition

$$\mathbf{Q}(z; N, k)\begin{pmatrix} z^{\#\Delta - k} & 0 \\ 0 & z^{k - \#\Delta}\end{pmatrix} = \mathbb{I} + O\left(\frac{1}{z}\right) \quad \text{as } z \to \infty.$$

Furthermore, at each node $x_{N,n}$, the matrix $\mathbf{Q}(z; N, k)$ has a simple pole. If n belongs to the complementary set
$$\nabla := \mathbb{Z}_N \setminus \Delta,$$
then the first column is analytic at $x_{N,n}$ and the pole is in the second column such that the residue satisfies the condition

$$\operatorname*{Res}_{z=x_{N,n}} \mathbf{Q}(z;N,k) = \lim_{z\to x_{N,n}} \mathbf{Q}(z;N,k) \begin{pmatrix} 0 & w_{N,n} \prod_{m\in\Delta}(x_{N,n}-x_{N,m})^2 \\ 0 & 0 \end{pmatrix}, \tag{1.44}$$

for $n \in \nabla$. If $n \in \Delta$, then the second column is analytic at $x_{N,n}$ and the pole is in the first column such that the residue satisfies the condition

$$\operatorname*{Res}_{z=x_{N,n}} \mathbf{Q}(z;N,k) = \lim_{z\to x_{N,n}} \mathbf{Q}(z;N,k) \begin{pmatrix} 0 & 0 \\ \dfrac{1}{w_{N,n}} \prod_{\substack{m\in\Delta \\ m\neq n}}(x_{N,n}-x_{N,m})^{-2} & 0 \end{pmatrix}, \tag{1.45}$$

for $n \in \Delta$. Thus the triangularity of the residue matrices has been reversed for nodes in $\Delta \subset X_N$.

The relation between the solution $\mathbf{P}(z; N, k)$ of Interpolation Problem 1.2 and the matrix $\mathbf{Q}(z; N, k)$ obtained therefrom by selective reversal of residue triangularity gives rise in a special case to a remarkable duality between pairs of weights $\{w_{N,n}\}$ defined on the same set of nodes and their corresponding families of discrete orthogonal polynomials that comes up in applications. Given nodes X_N and weights $\{w_{N,n}\}$, the dual polynomials arise by taking $\Delta = \mathbb{Z}_N$ in the change of variables (1.43) and then defining

$$\overline{\mathbf{P}}(z;N,\overline{k}) := \sigma_1 \mathbf{Q}(z;N,k)\sigma_1, \qquad \text{where } \overline{k} := N-k.$$

Here σ_1 is another Pauli matrix:

$$\sigma_1 := \begin{pmatrix} 0 & 1 \\ 1 & 0 \end{pmatrix}.$$

Thus we are reversing the triangularity at all of the nodes and swapping rows and columns of the resulting matrix. It is easy to check that $\overline{\mathbf{P}}(z; N, \overline{k})$ satisfies

$$\overline{\mathbf{P}}(z;N,\overline{k}) \begin{pmatrix} z^{-\overline{k}} & 0 \\ 0 & z^{\overline{k}} \end{pmatrix} = \mathbb{I} + O\left(\frac{1}{z}\right) \qquad \text{as } z \to \infty$$

and is a matrix with simple poles in the second column at all nodes such that

$$\operatorname*{Res}_{z=x_{N,n}} \overline{\mathbf{P}}(z;N,\overline{k}) = \lim_{z\to x_{N,n}} \overline{\mathbf{P}}(z;N,\overline{k}) \begin{pmatrix} 0 & \overline{w}_{N,n} \\ 0 & 0 \end{pmatrix}$$

holds for $n \in \mathbb{Z}_N$, where the *dual weights* $\{\overline{w}_{N,n}\}$ are defined by the identity

$$w_{N,n}\overline{w}_{N,n} \prod_{\substack{m=0 \\ m\neq n}}^{N-1} (x_{N,n}-x_{N,m})^2 = 1. \tag{1.46}$$

Comparing with Interpolation Problem 1.2, we see that $\overline{P}_{11}(z; N, \overline{k})$ is the monic orthogonal polynomial $\overline{\pi}_{N,\overline{k}}(z)$ of degree $\overline{k}$ associated with the dual weights $\{\overline{w}_{N,j}\}$ (and the same set of nodes X_N). In this sense, families of discrete orthogonal polynomials always come in dual pairs. An explicit relation between the dual polynomials comes from the representation of $\mathbf{P}(z; N, k)$ given by Proposition 1.3:

$$\begin{aligned}
\overline{\pi}_{N,\overline{k}}(z) &= \overline{P}_{11}(z;N,\overline{k}) \\
&= P_{22}(z;N,k) \prod_{n=0}^{N-1}(z-x_{N,n}) \\
&= \sum_{m=0}^{N-1} w_{N,m}\gamma_{N,k-1}^2 \pi_{N,k-1}(x_{N,m}) \prod_{\substack{n=0 \\ n\neq m}}^{N-1}(z-x_{N,n}).
\end{aligned} \tag{1.47}$$

INTRODUCTION
19

Since the left-hand side is a monic polynomial of degree $\overline{k} = N - k$ and the right-hand side is apparently a polynomial of degree $N-1$, equation (1.47) furnishes k relations involving the weights and the normalization constants $\gamma_{N,k}$.

In particular, if we evaluate (1.47) for $z = x_{N,l}$ for some $l \in \mathbb{Z}_N$, then only one term from the sum on the right-hand side survives, and we find

$$\overline{\pi}_{N,\overline{k}}(x_{N,l}) = \gamma_{N,k-1}^2 w_{N,l} \prod_{\substack{n=0 \\ n \neq l}}^{N-1} (x_{N,l} - x_{N,n}) \cdot \pi_{N,k-1}(x_{N,l}), \qquad (1.48)$$

an identity relating values of each discrete orthogonal polynomial and a corresponding dual polynomial at any given node. The identity (1.48) has also been derived by Borodin [Bor02].

Furthermore, by using (1.47) twice, along with the fact that $\overline{\overline{\pi}}_k(z) \equiv \pi_k(z)$ (*i.e.*, duality is an involution), we can obtain some additional identities involving the discrete orthogonal polynomials and their duals. By involution, (1.47) implies that

$$\pi_{N,k}(z) = \overline{\gamma}_{N,\overline{k}-1}^2 \gamma_{N,k}^2 \sum_{m=0}^{N-1} \pi_{N,k}(x_{N,m}) \prod_{\substack{n=0 \\ n \neq m}}^{N-1} \frac{z - x_{N,n}}{x_{N,m} - x_{N,n}}.$$

The sum on the right-hand side is the Lagrange interpolating polynomial of degree $N - 1$ (at most) that agrees with $\pi_{N,k}(z)$ at all N nodes. Of course this identifies the sum with $\pi_{N,k}(z)$ itself, and we therefore deduce the relation

$$\overline{\gamma}_{N,\overline{k}-1} = \frac{1}{\gamma_{N,k}}$$

between the leading coefficients of the discrete orthogonal polynomials and their duals.

◁ **Remark:** We want to point out that the notion of duality described here is different from that explained in [NikSU91]. The latter generally involves relationships between families of discrete orthogonal polynomials with two different sets of nodes of orthogonalization. For example, the Hahn polynomials are orthogonal on a lattice of equally spaced points, and the polynomials dual to the Hahn polynomials by the scheme of [NikSU91] are orthogonal on a quadratic lattice for which $x_{N,n} - x_{N,n-1}$ is proportional to n. However, the polynomials dual to the Hahn polynomials under the scheme described above are the associated Hahn polynomials, which are orthogonal on the same equally spaced nodes as the Hahn polynomials themselves. The notion of duality described above coincides with that described in [Bor02] and is also equivalent to the "hole-particle transformation" considered by Johansson [Joh02]. ▷

1.4.4 Overview of the key steps

The characterization of the discrete orthogonal polynomials in terms of Interpolation Problem 1.2 is the starting point for our asymptotic analysis. Our rigorous analysis of $\mathbf{P}(z; N, k)$ consists of three steps:

1. We introduce a change of variables, transforming $\mathbf{P}(z; N, k)$ into $\mathbf{X}(z)$, another matrix function of z. The transformation mediating between $\mathbf{P}(z; N, k)$ and $\mathbf{X}(z)$ is explicit and exact. The matrix $\mathbf{X}(z)$ is shown to satisfy a matrix Riemann-Hilbert problem that is equivalent to Interpolation Problem 1.2.

2. We construct an explicit model $\hat{\mathbf{X}}(z)$ for $\mathbf{X}(z)$ on the basis of formal asymptotics. We call $\hat{\mathbf{X}}(z)$ a *global parametrix* for $\mathbf{X}(z)$.

3. We compare $\mathbf{X}(z)$ to the global parametrix $\hat{\mathbf{X}}(z)$ by considering the error $\mathbf{E}(z) := \mathbf{X}(z)\hat{\mathbf{X}}(z)^{-1}$, which should be close to the identity matrix if the formally obtained global parametrix $\hat{\mathbf{X}}(z)$ is indeed a good approximation of $\mathbf{X}(z)$. We rigorously analyze $\mathbf{E}(z)$ by viewing its definition in terms of $\mathbf{X}(z)$ as another change of variables since $\hat{\mathbf{X}}(z)$ is known explicitly from step 2. This means that we may pose an equivalent Riemann-Hilbert problem for $\mathbf{E}(z)$. We prove that this Riemann-Hilbert problem

may be solved by a convergent Neumann series if N is sufficiently large. The series for $\mathbf{E}(z)$ is also an asymptotic series whose first term is the identity matrix, such that $\mathbf{E}(z) - \mathbb{I}$ is of order $1/N$ in a suitable precise sense. This gives an asymptotic formula for the unknown matrix $\mathbf{X}(z) = \mathbf{E}(z)\hat{\mathbf{X}}(z)$. Inverting the explicit change of variables from step 1 linking $\mathbf{X}(z)$ with $\mathbf{P}(z; N, k)$, we finally arrive at an asymptotic formula for $\mathbf{P}(z; N, k)$.

The first step in this process is the most crucial since the explicit transformation from $\mathbf{P}(z; N, k)$ to $\mathbf{X}(z)$ has to result in a problem that has been properly prepared for asymptotic analysis. The transformation is best presented as a composition of several subsequent transformations:

1(a). A transformation (1.43) is introduced from $\mathbf{P}(z; N, k)$ to a new unknown matrix $\mathbf{Q}(z; N, k)$, having the effect of moving poles at some of the nodes in X_N from the second column of $\mathbf{P}(z; N, k)$ to the first column of $\mathbf{Q}(z; N, k)$. This transformation turns out to be necessary in our approach to take into account subintervals of $[a, b]$ that are saturated with zeros of $\pi_{N,k}(z)$ in the sense that there is a zero between each pair of neighboring nodes (recall Proposition 1.1). The saturated regions are not known in advance but are detected by the equilibrium measure (see step 1(c) below).

1(b). The matrix $\mathbf{Q}(z; N, k)$ is transformed into $\mathbf{R}(z)$, a matrix that has, instead of polar singularities, a jump discontinuity across a contour in the complex z-plane along which $\mathbf{R}(z)$ takes continuous boundary values. To see how a pole may be removed at the cost of a jump across a contour, consider a point x_0 at which a matrix function $\mathbf{M}(z)$ is meromorphic, having a simple pole in the second column such that, for some given constant w_0,

$$\operatorname*{Res}_{z=x_0} \mathbf{M}(z) = \lim_{z \to x_0} \mathbf{M}(z) \begin{pmatrix} 0 & w_0 \\ 0 & 0 \end{pmatrix}. \tag{1.49}$$

If $f(z)$ is a scalar function analytic in the region $0 < |z - x_0| < \epsilon$ for some $\epsilon > 0$ having a simple pole at x_0 with residue w_0 (obviously there are many such functions and consequently significant freedom in making a choice), then we may try to define a new matrix function $\mathbf{N}(z)$ by choosing some positive $\delta < \epsilon$ sufficiently small and setting

$$\mathbf{N}(z) = \begin{cases} \mathbf{M}(z), & \text{for } |z - x_0| > \delta, \\ \mathbf{M}(z) \begin{pmatrix} 1 & -f(z) \\ 0 & 1 \end{pmatrix}, & \text{for } |z - x_0| < \delta. \end{cases} \tag{1.50}$$

It follows that the singularity of $\mathbf{N}(z)$ at $z = x_0$ is removable. Therefore $\mathbf{N}(z)$ may be considered to be analytic in the region $|z - x_0| < \delta$ and also at each point of the region $|z - x_0| > \delta$ where additionally $\mathbf{M}(z)$ is known to be analytic. In place of the residue condition (1.49), we now have a known jump discontinuity across the circle $|z - x_0| = \delta$ along which $\mathbf{N}(z)$ takes continuous boundary values from the inside (denoted $\mathbf{N}_+(z)$) and the outside (denoted $\mathbf{N}_-(z)$):

$$\mathbf{N}_+(z) = \mathbf{N}_-(z) \begin{pmatrix} 1 & -f(z) \\ 0 & 1 \end{pmatrix}, \qquad \text{for } |z - x_0| = \delta. \tag{1.51}$$

Obviously, the disc $|z - x_0| < \delta$ can be replaced by another domain D containing x_0. This technique of removing poles was first introduced in [DeiKKZ96].

The problem at hand is more complicated because the number of poles grows in the limit of interest; in this limit the poles accumulate on a fixed set, and thus it is not feasible to surround each with its own circle of fixed size. In [KamMM03] a generalization of the technique described above was developed precisely to allow for the simultaneous removal of a large number of poles in a way that is asymptotically advantageous as the number of poles increases. This generalization employs a single function $f(z)$ with simple poles at $x_{N,n}$, for $n = 0, \ldots, N-1$, having corresponding residues $w_{N,n}$, and makes the change of variables (1.50) in a common domain D containing all of the points $x_{N,0}, \ldots, x_{N,N-1}$. The essential asymptotic analysis is then related to the nature of the jump condition that generalizes (1.51) for z on the boundary of D. This jump condition can have different asymptotic properties in the limit $N \to \infty$ according to the placement of the boundary of D in the complex plane.

Further difficulties arise here because it turns out that the correct location for the boundary of D needed to facilitate the asymptotic analysis in the limit $N \to \infty$ coincides in part with the interval $[a, b]$ that contains the poles, and in the context of the method in [KamMM03] this leads to singularities both in the boundary values of the matrix unknown and also in the jump matrix relating the boundary values. These singularites are an obstruction to further analysis. Therefore the transformation we will introduce from $\mathbf{Q}(z; N, k)$ to $\mathbf{R}(z)$ uses a further variation of the pole removal technique developed in [Mil02] in which two different residue-interpolating functions $f_1(z)$ and $f_2(z)$ are used in respective disjoint domains D_1 and D_2 such that all of the poles $x_{N,n}$ are common boundary points of both domains. This version of the pole removal technique ultimately enables subsequent detailed analysis in the neighborhood of the interval $[a, b]$ in which $\mathbf{Q}(z; N, k)$ has poles.

1(c). $\mathbf{R}(z)$ is transformed into $\mathbf{S}(z)$ by a change of variables that is written explicitly in terms of the equilibrium measure. The equilibrium measure is the solution of a variational problem of logarithmic potential theory that is posed in terms of the functions $\rho^0(x)$ and $V(x)$ given on the interval $[a, b]$ and the constant $c \in (0, 1)$. The fundamental properties of the equilibrium measure are well known in general, and for particular cases of $\rho^0(x)$, $V(x)$, and c, it is not difficult to calculate the equilibrium measure explicitly. The purpose of introducing the equilibrium measure is that the variational problem it satisfies entails some constraints that impose strict inequalities on variational derivatives. These variational derivatives ultimately appear in the problem with a factor of N in certain exponents, and the inequalities lead to desirable exponential decay as $N \to \infty$.

The technique of preparing a matrix Riemann-Hilbert problem for subsequent asymptotic analysis with the introduction of an appropriate equilibrium measure first appeared in the paper [DeiVZ97] and was subsequently applied to the computation of asymptotics for orthogonal polynomials with continuous weights in [DeiKMVZ99a, DeiKMVZ99b]. The key quantity in all of these papers is the complex logarithmic potential of the equilibrium measure, the g-function. In order to apply these methods in the discrete weights context, we need to modify the relationship between the g-function and the equilibrium measure (see (4.5) and (4.7) below) to reflect the local reversal of triangularity described in 1(a) above. This amounts to a further generalization of the technique introduced in [DeiVZ97].

1(d). The final transformation explicitly relates $\mathbf{S}(z)$ to a matrix $\mathbf{X}(z)$. The matrix $\mathbf{S}(z)$ is apparently difficult to analyze in the neighborhood of subintervals of $[a, b]$ where constraints in the variational problem are not active and consequently exponential decay is not obvious. A model for this kind of situation is a matrix $\mathbf{M}(z)$ that takes continuous boundary values on an interval I of the real axis from above (denoted $\mathbf{M}_+(z)$) and below (denoted $\mathbf{M}_-(z)$) that satisfy a jump relation of the form

$$\mathbf{M}_+(z) = \mathbf{M}_-(z) \begin{pmatrix} e^{iN\theta(z)} & 1 \\ 0 & e^{-iN\theta(z)} \end{pmatrix},$$

where $\theta(z)$ is a real-analytic function that is strictly increasing for $z \in I$. This is therefore a rapidly oscillatory jump relation that has no obvious limit as $N \to \infty$. However, noting the algebraic factorization

$$\begin{pmatrix} e^{iN\theta(z)} & 1 \\ 0 & e^{-iN\theta(z)} \end{pmatrix} = \begin{pmatrix} 1 & 0 \\ e^{-iN\theta(z)} & 1 \end{pmatrix} \begin{pmatrix} 0 & 1 \\ -1 & 0 \end{pmatrix} \begin{pmatrix} 1 & 0 \\ e^{iN\theta(z)} & 1 \end{pmatrix}$$

and using the analyticity of $\theta(z)$, we may choose some sufficiently small $\epsilon > 0$ and define a new unknown by setting

$$\mathbf{N}(z) := \begin{cases} \mathbf{M}(z) \begin{pmatrix} 1 & 0 \\ -e^{iN\theta(z)} & 1 \end{pmatrix}, & \text{for } \Re(z) \in I \text{ and } 0 < \Im(z) < \epsilon, \\ \mathbf{M}(z) \begin{pmatrix} 1 & 0 \\ e^{-iN\theta(z)} & 1 \end{pmatrix}, & \text{for } \Re(z) \in I \text{ and } -\epsilon < \Im(z) < 0, \\ \mathbf{M}(z), & \text{otherwise}. \end{cases}$$

The matrix $\mathbf{N}(z)$ has jump discontinuities along the three parallel contours I, $I+i\epsilon$, and $I-i\epsilon$. If on any of these we indicate the boundary value taken by $\mathbf{N}(z)$ from above as $\mathbf{N}_+(z)$ and from below as $\mathbf{N}_-(z)$, then the oscillatory jump condition for $\mathbf{M}(z)$ in I is replaced by the three different formulae:

$$\mathbf{N}_+(z) = \mathbf{N}_-(z) \begin{pmatrix} 1 & 0 \\ e^{iN\theta(z)} & 1 \end{pmatrix}, \qquad z \in I + i\epsilon, \tag{1.52}$$

$$\mathbf{N}_+(z) = \mathbf{N}_-(z) \begin{pmatrix} 1 & 0 \\ e^{-iN\theta(z)} & 1 \end{pmatrix}, \qquad z \in I - i\epsilon, \tag{1.53}$$

$$\mathbf{N}_+(z) = \mathbf{N}_-(z) \begin{pmatrix} 0 & 1 \\ -1 & 0 \end{pmatrix}, \qquad z \in I. \tag{1.54}$$

The Cauchy-Riemann equations satisfied by the function $\theta(z)$ in I imply that $\Re(i\theta(z))$ is negative for $\Im(z) = \epsilon$ and positive for $\Im(z) = -\epsilon$. Thus the jump conditions (1.52)–(1.54) all have obvious asymptotics as $N \to \infty$.

The replacement of an oscillatory jump matrix by an exponentially decaying one on the basis of algebraic factorization is the essence of the steepest-descent method for Riemann-Hilbert problems first proposed in [DeiZ93]. Our transformation from $\mathbf{S}(z)$ to $\mathbf{X}(z)$ will be based on this key idea but will involve more complicated factorizations of both upper- and lower-triangular matrices.

These three steps of our analysis of the matrix $\mathbf{P}(z; N, k)$ solving Interpolation Problem 1.2 will be carried out in Chapters 4 and 5.

1.5 OUTLINE OF THE REST OF THE BOOK

Our main results are presented in Chapter 2 (for the discrete orthogonal polynomials themselves) and Chapter 3 (for corresponding applications). The subsequent chapters concern the proof of the main results: Chapters 4, 5, and 6 contain the proof of results stated in Chapter 2, and the results stated in Chapter 3 are proven in Chapter 7.

The detailed asymptotic behavior in the limit $N \to \infty$ of the discrete orthogonal polynomials in overlapping sets that cover the entire complex plane will be discussed in Chapter 2. After some important definitions and notation are established in §2.1 and §2.2, the results themselves will be given in in §2.3. In §2.4 we show how the general theory applies in some classical cases, specifically the Krawtchouk polynomials and two types of polynomials in the Hahn family. The equilibrium measures for the Hahn polynomials are also described in Theorem 2.17.

Further results of our analysis in the context of statistical ensembles associated with families of discrete orthogonal polynomials are discussed in Chapter 3. First, we introduce the notion of a discrete orthogonal polynomial ensemble in §3.1 and describe the ensembles associated with dual polynomials in §3.2. In §3.4, we discuss rhombus tilings of a hexagon as a specific application of discrete orthogonal polynomial ensembles and their duals. Our general results on the universality of various statistics in the limit $N \to \infty$ are explained in §3.3. The specific results implied by the general ones in the context of the hexagon tiling problem are described in §3.4.2. We also obtain new results on the continuum limit of the Toda lattice, which we explain in §3.5.

As mentioned above, Chapters 4 and 5 contain the complete asymptotic analysis of the matrix $\mathbf{P}(z; N, k)$ in the limit $N \to \infty$. This analysis is then used in Chapter 6 to establish the results presented in §2.3 and used again in Chapter 7 to establish the results presented in §3.3.

In Chapter 4 we describe all details of a sequence of algebraic transformations of the interpolation problem for the discrete orthogonal polynomials to arrive at a simpler Riemann-Hilbert problem to which a formal asymptotic analysis can be applied. For this purpose, we exploit a transformation from a Riemann-Hilbert problem with pole conditions (see Interpolation Problem 1.2) to a Riemann-Hilbert problem on a contour, a doubly constrained equilibrium measure, and hole-particle duality. This chapter is the technical core of the analysis of Riemann-Hilbert problems with pole conditions. Chapter 5 concerns the construction of a global parametrix and rigorous error estimates. By combining the calculations in Chapters 4 and 5, we prove the theorems stated in §2.3 in Chapter 6. Using the asymptotic analysis of the Riemann-Hilbert problem discussed in Chapters 4 and 5, we prove in Chapter 7 the theorems stated in §3.3.

Appendix A summarizes construction of the solution of a limiting Riemann-Hilbert problem by means of hyperelliptic function theory. Appendix B gives a proof of the determination of the equilibrium measure of the Hahn weight presented in §2.4. Finally, Appendix C contains a list of some important symbols used frequently throughout the book.

For the asymptotic results given here that correspond to theorems already stated in our announcement [BaiKMM03], we generally obtain significantly sharper error estimates. Since we published that paper, we have learned how to circumvent certain technical difficulties related to the continuum limit of the discrete orthogonality measures and the possibility of transition points where triangularity of residue matrices changes abruptly. In our opinion, these technical innovations do more than make the error estimates sharper; they also make the proofs more elegant.

1.6 RESEARCH BACKGROUND

The work described in this mongraph is connected with three different themes of current research.

First, in the context of approximation theory, there has been recent activity [DeiKMVZ99a, DeiKMVZ99b] in the study of polynomials orthogonal on the real axis with respect to *general* continuous varying weights and the corresponding large-degree pointwise asymptotics. The significance of the work [DeiKMVZ99a, DeiKMVZ99b] is that the method is not at all particular to any special classical formulae for weights; they are completely general. Thus a natural question is whether it is possible to further generalize the method in [DeiKMVZ99a, DeiKMVZ99b] to handle the discrete weights. However, it has turned out that discrete weights are of such a fundamentally different character than their continuous counterparts that this would require the development of new tools for asymptotic analysis. The setting for the work [DeiKMVZ99a, DeiKMVZ99b] is the characterization of the orthogonal polynomials in terms of the solution of a certain matrix-valued Riemann-Hilbert problem [FokIK91] with jump conditions on contours. For discrete weights, the corresponding Riemann-Hilbert problem is defined by constraints on residues of poles. Under the conditions that we consider in this work, each point mass added to the weight amounts to a pole in the matrix solution of the Riemann-Hilbert problem, so analyzing the asymptotics of an accumulation of poles becomes the main difficulty.

Second, there has been some recent progress [KamMM03, Mil02] in the integrable systems literature concerning the problem of computing asymptotics for solutions of integrable nonlinear partial differential equations (*e.g.*, the nonlinear Schrödinger equation) in the limit where the spectral data associated with the solution via the inverse scattering transform is made up of a large number of discrete eigenvalues. Significantly, inverse scattering theory also exploits much of the theory of matrix Riemann-Hilbert problems, and it turns out that the discrete eigenvalues appear as poles in the corresponding matrix-valued unknown. So, the methods recently developed in the context of inverse scattering actually suggest a general scheme by means of which an accumulation of poles in the matrix unknown can be analyzed.

Finally, a number of problems in probability theory have recently been identified that are in some sense solved in terms of discrete orthogonal polynomials, and certain statistical questions can be translated into corresponding questions about the asymptotic behavior of the polynomials. The particular problems we have in mind are related to statistics of random tilings of various shapes, to last-passage percolation models, and also to certain natural measures on sets of partitions. The joint probability distributions in these problems are examples of discrete orthogonal polynomial ensembles [Joh00, Joh01]. Roughly speaking, the analogy is that the relationship between universal asymptotic properties of discrete orthogonal polynomials and universal statistics for discrete orthogonal polynomial ensembles is the same as the relationship between universal asymptotic properties of polynomials orthogonal with respect to continuous weights and universal eigenvalue statistics of certain random matrix ensembles. The techniques required for computing asymptotics of discrete orthogonal polynomials with general weights have become available at just the time when questions that can be answered with these tools are appearing in the applied literature.

Chapter Two

Asymptotics of General Discrete Orthogonal Polynomials in the Complex Plane

In this chapter we state our results for the asymptotics of the discrete orthogonal polynomials subject to the conditions described in the introduction, in the limit $N \to \infty$. The asymptotic formulae we will present characterize the polynomials in terms of an equilibrium measure (described below in §2.1) and also the function theory of a hyperelliptic Riemann surface associated with the equilibrium measure (described below in §2.2). The results themselves will follow in §2.3.

2.1 THE EQUILIBRIUM ENERGY PROBLEM

2.1.1 The equilibrium measure

It has been recognized for some time (see [Rak96, DraS97], as well as the review article [KuiR98]), that the asymptotic behavior of discrete orthogonal polynomials in the limit $N \to \infty$ with $k/N \to c \in (0,1)$, and in particular the distribution of zeros in (a,b), is related to a constrained equilibrium problem for logarithmic potentials in an *external field* $\varphi(x)$ given by the formula

$$\varphi(x) := V(x) + \int_a^b \log|x-y|\rho^0(y)\,dy\,, \tag{2.1}$$

for $x \in (a,b)$. Under our assumptions about the weights, we can also view $\varphi(x)$ as being defined via a continuum limit:

$$\varphi(x) := -\lim_{N \to \infty} \frac{\log(w_{N,n})}{N}\,, \tag{2.2}$$

where $w_{N,n}$ is expressed in terms of $x_{N,n}$, which in turn is identified with x. Eliminating $\varphi(x)$ between (2.1) and (2.2) results in a more general version of the limiting statement (1.19). The external field $\varphi(x)$ we need here is analogous to the continuum limit of that usually encountered in the logarithmic potential theory of orthogonal polynomials [Sze91].

Here the field $\varphi(x)$ is a real-analytic function in the open interval (a,b) because $V(x)$ and $\rho^0(x)$ are (by assumption) real-analytic functions in a neighborhood of $[a,b]$. Unlike $V(x)$ and $\rho^0(x)$, however, the field $\varphi(x)$ does not extend analytically beyond the endpoints of (a,b) because of the condition (1.14).

Given $c \in (0,1)$ and $\varphi(x)$ as above, consider the quadratic functional

$$E_c[\mu] := c\int_a^b \int_a^b \log\frac{1}{|x-y|}\,d\mu(x)\,d\mu(y) + \int_a^b \varphi(x)\,d\mu(x) \tag{2.3}$$

of Borel measures μ on $[a,b]$. The subscript denotes the dependence of the energy functional on the parameter c. Let $\mu_{\min}^c$ be the measure that minimizes $E_c[\mu]$ over the class of measures satisfying the upper and lower constraints

$$0 \le \int_{x \in \mathcal{B}} d\mu(x) \le \frac{1}{c}\int_{x \in \mathcal{B}} \rho^0(x)\,dx\,, \tag{2.4}$$

for all Borel sets $\mathcal{B} \subset [a,b]$, and the normalization condition

$$\int_a^b d\mu(x) = 1\,. \tag{2.5}$$

The superscript on the minimizer indicates the value of the parameter c for which the energy functional (2.3) is minimized. The existence of a unique minimizer under the conditions enumerated in §1.3.1 follows from the Gauss-Frostman Theorem; see [SafT97] and [DraS97] for details. We will refer to the minimizer as the *equilibrium measure*.

That a variational problem plays a central role in asymptotic behavior is a familiar theme in the theory of orthogonal polynomials. The key new feature contributed by discreteness is the appearance of the upper constraint on the equilibrium measure (*i.e.*, the upper bound in (2.4)). Since the equilibrium measure gives the distribution of zeros of $\pi_{N,k}(z)$ in $[a,b]$, the upper constraint can be traced to the exclusion principle for zeros described in Proposition 1.1.

The theory of the "doubly constrained" variational problem we are considering is well established. In particular, the analytic properties we assume for $V(x)$ and $\rho^0(x)$ turn out to be unnecessary for the mere existence of the equilibrium measure. However, the analyticity of $V(x)$ and $\rho^0(x)$ provides additional regularity that we wish to exploit. In particular, we have the following result from the paper [Kui00].

Proposition 2.1 (Kuijlaars). *Let $V(x)$ and $\rho^0(x)$ be functions analytic in a complex neighborhood of $[a,b]$ with $\rho^0(x) > 0$ in $[a,b]$. Then the equilibrium measure $\mu^c_{\min}$ is continuously differentiable with respect to $x \in (a,b)$. Moreover, the derivative $d\mu^c_{\min}/dx$ is piecewise-analytic, with a finite number of points of nonanalyticity that may not occur at any x where both (strict) inequalities $d\mu^c_{\min}/dx(x) > 0$ and $d\mu^c_{\min}/dx(x) < \rho^0(x)/c$ hold.*

At a formal level, finding μ minimizing $E_c[\mu]$ subject to the normalization constraint (2.5) may be viewed as seeking a critical point of the modified functional

$$F_c[\mu] := E_c[\mu] - \ell_c \int_a^b d\mu(x),$$

where ℓ_c is a Lagrange multiplier. When $\mu = \mu^c_{\min}$ and ℓ_c is an appropriate associated real constant, variations of $F_c[\mu]$ vanish in subsets of $[a,b]$ where neither the upper nor the lower constraints are active. The Lagrange multiplier ℓ_c (the subscript indicates the dependence on the parameter c) is known in logarithmic potential theory as the *Robin constant*.

2.1.2 Simplifying assumptions on the equilibrium measure

For simplicity of exposition we want to exclude certain nongeneric phenomena that may occur even under conditions of analyticity of $V(x)$ and $\rho^0(x)$. Let $\underline{\mathcal{F}} \subset [a,b]$ denote the closed set of x-values where $d\mu^c_{\min}/dx(x) = 0$, and let $\overline{\mathcal{F}} \subset [a,b]$ denote the closed set of x-values where $d\mu^c_{\min}/dx(x) = \rho^0(x)/c$. We will make the following assumptions:

1. Each connected component of $\underline{\mathcal{F}}$ and $\overline{\mathcal{F}}$ has a nonempty interior. Therefore $\underline{\mathcal{F}}$ and $\overline{\mathcal{F}}$ are both finite unions of closed intervals with each interval containing more than one point. Note that this does not exclude the possibility of either $\underline{\mathcal{F}}$ or $\overline{\mathcal{F}}$ being empty.

2. For each open subinterval U of $(a,b) \setminus (\underline{\mathcal{F}} \cup \overline{\mathcal{F}})$ and each limit point $z_0 \in \underline{\mathcal{F}}$ of U, we have

$$\lim_{x \to z_0, x \in U} \frac{1}{\sqrt{|x-z_0|}} \frac{d\mu^c_{\min}}{dx}(x) = K, \qquad \text{with } 0 < K < \infty, \tag{2.6}$$

and for each limit point $z_0 \in \overline{\mathcal{F}}$ of U, we have

$$\lim_{x \to z_0, x \in U} \frac{1}{\sqrt{|x-z_0|}} \left[\frac{1}{c}\rho^0(x) - \frac{d\mu^c_{\min}}{dx}(x)\right] = K, \qquad \text{with } 0 < K < \infty. \tag{2.7}$$

Therefore the density of the equilibrium measure meets each constraint exactly like a square root.

3. A constraint is active at each endpoint: $\{a,b\} \subset \underline{\mathcal{F}} \cup \overline{\mathcal{F}}$.

It is difficult to translate these conditions on $\mu_{\min}^c$ into sufficient conditions on c, $V(x)$, and $\rho^0(x)$. However, there is a sense in which the conditions above are satisfied generically. By genericity, we mean that given $V(x)$ and $\rho^0(x)$, the set of values of c for which the conditions fail is discrete. For the analogous problem in the continuous-weights case, the conditions 1 and 2 above are proved to be generic in [KuiM00] and we expect the same of the discrete weights. For further arguments supporting the claim of the generic nature of these two conditions, see the discussion at the beginning of §5.1.2. However, the condition 3 above has no counterpart in continuous-weights cases. Nevertheless, for the classical discrete weights of the Krawtchouk and Hahn classes this condition indeed holds for all but a finite number of $c \in (0,1)$.

◁ **Remark:** Relaxing the condition that a constraint should be active at each endpoint requires specific local analysis near these two points. We expect that a constraint being active at each endpoint is a generic phenomenon in the sense that the opposite situation occurs only for isolated values of c. We know this statement to be true in all relevant classical cases. For the Krawtchouk polynomials only the values $c = p$ and $c = q = 1 - p$ correspond to an equilibrium measure that is not constrained at both endpoints (see [DraS00] and §2.4.1). The situation is similar for the Hahn polynomials, where only the values $c = c_A$ and $c = c_B$ defined by (2.73) are exceptional (see §2.4.2). While the values of c for which no constraint is active at an endpoint of (a,b) are exceptional, the behavior of the equilibrium measure near that endpoint at the exceptional values of c is again, in a sense, generic. In particular, we have the following result.

Proposition 2.2. *Suppose that $\rho^0(x)$ and $V(x)$ are analytic functions for $x \in (a,b)$ having, along with $V'(x)$, continuous extensions to $[a,b]$. Suppose also that one of the endpoints a or b is not contained in $\underline{\mathcal{F}} \cup \overline{\mathcal{F}}$. Then if $\rho^0(x)$ and $d\mu_{\min}^c/dx$ are Hölder-continuous at this endpoint with exponent $\nu > 0$, as $x \in (a,b)$ tends toward the endpoint,*

$$\rho^0(x) - 2c\frac{d\mu_{\min}^c}{dx}(x) \to 0.$$

Thus if neither constraint is active at an endpoint of (a,b), then at that endpoint the equilibrium measure takes on the average value of the upper and lower constraints.

Proof. Since for some neighborhood U of the endpoint no constraint is active in $(a,b) \cap U$, an Euler-Lagrange variational derivative of $E_c[\mu]$ satisfies an equilibrium condition (see (2.14)) at each point of $(a,b) \cap U$. Differentiating this condition with respect to x yields

$$\text{P. V.} \int_a^b \frac{f(y)\,dy}{x-y} + V'(x) \equiv 0, \quad \text{where } f(x) := \rho^0(x) - 2c\frac{d\mu_{\min}^c}{dx}(x),$$

which holds for all x in the interior of the subinterval. Removing the singularity from the principal value integral gives the identity

$$f(x)\log\left(\frac{x-a}{b-x}\right) + \int_a^b \frac{f(y)-f(x)}{x-y}\,dy + V'(x) \equiv 0, \tag{2.8}$$

where the integral is nonsingular by virtue of the Hölder condition and is uniformly bounded. If $K > 0$ is the Hölder constant for f, then we have

$$\left|\int_a^b \frac{f(y)-f(x)}{x-y}\,dy\right| \leq \int_a^b \left|\frac{f(y)-f(x)}{x-y}\right|\,dy$$

$$\leq K \int_a^b |x-y|^{\nu-1}\,dy$$

$$= \frac{K}{\nu}\left[(x-a)^\nu + (b-x)^\nu\right]$$

$$\leq \frac{K(b-a)^\nu}{2^{\nu-1}\nu}.$$

When we let x tend toward the endpoint of interest in (2.8), all terms but the first one on the left-hand side remain bounded. Hence it is necessary that the first term involving the logarithm be bounded as well; by the Hölder condition satisfied by $f(x)$ at the endpoint, it follows that $f(x)$ tends to zero as x tends toward the endpoint. □

This fact provides the key to modifications of the analysis we will present in Chapter 5 that are necessary to handle the exceptional values of $c \in (0,1)$. These modifications will be described elsewhere. ▷

2.1.3 Voids, bands, and saturated regions

Under the conditions enumerated in §1.3.1 and §2.1.2, the equilibrium measure $\mu^c_{\min}$ partitions (a,b) into three kinds of subintervals, with a finite number of each and each having a nonempty interior. The three types are defined as follows.

Definition 2.3 (Voids). *A void Γ is an open subinterval of $[a,b]$ of maximal length in which $\mu^c_{\min}(x) \equiv 0$, and thus the equilibrium measure realizes the lower constraint.*

Definition 2.4 (Bands). *A band I is an open subinterval of $[a,b]$ of maximal length where $\mu^c_{\min}(x)$ is a measure with a real-analytic density satisfying $0 < d\mu^c_{\min}/dx < \rho^0(x)/c$.*

Definition 2.5 (Saturated regions). *A saturated region Γ is an open subinterval of $[a,b]$ of maximal length in which $d\mu^c_{\min}/dx \equiv \rho^0(x)/c$, and thus the equilibrium measure realizes the upper constraint.*

Voids and saturated regions will also be called *gaps* when it is not necessary to distinguish between these two types of intervals. The closure of the union of all subintervals of the three types defined above is the interval $[a,b]$. From condition 1 in §2.1.2 above, bands cannot be adjacent to each other, and from condition 3 in §2.1.2, a band may not be adjacent to an endpoint of $[a,b]$. Thus subject to our assumptions, a band always has on each side either a void or a saturated region, and the equilibrium measure thus determines a set of numbers in (a,b),

$$a < \alpha_0 < \beta_0 < \alpha_1 < \beta_1 < \cdots < \alpha_G < \beta_G < b,$$

that are the endpoints of the bands. Thus the bands are open intervals of the form

$$I_j := (\alpha_j, \beta_j), \qquad \text{for } j = 0, \ldots, G.$$

The corresponding gaps are the intervals (a, α_0), (β_G, b), which we refer to as the *exterior gaps*, and

$$\Gamma_j := (\beta_{j-1}, \alpha_j), \qquad \text{for } j = 1, \ldots, G,$$

which we refer to as the *interior gaps*.

2.1.4 Quantities derived from the equilibrium measure

The variational derivative of $E_c[\mu]$ evaluated on the equilibrium measure $\mu = \mu^c_{\min}$ is given by

$$\frac{\delta E_c}{\delta \mu}(x) := -2c \int_a^b \log|x - y| \, d\mu^c_{\min}(y) + \varphi(x), \qquad (2.9)$$

and we may define an analytic logarithmic potential of the equilibrium measure by the formula

$$L_c(z) := c \int_a^b \log(z - x) \, d\mu^c_{\min}(x), \qquad \text{for } z \in \mathbb{C} \setminus (-\infty, b]. \qquad (2.10)$$

From any gap Γ we may introduce a function $\overline{L}_c^\Gamma(z)$ analytic in z for $\Re(z) \in \Gamma$ and $|\Im(z)|$ sufficiently small that satisfies

$$\overline{L}_c^\Gamma(z) = c \int_a^b \log|z - x| \, d\mu^c_{\min}(x), \qquad \text{for } z \in \Gamma. \qquad (2.11)$$

And from any band I we may introduce a function $\overline{L}_c^I(z)$ analytic in z for $\Re(z) \in I$ and $|\Im(z)|$ sufficiently small that satisfies

$$\overline{L}_c^I(z) = c \int_a^b \log|z - x| \, d\mu^c_{\min}(x), \qquad \text{for } z \in I. \qquad (2.12)$$

Recall the Lagrange multiplier ℓ_c. If Γ is a void, then admissible variations of $\mu_{\min}^c$ are positive, and a simple variational calculation shows that for $x \in \Gamma$ we have the strict inequality

$$\frac{\delta E_c}{\delta \mu}(x) > \ell_c. \tag{2.13}$$

Thus for each void Γ we may introduce a positive function having an analytic extension from the interior:

$$\xi_\Gamma(x) := \frac{\delta E_c}{\delta \mu}(x) - \ell_c.$$

In a band I, variations of the equilibrium measure in I are free (*i.e.*, may be of either sign). Thus for $x \in I$, we have the equilibrium condition

$$\frac{\delta E_c}{\delta \mu}(x) \equiv \ell_c. \tag{2.14}$$

For each band I, we may introduce two positive functions having analytic extensions from the interior:

$$\psi_I(x) := \frac{d\mu_{\min}^c}{dx}(x) \quad \text{and} \quad \overline{\psi}_I(x) := \frac{1}{c}\rho^0(x) - \frac{d\mu_{\min}^c}{dx}(x). \tag{2.15}$$

If Γ is a saturated region, then variations of the equilibrium measure in Γ are strictly negative, and for $x \in \Gamma$, we have the strict variational inequality

$$\frac{\delta E_c}{\delta \mu}(x) < \ell_c. \tag{2.16}$$

It follows that for each saturated region Γ we may introduce a positive function having an analytic extension from the interior:

$$\xi_\Gamma(x) := \ell_c - \frac{\delta E_c}{\delta \mu}(x).$$

In addition to the functions $\overline{L}_c^\Gamma(x)$ and $\xi_\Gamma(x)$ that extend analytically from each gap Γ, and the functions $\overline{L}_c^I(x)$, $\psi_I(x)$, and $\overline{\psi}_I(x)$ that extend analytically from each band I, we may define for each band endpoint a function analytic in a neighborhood of this point. If $z = \alpha$ is a left band edge separating a void Γ (for real $z < \alpha$) from a band I (for real $z > \alpha$), then according to the generic assumption (2.6) in §2.1.2, the function defined by

$$\tau_\Gamma^{\nabla,L}(z) := \left(2\pi N c \int_\alpha^z \psi_I(x)\,dx\right)^{2/3}, \quad \tau_\Gamma^{\nabla,L}(z) > 0 \text{ for } z > \alpha, \tag{2.17}$$

extends to a neighborhood of $z = \alpha$ as an invertible conformal mapping. If $z = \beta$ is a right band edge separating a void Γ (for real $z > \beta$) from a band I (for real $z < \beta$), then (2.6) implies that the function defined by

$$\tau_\Gamma^{\nabla,R}(z) := \left(-2\pi N c \int_\beta^z \psi_I(x)\,dx\right)^{2/3}, \quad \tau_\Gamma^{\nabla,R}(z) > 0 \text{ for } z < \beta, \tag{2.18}$$

extends to a neighborhood of $z = \beta$ as an invertible conformal mapping. If $z = \alpha$ is a left band edge separating a saturated region Γ (for real $z < \alpha$) from a band I (for real $z > \alpha$), then according to the generic assumption (2.7) in §2.1.2, the function defined by

$$\tau_\Gamma^{\Delta,L}(z) := \left(2\pi N c \int_\alpha^z \overline{\psi}_I(x)\,dx\right)^{2/3}, \quad \tau_\Gamma^{\Delta,L}(z) > 0 \text{ for } z > \alpha, \tag{2.19}$$

extends to a neighborhood of $z = \alpha$ as an invertible conformal mapping. If $z = \beta$ is a right band edge separating a saturated region Γ (for real $z > \beta$) from a band I (for real $z < \beta$), then (2.7) implies that the function defined by

$$\tau_\Gamma^{\Delta,R}(z) := \left(-2\pi N c \int_\beta^z \overline{\psi}_I(x)\,dx\right)^{2/3}, \quad \tau_\Gamma^{\Delta,R}(z) > 0 \text{ for } z < \beta, \tag{2.20}$$

extends to a neighborhood of $z = \beta$ as an invertible conformal mapping.

For later use, it will be useful to define a real constant θ_Γ corresponding to each gap. For each void $\Gamma = \Gamma_j$ surrounded by bands I_{j-1} and I_j, we define a constant by

$$\theta_{\Gamma_j} := -2\pi c \int_{\alpha_j}^{b} \frac{d\mu_{\min}^c}{dx}(x)\, dx\,. \tag{2.21}$$

Similarly, for each saturated region $\Gamma = \Gamma_j$ surrounded by bands I_{j-1} and I_j (necessarily an interior gap), we define a constant by

$$\theta_{\Gamma_j} := 2\pi c \int_{\alpha_j}^{b} \left[\frac{\rho^0(x)}{c} - \frac{d\mu_{\min}^c}{dx}(x)\right] dx\,. \tag{2.22}$$

There is no difficulty using the same symbol θ_Γ on the left-hand side of these two definitions because a given gap $\Gamma = \Gamma_j$ is either a void or a saturated region but cannot be both. For the gaps $\Gamma = (a, \alpha_0)$ and $\Gamma = (\beta_G, b)$, which exist according to the genericity assumptions stated in §2.1.2, it will also be useful to define associated constants θ_Γ. Whether each of these gaps is a void or a saturated region, we define

$$\theta_{(a,\alpha_0)} := -2\pi c \quad \text{and} \quad \theta_{(\beta_G,b)} := 0\,. \tag{2.23}$$

2.1.5 Dual equilibrium measures

According to (1.46), if $V_N(x)$ is associated with the weights $\{w_{N,n}\}$ and if $\overline{V}_N(x)$ is associated with the dual weights $\{\overline{w}_{N,n}\}$, then we have the simple identity $\overline{V}_N(x) = -V_N(x)$. This leads to the useful fact that knowing the equilibrium measure for one family of discrete orthogonal polynomials is equivalent to knowing the equilibrium measure for the dual discrete orthogonal polynomials. We have the following specific result.

Proposition 2.6. *Let $E_c[\mu; V, \rho^0]$ be the energy functional (2.3) with external field $\varphi(x)$ given by (2.1) in terms of analytic functions $V(x)$ and $\rho^0(x)$, and let $\ell_c[V, \rho^0]$ denote the corresponding Lagrange multiplier. Let $P(c, V, \rho^0)$ denote the problem of finding the measure μ on (a,b) minimizing $E_c[\mu; V, \rho^0]$, subject to conditions (2.4) and (2.5). If for all $c \in (0,1)$, $\mu_{\min}^c$ is the solution of the problem $P(c, V, \rho^0)$, then the measure $\bar{\mu}_{\min}^{1-c}$ with density*

$$\frac{d\bar{\mu}_{\min}^{1-c}}{dx}(x) := \frac{1}{1-c}\left(\rho^0(x) - c\frac{d\mu_{\min}^c}{dx}(x)\right) \tag{2.24}$$

is the solution of the problem $P(1-c, -V, \rho^0)$. Also,

$$\left.\frac{\delta E_{1-c}[\bar{\mu}; -V, \rho^0]}{\delta\bar{\mu}}\right|_{\bar{\mu}_{\min}^{1-c}} = -\left.\frac{\delta E_c[\mu; V, \rho^0]}{\delta\mu}\right|_{\mu_{\min}^c} \quad \text{and} \quad \ell_{1-c}[-V, \rho^0] = -\ell_c[V, \rho^0]\,. \tag{2.25}$$

Proof. Clearly, the measure with density given on (a,b) by (2.24) satisfies both conditions (2.4) and (2.5). A direct calculation then shows that $E_{1-c}[\bar{\mu}; -V, \rho^0]$, when considered as a functional of μ by the relation

$$\frac{d\bar{\mu}}{dx}(x) = \frac{1}{1-c}\left(\rho^0(x) - c\frac{d\mu}{dx}(x)\right)\,,$$

is linearly related to the functional $E_c[\mu; V, \rho^0]$:

$$E_{1-c}[\bar{\mu}; -V, \rho^0] = \frac{c}{1-c}E_c[\mu; V, \rho^0] - \frac{1}{1-c}\int_a^b V(x)\rho^0(x)\, dx\,.$$

Since c and $1-c$ are both positive, we obtain (2.24). The proof of (2.25) is similar. □

2.2 ELEMENTS OF HYPERELLIPTIC FUNCTION THEORY

Let the analytic function $R(z)$ be defined for $z \in \mathbb{C} \setminus \cup_j I_j$ to satisfy

$$R(z)^2 = \prod_{j=0}^{G}(z-\alpha_j)(z-\beta_j) \quad \text{and} \quad R(z) \sim z^{G+1} \text{ as } z \to \infty, \tag{2.26}$$

and for z in the same domain, define

$$h'(z) := \frac{1}{2\pi i R(z)} \int_{\cup_j I_j} \frac{\eta'(x) R_+(x)}{x-z} dx + \frac{1}{R(z)}\left[\kappa z^G + \sum_{m=0}^{G-1} f_m z^m\right], \tag{2.27}$$

where $R_+(x)$ denotes the boundary value taken by $R(z)$ from the upper half-plane and where the constants f_m, $m = 0, \ldots, G-1$, are chosen (uniquely—see Appendix A) so that

$$\int_{\Gamma_j} h'(z)\, dz = 0, \quad \text{for } j = 1, \ldots, G. \tag{2.28}$$

Then we define a function for $z \in \mathbb{C} \setminus (-\infty, \beta_G]$ by the integral

$$h(z) := \kappa \log(z) + \int_z^\infty \left[\frac{\kappa}{s} - h'(s)\right] ds, \tag{2.29}$$

where the path of integration lies in $\mathbb{C} \setminus (-\infty, \beta_G]$. Furthermore, we define a constant γ by

$$\gamma := \eta(\beta_G) - 2h(\beta_G). \tag{2.30}$$

The combination $N\ell_c + \gamma$ plays an important role in what follows. Since γ remains bounded as $N \to \infty$, we may interpret γ as a higher-order correction to the scaled Robin constant $N\ell_c$.

Suppose that $G > 0$. It may be verified that for z in any interior gap Γ_j, the difference of boundary values taken by $h(z)$ depends on j but is independent of z. Thus there are constants c_j, $j = 1, \ldots, G$, such that

$$h_+(z) - h_-(z) := \lim_{\epsilon \downarrow 0} h(z+i\epsilon) - h(z-i\epsilon) = ic_j, \quad \text{for } z \in \Gamma_j. \tag{2.31}$$

Moreover, it can be checked directly that the constants c_j are real and linear in κ, so that we may write

$$c_j = c_{j,0} + \omega_j \kappa,$$

for some other real constants $c_{j,0}$ and ω_j that are independent of κ. We define a vector $\mathbf{r}$ with components

$$r_j := N\theta_{\Gamma_j} - c_{j,0}, \tag{2.32}$$

for $j = 1, \ldots, G$, and a vector $\mathbf{\Omega}$ with components ω_j, for $j = 1, \ldots, G$.

The function $iR_+(z)$ may be analytically continued from any band I to the complex plane with the real intervals $(-\infty, \alpha_0]$ and $[\beta_G, \infty)$ and the closures of the interior gaps $\Gamma_1, \ldots, \Gamma_G$ deleted. We call this analytic continuation $y(z)$, and for z in this domain of definition, we introduce a vector function $\mathbf{m}(z)$ having components $m_p(z) := z^{p-1}/y(z)$, for $p = 1, 2, \ldots, G$. Next, a constant $G \times G$ matrix $\mathbf{A} = (\mathbf{a}^{(1)}, \mathbf{a}^{(2)}, \ldots, \mathbf{a}^{(G)})$ is defined by insisting that the linear equations

$$\mathbf{A} \int_{\beta_{j-1}}^{\alpha_j} \mathbf{m}_-(z)\, dz = \pi i \mathbf{e}^{(j)}, \quad \text{for } j = 1, \ldots, G, \tag{2.33}$$

are satisfied, where $\mathbf{m}_-(z)$ denotes the boundary value taken on the real axis from the lower half-plane and $\mathbf{e}^{(j)}$ is column j of the $G \times G$ identity matrix. This determines vectors $\mathbf{b}^{(j)}$ by the definition

$$\mathbf{b}^{(j)} := -2\mathbf{A} \sum_{m=1}^{j} \int_{\alpha_{m-1}}^{\beta_{m-1}} \mathbf{m}(z)\, dz, \tag{2.34}$$

and we obtain a second $G \times G$ constant *Riemann matrix* from these column vectors by setting $\mathbf{B} := (\mathbf{b}^{(1)}, \mathbf{b}^{(2)}, \ldots, \mathbf{b}^{(G)})$. A *Riemann constant vector* $\mathbf{k}$ may now be defined by the formula

$$\mathbf{k} := \begin{cases} \pi i \sum_{j \text{ odd}} \mathbf{e}^{(j)} + \dfrac{1}{2}\sum_{j=1}^{G} \mathbf{b}^{(j)}, & G \text{ odd}, \\[2mm] \pi i \sum_{j \text{ even}} \mathbf{e}^{(j)} + \dfrac{1}{2}\sum_{j=1}^{G} \mathbf{b}^{(j)}, & G \text{ even}. \end{cases} \tag{2.35}$$

The matrix $\mathbf{B}$ is real, symmetric, and negative-definite, so we may use it to define a *Riemann theta function* for $\mathbf{w} \in \mathbb{C}^G$ by the Fourier series

$$\Theta(\mathbf{w}) := \sum_{\mathbf{n} \in \mathbb{Z}^G} t_\mathbf{n} e^{\mathbf{n}^T \mathbf{w}}, \quad \text{with Fourier coefficients } t_\mathbf{n} := \exp\left(\frac{1}{2}\mathbf{n}^T \mathbf{B} \mathbf{n}\right). \tag{2.36}$$

Next, for $z \in \mathbb{C} \setminus \mathbb{R}$, define the *Abel-Jacobi mapping* by setting

$$\mathbf{w}(z) := \int_{\alpha_0}^{z} \mathbf{Am}(s)\, ds, \tag{2.37}$$

where the path of integration lies in the half-plane $\Im(s) = \Im(z)$ but is otherwise arbitrary. As special cases we set

$$\mathbf{w}_+(\infty) := \lim_{\substack{z \to \infty \\ \Im(z) > 0}} \mathbf{w}(z) \quad \text{and} \quad \mathbf{w}_-(\infty) := \lim_{\substack{z \to \infty \\ \Im(z) < 0}} \mathbf{w}(z). \tag{2.38}$$

Let $\lambda(z)$ be defined in the same domain as $y(z)$ by

$$\lambda(z)^4 = \prod_{j=0}^{G} \frac{z - \alpha_j}{z - \beta_j} \quad \text{and} \quad \lambda(z) \to 1 \text{ as } z \to \infty \text{ with } \Im(z) > 0. \tag{2.39}$$

In terms of $\lambda(z)$ we define two functions in the same domain by setting

$$u(z) := \frac{1}{2}\left[\lambda(z) + \frac{1}{\lambda(z)}\right] \quad \text{and} \quad v(z) := \frac{1}{2i}\left[\lambda(z) - \frac{1}{\lambda(z)}\right]. \tag{2.40}$$

The polynomial equation

$$\prod_{j=0}^{G}(x - \alpha_j) - \prod_{j=0}^{G}(x - \beta_j) = 0 \tag{2.41}$$

of degree G has exactly one root $x = x_j$ in each interior gap Γ_j, for $j = 1, \ldots, G$. Denoting the boundary values of $\mathbf{w}(z)$ taken on the real axis from the upper and lower half-planes by $\mathbf{w}_+(z)$ and $\mathbf{w}_-(z)$, respectively, we define two vectors by setting

$$\mathbf{q}_u := \sum_{j=1}^{G} \mathbf{w}_-(x_j) + \mathbf{k} \quad \text{and} \quad \mathbf{q}_v := \sum_{j=1}^{G} \mathbf{w}_+(x_j) + \mathbf{k}. \tag{2.42}$$

In terms of these ingredients we may now define two functions that turn out to extend analytically to the domain $\mathbb{C} \setminus (-\infty, \beta_G]$. If $G > 0$, set

$$W(z) := \begin{cases} u(z)e^{h(z)} \dfrac{\Theta(\mathbf{w}_+(\infty) - \mathbf{q}_u)}{\Theta(\mathbf{w}_+(\infty) - \mathbf{q}_u - i\mathbf{r} + i\kappa\boldsymbol{\Omega})} \dfrac{\Theta(\mathbf{w}(z) - \mathbf{q}_u - i\mathbf{r} + i\kappa\boldsymbol{\Omega})}{\Theta(\mathbf{w}(z) - \mathbf{q}_u)}, & \Im(z) > 0, \\ -v(z)e^{h(z)} \dfrac{\Theta(\mathbf{w}_-(\infty) - \mathbf{q}_v)}{\Theta(\mathbf{w}_-(\infty) - \mathbf{q}_v + i\mathbf{r} - i\kappa\boldsymbol{\Omega})} \dfrac{\Theta(\mathbf{w}(z) - \mathbf{q}_v + i\mathbf{r} - i\kappa\boldsymbol{\Omega})}{\Theta(\mathbf{w}(z) - \mathbf{q}_v)}, & \Im(z) < 0, \end{cases} \tag{2.43}$$

and

$$Z(z) := \begin{cases} iv(z)e^{-h(z)} \dfrac{\Theta(\mathbf{w}_-(\infty) - \mathbf{q}_v)}{\Theta(\mathbf{w}_-(\infty) - \mathbf{q}_v + i\mathbf{r} - i\kappa\boldsymbol{\Omega})} \dfrac{\Theta(\mathbf{w}(z) - \mathbf{q}_v + i\mathbf{r} - i\kappa\boldsymbol{\Omega})}{\Theta(\mathbf{w}(z) - \mathbf{q}_v)}, & \Im(z) > 0, \\ iu(z)e^{-h(z)} \dfrac{\Theta(\mathbf{w}_+(\infty) - \mathbf{q}_u)}{\Theta(\mathbf{w}_+(\infty) - \mathbf{q}_u - i\mathbf{r} + i\kappa\boldsymbol{\Omega})} \dfrac{\Theta(\mathbf{w}(z) - \mathbf{q}_u - i\mathbf{r} + i\kappa\boldsymbol{\Omega})}{\Theta(\mathbf{w}(z) - \mathbf{q}_u)}, & \Im(z) < 0. \end{cases} \tag{2.44}$$

And if $G = 0$, set

$$W(z) := \begin{cases} u(z)e^{h(z)}, & \Im(z) > 0, \\ -v(z)e^{h(z)}, & \Im(z) < 0, \end{cases} \tag{2.45}$$

and

$$Z(z) := \begin{cases} iv(z)e^{-h(z)}, & \Im(z) > 0, \\ iu(z)e^{-h(z)}, & \Im(z) < 0. \end{cases} \tag{2.46}$$

Finally, for any gap Γ we may define two corresponding functions $H_\Gamma^\pm(z)$ in terms of $W(z)$ and $Z(z)$:

$$H_\Gamma^\pm(z) := \frac{W(z)}{\sqrt{2}} e^{(\gamma - \eta(z) - iN\operatorname{sgn}(\Im(z))\theta_\Gamma)/2} \pm \frac{Z(z)}{\sqrt{2}} e^{-(\gamma - \eta(z) - iN\operatorname{sgn}(\Im(z))\theta_\Gamma)/2}. \tag{2.47}$$

2.3 RESULTS ON ASYMPTOTICS OF DISCRETE ORTHOGONAL POLYNOMIALS

Subject to the basic assumptions described in §1.3.1 and the simplifying assumptions described in §2.1.2, we have the following results, the proofs of which will be given in Chapter 6.

Theorem 2.7 (Outer asymptotics of $\pi_{N,k}(z)$). *Let K be a closed set with $K \cap [a,b] = \emptyset$. Then there exists a constant $C_K > 0$ such that*

$$\pi_{N,k}(z) = e^{NL_c(z)}\left[W(z) + \varepsilon_N(z)\right],$$

where the estimate

$$\sup_{z \in K} |\varepsilon_N(z)| \leq \frac{C_K}{N}$$

holds for sufficiently large N, and $W(z)$ defined by (2.43) or (2.45) is a function that is nonvanishing and uniformly bounded in K independently of N. Furthermore, the product $e^{NL_c(z)}W(z)$ is analytic for $z \in \mathbb{C} \setminus [a,b]$.

Theorem 2.8 (Asymptotics of leading coefficients and recurrence coefficients). *If there is only a single band of unconstrained support of the equilibrium measure $\mu_{\min}^c$ in $[a,b]$, with endpoints $\alpha_0 < \beta_0$, then*

$$\gamma_{N,k}^2 = \frac{4}{\beta_0 - \alpha_0}e^{N\ell_c + \gamma}\left(1 + \varepsilon_N^{(1)}\right),$$

$$\gamma_{N,k-1}^2 = \frac{\beta_0 - \alpha_0}{4}e^{N\ell_c + \gamma}\left(1 + \varepsilon_N^{(2)}\right),$$

$$b_{N,k-1} = \frac{\beta_0 - \alpha_0}{4}\left(1 + \varepsilon_N^{(3)}\right),$$

and

$$a_{N,k} = \frac{\beta_0 + \alpha_0}{2} + \varepsilon_N^{(4)},$$

where there is a constant $C > 0$ such that the estimates $|\varepsilon_N^{(m)}| \leq C/N$, $m = 1, 2, 3, 4$, all hold for sufficiently large N. More generally, if for some $G > 0$ there are $G + 1$ disjoint bands with endpoints $\alpha_0 < \beta_0 < \alpha_1 < \beta_1 < \cdots < \alpha_G < \beta_G$, then

$$\gamma_{N,k}^2 = \frac{4e^{N\ell_c + \gamma}}{\displaystyle\sum_{j=0}^{G}(\beta_j - \alpha_j)} \frac{\Theta(\mathbf{w}_-(\infty) - \mathbf{q}_v + i\mathbf{r} - i\kappa\mathbf{\Omega})\Theta(\mathbf{w}_+(\infty) - \mathbf{q}_v)}{\Theta(\mathbf{w}_+(\infty) - \mathbf{q}_v + i\mathbf{r} - i\kappa\mathbf{\Omega})\Theta(\mathbf{w}_-(\infty) - \mathbf{q}_v)}\left(1 + \varepsilon_N^{(1)}\right),$$

$$\gamma_{N,k-1}^2 = \frac{e^{N\ell_c + \gamma}}{4}\left[\sum_{j=0}^{G}(\beta_j - \alpha_j)\right]\frac{\Theta(\mathbf{w}_+(\infty) - \mathbf{q}_v - i\mathbf{r} + i\kappa\mathbf{\Omega})\Theta(\mathbf{w}_-(\infty) - \mathbf{q}_v)}{\Theta(\mathbf{w}_-(\infty) - \mathbf{q}_v - i\mathbf{r} + i\kappa\mathbf{\Omega})\Theta(\mathbf{w}_+(\infty) - \mathbf{q}_v)}\left(1 + \varepsilon_N^{(2)}\right),$$

$$b_{N,k-1} = \frac{1}{4}\left[\sum_{j=0}^{G}(\beta_j - \alpha_j)\right]\frac{\Theta(\mathbf{w}_-(\infty) - \mathbf{q}_v)}{\Theta(\mathbf{w}_+(\infty) - \mathbf{q}_v)}$$

$$\cdot\sqrt{\frac{\Theta(\mathbf{w}_+(\infty) - \mathbf{q}_v - i\mathbf{r} + i\kappa\mathbf{\Omega})\Theta(\mathbf{w}_+(\infty) - \mathbf{q}_v + i\mathbf{r} - i\kappa\mathbf{\Omega})}{\Theta(\mathbf{w}_-(\infty) - \mathbf{q}_v - i\mathbf{r} + i\kappa\mathbf{\Omega})\Theta(\mathbf{w}_-(\infty) - \mathbf{q}_v + i\mathbf{r} - i\kappa\mathbf{\Omega})}}\left(1 + \varepsilon_N^{(3)}\right),$$

and

$$a_{N,k} = \frac{i\mathbf{a}^{(G)} \cdot \nabla\Theta(\mathbf{w}_+(\infty) + \mathbf{q}_v - i\mathbf{r} + i\kappa\mathbf{\Omega})}{\Theta(\mathbf{w}_+(\infty) + \mathbf{q}_v - i\mathbf{r} + i\kappa\mathbf{\Omega})} - \frac{i\mathbf{a}^{(G)} \cdot \nabla\Theta(\mathbf{w}_+(\infty) + \mathbf{q}_v)}{\Theta(\mathbf{w}_+(\infty) + \mathbf{q}_v)}$$
$$+ \frac{i\mathbf{a}^{(G)} \cdot \nabla\Theta(\mathbf{w}_+(\infty) - \mathbf{q}_v + i\mathbf{r} - i\kappa\mathbf{\Omega})}{\Theta(\mathbf{w}_+(\infty) - \mathbf{q}_v + i\mathbf{r} - i\kappa\mathbf{\Omega})} - \frac{i\mathbf{a}^{(G)} \cdot \nabla\Theta(\mathbf{w}_+(\infty) - \mathbf{q}_v)}{\Theta(\mathbf{w}_+(\infty) - \mathbf{q}_v)}$$
$$+ \frac{1}{2} \frac{\sum_{j=0}^{G}(\beta_j^2 - \alpha_j^2)}{\sum_{j=0}^{G}(\beta_j - \alpha_j)} + \varepsilon_N^{(4)},$$

where $\mathbf{a}^{(G)} \cdot \nabla\Theta$ denotes the directional derivative of $\Theta(\cdot)$ in the direction of $\mathbf{a}^{(G)}$ in $\mathbb{C}^G$ and where there is a constant $C > 0$ such that the estimates $|\varepsilon_N^{(m)}| \leq C/N$, $m = 1, 2, 3, 4$, all hold for sufficiently large N.

For an interval $J \subset [a,b]$ and any $\delta > 0$, define the compact set

$$K_J^\delta := \bigcup_{w \in J} \{z \in \mathbb{C} \text{ such that } |z-w| \leq \delta\}. \tag{2.48}$$

Theorem 2.9 (Asymptotics of $\pi_{N,k}(z)$ in voids). *Let $J \subset [a,b]$ be a closed interval and let Γ be a void. If $\Gamma = (a, \alpha_0)$, then assume that $J \subset [a, \alpha_0)$. If $\Gamma = (\beta_G, b)$, then assume that $J \subset (\beta_G, b]$. Finally, if $\Gamma = \Gamma_j = (\beta_{j-1}, \alpha_j)$ for some $j = 1, \ldots, G$, then assume that $J \subset \Gamma$. There is a positive δ and a constant $C_J^\delta > 0$ such that for $z \in K_J^\delta$ defined by (2.48), we have*

$$\pi_{N,k}(z) = e^{N\overline{L}_c^\Gamma(z)} \left[A_\Gamma^\nabla(z) + \varepsilon_N(z)\right],$$

where the estimate

$$\sup_{z \in K_J^\delta} |\varepsilon_N(z)| \leq \frac{C_J^\delta}{N} \tag{2.49}$$

holds for sufficiently large N and

$$A_\Gamma^\nabla(z) := e^{N(L_c(z) - \overline{L}_c^\Gamma(z))} W(z),$$

with $W(z)$ given by (2.43) or (2.45), is a function that is real-analytic and uniformly bounded in K_J^δ independently of N. If Γ is adjacent to either endpoint, $z = a$ or $z = b$, then $A_\Gamma^\nabla(z)$ does not vanish in K_J^δ. Otherwise, $A_\Gamma^\nabla(z)$ has at most one (real) zero in K_J^δ.

The possible lone zero of $A_\Gamma^\nabla(z)$ in the void Γ is analogous to a *spurious zero* in approximation theory. The motion of a spurious zero through an interior gap Γ as parameters (like the degree k) are varied corresponds to the spontaneous emission of a zero from one band and its subsequent capture by an adjacent band separated by a void. At most one zero can be in transit in Γ for each choice of parameters.

◁ **Remark:** Note that if the void Γ contains no spurious zero for N sufficiently large, then Theorem 2.9 implies the existence of the limit, as $N \to \infty$, of $\pi_{N,k+1}(z)/\pi_{N,k}(z)$ uniformly for $z \in K_J^\delta$. In particular, this limit exists everywhere outside the support of the equilibrium measure if $G = 0$. The existence of such ratio asymptotics in voids was postulated (and a necessary formula for the limit, should it exist, was obtained in terms of the equilibrium measure) by Aptekarev and Van Assche under slightly more general conditions than we have considered. See [AptV01]. ▷

For z in the domain of analyticity of $\rho^0(z)$, we set

$$\theta^0(z) := 2\pi \int_z^b \rho^0(s)\, ds. \tag{2.50}$$

Theorem 2.10 (Asymptotics of $\pi_{N,k}(z)$ in saturated regions). *Let $J \subset \Gamma \subset [a,b]$ be a closed interval and let Γ be a saturated region. There is a positive δ and there are constants $C_J^\delta > 0$, $D_J^\delta > 0$, and $E_J^\delta > 0$ such that for $z \in K_J^\delta$ defined by (2.48) we have*

$$\pi_{N,k}(z) = e^{N\overline{L}_c^\Gamma(z)} \left[\left(A_\Gamma^\Delta(z) + \varepsilon_N(z) \right) \cos\left(\frac{N\theta^0(z)}{2} \right) + \delta_N(z) \right], \qquad (2.51)$$

where the estimates

$$\sup_{z \in K_J^\delta} |\varepsilon_N(z)| \leq \frac{C_J^\delta}{N} \qquad \text{and} \qquad \sup_{z \in K_J^\delta} |\delta_N(z)| \leq D_J^\delta e^{-NE_J^\delta} \qquad (2.52)$$

hold for sufficiently large N, and

$$A_\Gamma^\Delta(z) := 2 e^{N(L_c(z) - \overline{L}_c^\Gamma(z))} e^{-iN \operatorname{sgn}(\Im(z))\theta^0(z)/2} W(z),$$

with $W(z)$ given by (2.43) or (2.45), is a function that is real-analytic and uniformly bounded in K_J^δ independently of N. If Γ is adjacent to either endpoint, $z = a$ or $z = b$, then $A_\Gamma^\Delta(z)$ does not vanish in K_J^δ. Otherwise, $A_\Gamma^\Delta(z)$ has at most one (real) zero in K_J^δ.

When a saturated region meets an endpoint of $[a,b]$ (*i.e.*, when it is an exterior gap), we say that there is a *hard edge* at that endpoint. This terminology is borrowed from random matrix theory, where it refers to an ensemble of matrices all of which share a certain common bound on their spectra. For example, random Wishart matrices of the form $\mathbf{W} = \mathbf{X}^T \mathbf{X}$ for some real matrix $\mathbf{X}$ necessarily have nonnegative spectra, and for certain types of matrices $\mathbf{X}$ the asymptotic density of eigenvalues z of $\mathbf{W}$ can have a jump discontinuity at $z = 0$, being identically zero for $z < 0$ and strictly positive for $z > 0$ however small. Thus $z = 0$ is a hard edge for the spectrum in Wishart random matrix ensembles. We will see in Chapter 3 that the density of the scaled equilibrium measure $\mu_{\min}^c/c$ plays the same role in certain discrete random processes that the asymptotic density of eigenvalues plays in random matrix theory. Since the upper constraint always corresponds to a strictly positive density, and since the support of the equilibrium measure is a subset of $[a,b]$, an active upper constraint at either $z = a$ or $z = b$ implies a jump discontinuity in the density of the equilibrium measure at the corresponding endpoint, which explains our terminology.

Theorem 2.11 (Asymptotics of $\pi_{N,k}(z)$ near hard edges). *Suppose either that $\Gamma = (a, \alpha_0)$ is a saturated region and $J = [a,t]$ for some $t \in \Gamma$, or that $\Gamma = (\beta_G, b)$ is a saturated region and $J = [t,b]$ for some $t \in \Gamma$ (in both cases Γ is an exterior gap). Set*

$$\zeta := \begin{cases} N \int_a^z \rho^0(s)\,ds, & \text{if } a \in J \\ N \int_z^b \rho^0(s)\,ds, & \text{if } b \in J. \end{cases}$$

There is a positive δ and there are constants $C_J^\delta > 0$, $D_J^\delta > 0$, and $E_J^\delta > 0$ such that for $z \in K_J^\delta$ defined by (2.48) we have

$$\pi_{N,k}(z) = e^{N\overline{L}_c^\Gamma(z)} \zeta^{-\zeta} \left[\left(\tilde{A}_\Gamma^\Delta(z) + \tilde{\varepsilon}_N(z) \right) \frac{\Gamma(1/2 + \zeta) \cos(\pi\zeta)}{\sqrt{2\pi} e^{-\zeta}} + \zeta^\zeta \delta_N(z) \right], \qquad (2.53)$$

where $\tilde{\varepsilon}_N(z)$ and $\zeta^\zeta \delta_N(z)$ extend from $\zeta > 0$ as functions analytic in K_J^δ such that the estimates

$$\sup_{z \in K_J^\delta} |\tilde{\varepsilon}_N(z)| \leq \frac{C_J^\delta}{N} \qquad \text{and} \qquad \sup_{z \in K_J^\delta} |\delta_N(z)| \leq D_J^\delta e^{-NE_J^\delta}$$

hold for sufficiently large N and where

$$\tilde{A}_\Gamma^\Delta(z) := 2 e^{N(L_c(z) - \overline{L}_c^\Gamma(z))} e^{-iN\pi \operatorname{sgn}(\Im(\zeta))\zeta} W(z),$$

with $W(z)$ given by (2.43) or (2.45), is a function that is real-analytic, nonvanishing, and uniformly bounded in K_J^δ independently of N. Finally, note that $e^{N\overline{L}_c^\Gamma(z)} \zeta^{-\zeta}$ extends from $\zeta > 0$ as an analytic function in K_J^δ and that $\delta_N(z)$ represents exactly the same function as in Theorem 2.10.

◁ **Remark:** The fact that the asymptotic formulae presented in Theorem 2.11 are in terms of the Euler gamma function is directly related to the discrete nature of the weights. In a sense, the poles of the function $\Gamma(1/2 + \zeta)$ are "shadows" of the poles of the matrix $\mathbf{P}(z; N, k)$ solving Interpolation Problem 1.2. ▷

Theorem 2.12 (Exponential confinement of zeros in saturated regions). *Let $J \subset [a, b]$ be a closed interval and let Γ be a saturated region. If $\Gamma = (a, \alpha_0)$, then assume that $J \subset [a, \alpha_0)$. If $\Gamma = (\beta_G, b)$, then assume that $J \subset (\beta_G, b]$. Finally, if $\Gamma = \Gamma_j = (\beta_{j-1}, \alpha_j)$ for some $j = 1, \ldots, G$, then assume that $J \subset \Gamma$. There are positive constants D_J, E_J, and N_0 such that for every node $x_{N,n} \in X_N \cap J$ there exists a zero z_0 of the monic discrete orthogonal polynomial $\pi_{N,k}(z)$ with*

$$|z_0 - x_{N,n}| \leq D_J e^{-N E_J}, \qquad \text{whenever } N > N_0. \tag{2.54}$$

Moreover:

1. *If $\Gamma = (a, \alpha_0)$, so that $\min J \geq a$, then for each node $x_{N,n} \in J \cap X_N$ there is a zero z_0 of $\pi_{N,k}(z)$ such that*

$$x_{N,n} < z_0 < x_{N,n} + D_J e^{-N E_J}, \qquad \text{whenever } N > N_0, \tag{2.55}$$

and for each zero $z_0 \in J$ of $\pi_{N,k}(z)$ there is a node $x_{N,n} \in X_N$ such that (2.55) holds.

2. *If $\Gamma = (\beta_G, b)$, so that $\max J \leq b$, then for each node $x_{N,n} \in J \cap X_N$ there is a zero z_0 of $\pi_{N,k}(z)$ such that*

$$x_{N,n} - D_J e^{-N E_J} < z_0 < x_{N,n}, \qquad \text{whenever } N > N_0, \tag{2.56}$$

and for each zero $z_0 \in J$ of $\pi_{N,k}(z)$ there is a node $x_{N,n} \in X_N$ such that (2.56) holds.

3. *If $\Gamma = \Gamma_j = (\beta_{j-1}, \alpha_j)$ for some $j = 1, \ldots, G$, then exactly one of the following two mutually exclusive possibilities holds:*

   (a) *There is a node $x_{N,m} \in \Gamma \cap X_N$ such that $\pi_{N,k}(x_{N,m}) = 0$. For each node $x_{N,n} \in J \cap X_N$ with $x_{N,n} > x_{N,m}$, there is a zero z_0 of $\pi_{N,k}(z)$ such that (2.55) holds, and for each zero $z_0 \in J$ of $\pi_{N,k}(z)$ with $z_0 > x_{N,m}$, there is a node $x_{N,n} \in X_N$ such that (2.55) holds. For each node $x_{N,n} \in J \cap X_N$ with $x_{N,n} < x_{N,m}$, there is a zero z_0 of $\pi_{N,k}(z)$ such that (2.56) holds, and for each zero $z_0 \in J$ of $\pi_{N,k}(z)$ with $z_0 < x_{N,m}$, there is a node $x_{N,n} \in X_N$ such that (2.56) holds.*

   (b) *There is a consecutive pair of nodes $x_{N,m} \in \Gamma \cap X_N$ and $x_{N,m+1} \in \Gamma \cap X_N$ such that*
   - *For each node $x_{N,n} \in J \cap X_N$ with $x_{N,n} \geq x_{N,m+1}$, there is a zero z_0 of $\pi_{N,k}(z)$ such that (2.55) holds, and for each zero $z_0 \in J$ of $\pi_{N,k}(z)$ with $z_0 \geq x_{N,m+1}$, there is a node $x_{N,n} \in X_N$ such that (2.55) holds.*
   - *For each node $x_{N,n} \in J \cap X_N$ with $x_{N,n} \leq x_{N,m}$, there is a zero z_0 of $\pi_{N,k}(z)$ such that (2.56) holds, and for each zero $z_0 \in J$ of $\pi_{N,k}(z)$ with $z_0 \leq x_{N,m}$, there is a node $x_{N,n} \in X_N$ such that (2.56) holds.*
   - *There is at most one zero z_0 of $\pi_{N,k}(z)$ in the closed interval $[x_{N,m}, x_{N,m+1}]$, and if it exists, then $z_0 \in (x_{N,m}, x_{N,m+1})$.*

Note that in case 3(b), if there is a zero z_0 of $\pi_{N,k}(z)$ with $x_{N,m} < z_0 < x_{N,m+1}$, there need not be any node $x_{N,n} \in X_N$ such that (2.54) holds. This particular zero, and only this one, is not necessarily exponentially close to any node.

We refer to the node $x_{N,m}$ in 3(a) and the interval $[x_{N,m}, x_{N,m+1}]$ in 3(b), both of which serve to separate the two directions of perturbation of the zeros of $\pi_{N,k}(z)$ from the nodes, as *defects*, and to the zero possibly carried by the defect in 3(b), as a *spurious zero*. The remaining zeros correspond in a one-to-one fashion with the nodes; we refer to them as *Hurwitz zeros* by analogy with the approximation theory literature. See Figure 2.1.

◁ **Remark:** It should perhaps be stressed that in principle there is nothing that prevents a zero of $\pi_{N,k}(z)$ from coinciding *exactly* with one of the nodes $x_{N,n} \in X_N$. Indeed, this is the case in 3(a) above. However,

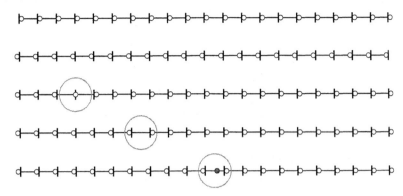

Figure 2.1 *First line: the pattern of zeros of $\pi_{N,k}(z)$ (small circles) and nodes (vertical segments) in a saturated region adjacent to the left endpoint $z = a$. Second line: same as above, but for a saturated region adjacent to the right endpoint $z = b$. Third line: a pattern of zeros and nodes in a saturated region between two bands; there is a single defect (circled) corresponding to a zero of $\pi_{N,k}(z)$ occurring exactly at a node. Fourth line: a pattern as above in which the defect (circled) is an interval $[x_{N,m}, x_{N,m+1}]$ that does not carry a spurious zero. Fifth line: same as above, but a case in which the defect (circled) carries a spurious zero (small shaded circle).*

Theorem 2.12 shows that in a saturated region adjacent to an endpoint $z = a$ or $z = b$ (*i.e.*, a saturated region that is also an exterior gap) all zeros become asymptotically distinct from (yet paradoxically converge rapidly to) nodes as $N \to \infty$. In a saturated region lying between two bands (*i.e.*, a saturated region that is also an interior gap) it is asymptotically possible for only a single zero to coincide exactly with a node. ▷

The precise location of a defect within a saturated region depends on all the parameters of the problem, and in some circumstances it may make sense for a parameter, say, appearing in the function $V(x)$ defined in (1.17), to be continuously varied. An example of such a parameter is the parameter T parametrizing the Toda flow. This is interesting because it can imply corresponding dynamics of the defects and any spurious zeros they may carry. If continuous deformation of a parameter leads to deformation of the phase vector $\mathbf{r} - \kappa\mathbf{\Omega}$, then the defect will move continuously through the saturated region Γ as well.

How does a defect move? If there is no spurious zero, then a defect $[x_{N,m}, x_{N,m+1}]$ can move to the right to become a defect $[x_{N,m+1}, x_{N,m+2}]$, as the zero of $\pi_{N,k}(z)$ just to the right of the node $x_{N,m+1}$ moves continuously to the left through the node. Then the same process occurs near the node $x_{N,m+2}$, and so on. Thus a defect without a spurious zero moves to the right by a process in which Hurwitz zeros move to the left an exponentially small amount, passing through the corresponding nodes, one after the other. During the continuous motion of a defect without a spurious zero, the situation described in 3(a) above occurs only at isolated values of the deformation parameter on which the phase vector $\mathbf{r} - \kappa\mathbf{\Omega}$ continuously depends. See the left diagram in Figure 2.2.

If the defect $[x_{N,m}, x_{N,m+1}]$ contains a spurious zero, then the motion of the defect to the right occurs by a change-of-identity process in which the spurious zero moves to the right through the defect toward $x_{N,m+1}$, and when it is exponentially close to $x_{N,m+1}$, it becomes a Hurwitz zero and the previously Hurwitz zero just to the right of $x_{N,m+1}$ becomes a spurious zero belonging to the new defect $[x_{N,m+1}, x_{N,m+2}]$. Thus a defect carrying a spurious zero moves to the right by a process in which zeros move to the right by an amount proportional to $1/N$, one after the other. During the continuous motion of a defect containing a spurious zero, the situation described in 3(a) above never occurs at all. See the right diagram in Figure 2.2.

In fact, a defect carrying a spurious zero that reaches an endpoint of Γ under deformation will generally be reflected back into Γ as a defect without a spurious zero. Thus as parameters are deformed, a defect may oscillate back and forth within a saturated region and act like a conveyor belt, carrying a spurious zero from one band to the next and returning empty to pick up the next zero. This phenomenon will be illustrated concretely with the Toda flow in §3.5.3.

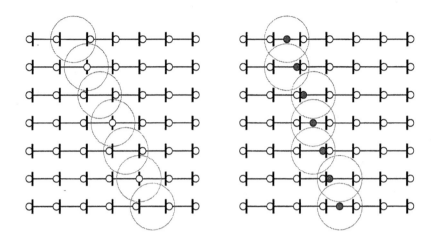

Figure 2.2 *Left: the motion of a defect (circled) that does not carry a spurious zero. Right: the motion of a defect (circled) carrying a spurious zero, which exchanges its identity with a Hurwitz zero in each step.*

Theorem 2.13 (Asymptotics of $\pi_{N,k}(z)$ in bands). *Let J be a closed interval and let $I = (\alpha_j, \beta_j)$ be a band. Assume that $J \subset I$. There is a positive δ and a constant $C_J > 0$ such that for $z \in K_J^\delta$ defined by (2.48), we have*

$$\pi_{N,k}(z) = e^{N\overline{L}_c^I(z)} \left[A_I(z) \cos\left(\Phi_I(z) + N\pi c \mu_{\min}^c([x,b]) - N\pi c \int_x^z \psi_I(s)\, ds \right) + \varepsilon_I(z) \right],$$

where x is any point (or endpoint) of I and $\varepsilon_I(z)$ satisfies the estimate

$$\sup_{z \in K_J^\delta} |\varepsilon_I(z)| \leq \frac{C_J}{N}. \tag{2.57}$$

Here $A_I(z)$ and $\Phi_I(z)$ are real-analytic functions of z that are uniformly bounded independently of N in K_J^δ. Moreover, $A_I(z)$ is strictly positive for real z. For real z, the identity

$$W_+(z) = \frac{1}{2} A_I(z) e^{i\Phi_I(z)}$$

holds, where $W(z)$ is defined by (2.43) or (2.45) and $W_+(z)$ indicates the boundary value taken from the upper half-plane. This relation serves as a definition of the analytic functions $A_I(z)$ and $\Phi_I(z)$.

Theorem 2.14 (Asymptotic description of the zeros of $\pi_{N,k}(z)$ in bands). *Let J be a closed interval and let $I = (\alpha_j, \beta_j)$ be a band. Assume that $J \subset I$. Then the zeros of $\pi_{N,k}(z)$ in J correspond in a one-to-one fashion with those of the model function*

$$C_I(z) := \cos\left(\Phi_I(z) + N\pi c \mu_{\min}^c([x,b]) - N\pi c \int_x^z \psi_I(s)\, ds \right),$$

where $\Phi_I(z)$ is as in the statement of Theorem 2.13 above and x is any point (or endpoint) of I. Moreover, there exists a constant $D_J > 0$ such that if N is sufficiently large, each pair of corresponding zeros z_0 of $\pi_{N,k}(z)$ and $\tilde{z}_0$ of $C_I(z)$ in J satisfies the estimate

$$|z_0 - \tilde{z}_0| \leq \frac{D_J}{N^2}.$$

Before we state the next results, we point out that the function $\overline{L}_c^I(z)$ defined for each band $I = (\alpha, \beta)$ (see (2.12)) may be considered to be analytic in a complex neighborhood of the closed interval $[\alpha, \beta]$. In particular, $\overline{L}_c^I(z)$ is analytic in a neighborhood of each endpoint of the band. The analytic continuation to a neighborhood U_α of $z = \alpha$ is accomplished by the identity

$$\overline{L}_c^I(z) = \overline{L}_c^\Gamma(z) + \frac{1}{2N}\left(-\tau_\Gamma^{\nabla,L}(z)\right)^{3/2}, \qquad \text{for } z \in U_\alpha \text{ with } \Im(z) \neq 0,$$

if the adjacent gap Γ is a void, and by the identity

$$\overline{L}_c^I(z) = \overline{L}_c^\Gamma(z) - \frac{1}{2N}\left(-\tau_\Gamma^{\Delta,L}(z)\right)^{3/2}, \qquad \text{for } z \in U_\alpha \text{ with } \Im(z) \neq 0,$$

if the adjacent gap Γ is a saturated region. Similarly, the analytic continuation to a neighborhood U_β of $z = \beta$ is accomplished by the identity

$$\overline{L}_c^I(z) = \overline{L}_c^\Gamma(z) + \frac{1}{2N}\left(-\tau_\Gamma^{\nabla,R}(z)\right)^{3/2}, \qquad \text{for } z \in U_\beta \text{ with } \Im(z) \neq 0,$$

if the adjacent gap Γ is a void, and by the identity

$$\overline{L}_c^I(z) = \overline{L}_c^\Gamma(z) - \frac{1}{2N}\left(-\tau_\Gamma^{\Delta,R}(z)\right)^{3/2}, \qquad \text{for } z \in U_\beta \text{ with } \Im(z) \neq 0,$$

if the adjacent gap Γ is a saturated region.

Theorem 2.15 (Asymptotics of $\pi_{N,k}(z)$ near band/void edges). *Let $z = \alpha$ be the left endpoint of a band I and suppose that a void Γ lies immediately to the left of $z = \alpha$. There exist constants $r > 0$ and $C > 0$ such that when $|z - \alpha| \leq r$,*

$$\pi_{N,k}(z) = e^{N\overline{L}_c^I(z)}\left[N^{1/6}\left(A_\Gamma^{\nabla,L}(z) + \varepsilon_A(z)\right)\mathrm{Ai}\left(-\left(\frac{3}{4}\right)^{2/3}\tau_\Gamma^{\nabla,L}(z)\right)\right.$$
$$\left. + N^{-1/6}\left(B_\Gamma^{\nabla,L}(z) + \varepsilon_B(z)\right)\mathrm{Ai}'\left(-\left(\frac{3}{4}\right)^{2/3}\tau_\Gamma^{\nabla,L}(z)\right)\right], \quad (2.58)$$

where the estimates

$$\sup_{|z-\alpha|\leq r}|\varepsilon_A(z)| \leq \frac{C}{N} \qquad \text{and} \qquad \sup_{|z-\alpha|\leq r}|\varepsilon_B(z)| \leq \frac{C}{N}$$

both hold for all N sufficiently large and where the leading coefficient functions defined by

$$A_\Gamma^{\nabla,L}(z) := \left(\frac{3}{4}\right)^{1/6}\sqrt{2\pi}e^{(\eta(z)-\gamma)/2}H_\Gamma^-(z) \cdot N^{-1/6}\left(-\tau_\Gamma^{\nabla,L}(z)\right)^{1/4},$$

$$B_\Gamma^{\nabla,L}(z) := -\left(\frac{3}{4}\right)^{-1/6}\sqrt{2\pi}e^{(\eta(z)-\gamma)/2}H_\Gamma^+(z) \cdot N^{1/6}\left(-\tau_\Gamma^{\nabla,L}(z)\right)^{-1/4}$$

are real-analytic functions for $|z - \alpha| \leq r$ that remain uniformly bounded in this disc as $N \to \infty$. Furthermore, we may also write

$$\pi_{N,k}(z) = e^{N\overline{L}_c^I(z)}\left[N^{1/6}A_\Gamma^{\nabla,L}(\alpha)\mathrm{Ai}\left(-\left(\frac{3}{4}\right)^{2/3}\tau_\Gamma^{\nabla,L}(z)\right) + \delta(z)\right], \quad (2.59)$$

where the estimate

$$\sup_{|z-\alpha|\leq rN^{-2/3}}|\delta(z)| \leq \frac{C}{N^{1/6}}$$

holds for all sufficiently large N.

Let $z = \beta$ be the right endpoint of a band I and suppose that a void Γ lies immediately to the right of $z = \beta$. There exist constants $r > 0$ and $C > 0$ such that when $|z - \beta| \leq r$,

$$\pi_{N,k}(z) = e^{N\overline{L}_c^I(z)}\left[N^{1/6}\left(A_\Gamma^{\nabla,R}(z) + \varepsilon_A(z)\right)\mathrm{Ai}\left(-\left(\frac{3}{4}\right)^{2/3}\tau_\Gamma^{\nabla,R}(z)\right)\right.$$
$$\left. + N^{-1/6}\left(B_\Gamma^{\nabla,R}(z) + \varepsilon_B(z)\right)\mathrm{Ai}'\left(-\left(\frac{3}{4}\right)^{2/3}\tau_\Gamma^{\nabla,R}(z)\right)\right], \quad (2.60)$$

where the estimates
$$\sup_{|z-\beta|\leq r}|\varepsilon_A(z)| \leq \frac{C}{N} \quad \text{and} \quad \sup_{|z-\beta|\leq r}|\varepsilon_B(z)| \leq \frac{C}{N}$$
both hold for all N sufficiently large and where the leading coefficient functions defined by
$$A_\Gamma^{\nabla,R}(z) := \left(\frac{3}{4}\right)^{1/6}\sqrt{2\pi}e^{(\eta(z)-\gamma)/2}H_\Gamma^+(z)\cdot N^{-1/6}\left(-\tau_\Gamma^{\nabla,R}(z)\right)^{1/4},$$
$$B_\Gamma^{\nabla,R}(z) := -\left(\frac{3}{4}\right)^{-1/6}\sqrt{2\pi}e^{(\eta(z)-\gamma)/2}H_\Gamma^-(z)\cdot N^{1/6}\left(-\tau_\Gamma^{\nabla,R}(z)\right)^{-1/4},$$
are real-analytic functions for $|z-\beta|\leq r$ that remain uniformly bounded in this disc as $N\to\infty$. Furthermore, we may also write
$$\pi_{N,k}(z) = e^{N\overline{L}_c^I(z)}\left[N^{1/6}A_\Gamma^{\nabla,R}(\beta)\text{Ai}\left(-\left(\frac{3}{4}\right)^{2/3}\tau_\Gamma^{\nabla,R}(z)\right) + \delta(z)\right], \tag{2.61}$$
where the estimate
$$\sup_{|z-\beta|\leq rN^{-2/3}}|\delta(z)| \leq \frac{C}{N^{1/6}}$$
holds for all sufficiently large N.

Note that these asymptotic formulae are similar in nature to the corresponding asymptotic formulae found in [DeiKMVZ99b] for polynomials orthogonal with respect to analytic weights on the whole real line. On the other hand, there is no analogue of a saturated region for continuous weights. The asymptotic behavior near the edge between a band I and a saturated region Γ involves the Airy function $\text{Bi}(\cdot)$, as well as $\text{Ai}(\cdot)$, and is the subject of the next theorem.

Theorem 2.16 (Asymptotics of $\pi_{N,k}(z)$ near band/saturated region edges). *Let $z=\alpha$ be the left endpoint of a band I and suppose that a saturated region Γ lies immediately to the left of $z=\alpha$. There exist constants $r>0$ and $C>0$ such that when $|z-\alpha|\leq r$,*
$$\pi_{N,k}(z) = e^{N\overline{L}_c^I(z)}\left[N^{1/6}\left(A_\Gamma^{\Delta,L}(z) + \varepsilon_A(z)\right)F_A^L(z) + N^{-1/6}\left(B_\Gamma^{\Delta,L}(z) + \varepsilon_B(z)\right)F_B^L(z)\right], \tag{2.62}$$
with
$$F_A^L(z) := \cos\left(\frac{N\theta^0(z)}{2}\right)\text{Bi}\left(-\left(\frac{3}{4}\right)^{2/3}\tau_\Gamma^{\Delta,L}(z)\right) - \sin\left(\frac{N\theta^0(z)}{2}\right)\text{Ai}\left(-\left(\frac{3}{4}\right)^{2/3}\tau_\Gamma^{\Delta,L}(z)\right),$$
$$F_B^L(z) := \cos\left(\frac{N\theta^0(z)}{2}\right)\text{Bi}'\left(-\left(\frac{3}{4}\right)^{2/3}\tau_\Gamma^{\Delta,L}(z)\right) - \sin\left(\frac{N\theta^0(z)}{2}\right)\text{Ai}'\left(-\left(\frac{3}{4}\right)^{2/3}\tau_\Gamma^{\Delta,L}(z)\right), \tag{2.63}$$
where the estimates
$$\sup_{|z-\alpha|\leq r}|\varepsilon_A(z)| \leq \frac{C}{N} \quad \text{and} \quad \sup_{|z-\alpha|\leq r}|\varepsilon_B(z)| \leq \frac{C}{N}$$
both hold for all N sufficiently large and where the leading coefficient functions defined by
$$A_\Gamma^{\Delta,L}(z) := \left(\frac{3}{4}\right)^{1/6}\sqrt{2\pi}e^{(\eta(z)-\gamma)/2}H_\Gamma^-(z)\cdot N^{-1/6}\left(-\tau_\Gamma^{\Delta,L}(z)\right)^{1/4},$$
$$B_\Gamma^{\Delta,L}(z) := \left(\frac{3}{4}\right)^{-1/6}\sqrt{2\pi}e^{(\eta(z)-\gamma)/2}H_\Gamma^+(z)\cdot N^{1/6}\left(-\tau_\Gamma^{\Delta,L}(z)\right)^{-1/4},$$
are real-analytic functions for $|z-\alpha|\leq r$ that remain uniformly bounded in this disc as $N\to\infty$. Furthermore, we may also write
$$\pi_{N,k}(z) = e^{N\overline{L}_c^I(z)}\left[N^{1/6}A_\Gamma^{\Delta,L}(\alpha)F_A^L(z) + \delta(z)\right], \tag{2.64}$$

where the estimate
$$\sup_{|z-\alpha|\leq rN^{-2/3}} |\delta(z)| \leq \frac{C}{N^{1/6}}$$
holds for all sufficiently large N.

Let $z = \beta$ be the right endpoint of a band I and suppose that a saturated region Γ lies immediately to the right of $z = \beta$. There exist constants $r > 0$ and $C > 0$ such that when $|z - \beta| \leq r$,
$$\pi_{N,k}(z) = e^{N\overline{L}_c^I(z)} \left[N^{1/6} \left(A_\Gamma^{\Delta,R}(z) + \varepsilon_A(z) \right) F_A^R(z) + N^{-1/6} \left(B_\Gamma^{\Delta,R}(z) + \varepsilon_B(z) \right) F_B^R(z) \right], \tag{2.65}$$
with
$$\begin{aligned} F_A^R(z) &:= \cos\left(\frac{N\theta^0(z)}{2}\right) \operatorname{Bi}\left(-\left(\frac{3}{4}\right)^{2/3} \tau_\Gamma^{\Delta,R}(z)\right) + \sin\left(\frac{N\theta^0(z)}{2}\right) \operatorname{Ai}\left(-\left(\frac{3}{4}\right)^{2/3} \tau_\Gamma^{\Delta,R}(z)\right), \\ F_B^R(z) &:= \cos\left(\frac{N\theta^0(z)}{2}\right) \operatorname{Bi}'\left(-\left(\frac{3}{4}\right)^{2/3} \tau_\Gamma^{\Delta,R}(z)\right) + \sin\left(\frac{N\theta^0(z)}{2}\right) \operatorname{Ai}'\left(-\left(\frac{3}{4}\right)^{2/3} \tau_\Gamma^{\Delta,R}(z)\right), \end{aligned} \tag{2.66}$$
where the estimates
$$\sup_{|z-\beta|\leq r} |\varepsilon_A(z)| \leq \frac{C}{N} \quad \text{and} \quad \sup_{|z-\beta|\leq r} |\varepsilon_B(z)| \leq \frac{C}{N}$$
both hold for all N sufficiently large and where the leading coefficient functions defined by
$$\begin{aligned} A_\Gamma^{\Delta,R}(z) &:= \left(\frac{3}{4}\right)^{1/6} \sqrt{2\pi} e^{(\eta(z)-\gamma)/2} H_\Gamma^-(z) \cdot N^{-1/6} \left(-\tau_\Gamma^{\Delta,R}(z)\right)^{1/4}, \\ B_\Gamma^{\Delta,R}(z) &:= -\left(\frac{3}{4}\right)^{-1/6} \sqrt{2\pi} e^{(\eta(z)-\gamma)/2} H_\Gamma^+(z) \cdot N^{1/6} \left(-\tau_\Gamma^{\Delta,R}(z)\right)^{-1/4} \end{aligned}$$
are real-analytic functions for $|z - \beta| \leq r$ that remain uniformly bounded in this disc as $N \to \infty$. Furthermore, we may also write
$$\pi_{N,k}(z) = e^{N\overline{L}_c^I(z)} \left[N^{1/6} A_\Gamma^{\Delta,R}(\beta) F_A^R(z) + \delta(z) \right], \tag{2.67}$$
where the estimate
$$\sup_{|z-\beta|\leq rN^{-2/3}} |\delta(z)| \leq \frac{C}{N^{1/6}}$$
holds for all sufficiently large N.

With the proper choice of the closed set K in Theorem 2.7, we see that the whole complex z-plane has been covered with overlapping closed sets, in each of which there is an associated asymptotic formula for $\pi_{N,k}(z)$ with rigorous error bounds.

2.4 EQUILIBRIUM MEASURES FOR SOME CLASSICAL DISCRETE ORTHOGONAL POLYNOMIALS

Since the asymptotic behavior of the discrete orthogonal polynomials is determined by the equilibrium measure $\mu_{\min}^c$ corresponding to the functions $\rho^0(x)$, $V(x)$, the interval $[a,b]$, and the constant c, it will be useful to demonstrate that the results stated in §2.3 can be made effective by a concrete calculation of the equilibrium measure. We consider below two classical cases. The equilibrium measure for the Krawtchouk polynomials was obtained by Dragnev and Saff in [DraS00]. The equilibrium measure for the Hahn polynomials has not appeared in the literature before (to our knowledge), and we present it below as well.

2.4.1 The Krawtchouk polynomials

The Krawtchouk polynomials [AbrS65] are orthogonal on a finite set of equally spaced nodes in the interval $(0,1)$:
$$x_{N,n} := \frac{2n+1}{2N}, \qquad \text{for } n = 0,1,2,\ldots,N-1\,.$$

The analytic probability density on $(0,1)$ is then given simply by $\rho^0(x) \equiv 1$. The corresponding weights are given by
$$w_{N,n}^{\text{Kraw}}(p,q) := \frac{N^{N-1}\sqrt{pq}}{q^N \Gamma(N)} \binom{N-1}{n} p^n q^{N-1-n}\,,$$

where p and q are positive parameters. The first factor that depends only on N, p, and q is not present in the classical formula [AbrS65] for the weights; we include it for convenience since the lattice spacing for our nodes is $1/N$ rather than being fixed. In any case, since
$$\prod_{\substack{m=0 \\ m \neq n}}^{N-1}(x_{N,n} - x_{N,m}) = \frac{(-1)^{N-1-n}}{N^{N-1}} n!(N-1-n)!\,,$$

the weights may also be written in the form
$$w_{N,n}^{\text{Kraw}}(p,q) = e^{-N V_N^{\text{Kraw}}(x_{N,n};l)} \prod_{\substack{m=0 \\ m \neq n}}^{N-1} |x_{N,n} - x_{N,m}|^{-1}\,,$$

where
$$V_N^{\text{Kraw}}(x;l) := lx\,, \qquad \text{with } l := \log \frac{q}{p}\,.$$

Note that in this case the function $V_N^{\text{Kraw}}(x;l)$ is coincidentally independent of N, so that $V^{\text{Kraw}}(x;l) = lx$ and $\eta^{\text{Kraw}}(x;l) \equiv 0$. These weights are therefore of the required form (see (1.16)) for our analysis. Since for the dual family of discrete orthogonal polynomials we should simply take the opposite sign of the function $V_N(x)$, we see that the polynomials dual to the Krawtchouk polynomials with parameter l are again Krawtchouk polynomials with parameter $-l$. A number of different involutions of the primitive parameters p and q correspond to changing the sign of l. For example, one could have $p \leftrightarrow 1/p$ and $q \leftrightarrow 1/q$, or simply $p \leftrightarrow q$. The latter involution is consistent with the typical assumption that $0 \leq p \leq 1$ and $p + q = 1$.

For the typical case when $0 \leq p \leq 1$ and $p + q = 1$, the above self-duality of the Krawtchouk polynomials implies that it is in fact sufficient to consider $0 \leq p \leq 1/2$. This fact was used in the paper [DraS00], where the equilibrium measure was explicitly constructed for all p in this range and for all $c \in (0,1)$. To summarize the results, it has been shown that there is a single band $I \subset [0,1]$, with endpoints $\alpha = \alpha(p,c) < \beta(p,c) = \beta$, for which there are explicit formulae. The behavior of the equilibrium measure in $(0,1) \setminus I$ depends on the relationship between c and p in the following way:

- If $0 \leq c < p$: The intervals $(0,\alpha)$ and $(\beta,1)$ are both voids.

- If $p < c < q$: The interval $(0,\alpha)$ is a saturated region, and the interval $(\beta,1)$ is a void.

- If $q < c \leq 1$: The intervals $(0,\alpha)$ and $(\beta,1)$ are both saturated regions.

This information supports our argument that the situation of having a constraint active at both endpoints of the interval is generic with respect to small perturbations of c. The borderline cases of $c = p$ and $c = q$ are also interesting. In the paper [DraS00] it is shown that
$$\alpha \to 0 \text{ as } c \to p \qquad \text{and} \qquad \beta \to 1 \text{ as } c \to q\,,$$

and for $c = p$ the density $d\mu_{\min}^c/dx$ of the equilibrium measure is equal to the average of the constraints at $x = 0$, while for $c = q$ it is equal to the average of the constraints at $x = 1$. These are thus both special cases of the general result stated in Proposition 2.2.

There exists an integral representation for the Krawtchouk polynomials, and an exhaustive asymptotic analysis of the polynomials has been carried out using this formula and the classical method of steepest descent; see [IsmS98]. Our formulae for the leading-order terms agree with those in [IsmS98] in the interior of all bands, voids, and saturated regions (the steepest-descent analysis is carried out with z held fixed away from all band edges and from the endpoints of the interval of accumulation of the nodes). The relative error obtained in [IsmS98] is typically of the order $O(N^{-1/2})$, although it is stated that under some circumstances this can be improved to $O(N^{-1})$ in some voids and saturated regions. The relative error estimates associated with the asymptotic formulae presented in §2.3 thus generally sharpen those in [IsmS98] in regions where the $O(N^{-1/2})$ relative error bound is obtained. It should be noted that while integral representations like that analyzed in [IsmS98] are not available for more general (nonclassical) discrete orthogonal polynomials, the methods to be developed in Chapters 4 and 5 that lead to the general theorems stated in §2.3 apply in the absence of any such representation.

An interesting fact is that the Krawtchouk weights are invariant under the Toda flow for each N. Indeed, from (1.32) for $p = 1$ (the Toda flow index p, not the Krawtchouk parameter p), we should consider the exponentially deformed Krawtchouk weights:

$$w_{N,n}^{\text{Kraw}}(p,q,t) := w_{N,n}^{\text{Kraw}}(p,q)e^{2x_{N,n}t}.$$

An elementary observation in the case $p + q = 1$ is that

$$w_{N,n}^{\text{Kraw}}(p_0, 1-p_0)e^{2x_{N,n}t} = w_{N,n}^{\text{Kraw}}(p(t), 1-p(t)),$$

where

$$p(t) := \frac{p_0}{p_0 + (1-p_0)e^{-2T}} \quad \text{and} \quad T = \frac{t}{N}. \tag{2.68}$$

Thus as t varies from $-\infty$ to $+\infty$, $p(t)$ increases monotonically under the Toda flow from $p(-\infty) = 0$ to $p(+\infty) = 1$. This interesting observation can be found in [AptV01] and [KuiM01]. Remarkably, this construction also arises naturally in the context of geometric quantization of the Riemann sphere (the quantum theory of spin); see [BloGPU03].

2.4.2 The Hahn and associated Hahn polynomials

Now we consider a semi-infinite lattice of equally spaced nodes

$$x_{N,n} := \frac{2n+1}{2N}, \quad \text{for } n = 0, 1, 2, \ldots, \tag{2.69}$$

and consider a corresponding three-parameter family of weights [AbrS65],

$$w_{N,n}(b,c,d) := \frac{N^{N-1}}{\Gamma(N)} \cdot \frac{\Gamma(b)\Gamma(c+n)\Gamma(d+n)}{\Gamma(n+1)\Gamma(b+n)\Gamma(c)\Gamma(d)}, \tag{2.70}$$

where b, c, and d are real parameters. The prefactor depending only on N is included as a convenient normalization factor that takes into account the fact that the lattice spacing in (2.69) is $1/N$.

Although the measure corresponding to the weight function (2.70) is supported on an infinite set, there are always only a finite number of orthogonal polynomials. For example, if one takes the parameters b, c, and d to be positive, then Stirling's formula shows that the weight decays for large n only if a certain inequality is satisfied among b, c, and d and that it decays only algebraically, like n^{-p} with the power p depending on b, c, and d. Therefore, for positive parameters, the weight function (2.70) has only a finite number of finite moments, and consequently only a finite number of powers of n may be orthogonalized.

We consider here a different way of arriving at a finite family of orthogonal polynomials starting from (2.70). If one takes a limit in the parameter space, letting the parameter c in (2.70) tend toward the negative integer $1 - N$, then one finds

$$w_{N,n}(b, 1-N, d) := \lim_{c \to 1-N} w_{N,n}(b,c,d) = \begin{cases} \dfrac{N^{N-1}}{\Gamma(N)} \binom{N-1}{n} \cdot (-1)^n \cdot \dfrac{\Gamma(b)\Gamma(d+n)}{\Gamma(b+n)\Gamma(d)}, & \text{if } n \in \mathbb{Z}_N, \\ 0, & \text{if } n \geq N. \end{cases}$$

The limiting weights are thus supported on $\mathbb{Z}_N$ rather than on an infinite lattice, and according to (2.69) the N nodes $x_{N,0} < \cdots < x_{N,N-1}$ are equally spaced with spacing $1/N$ and $x_{N,0} = 1/(2N)$. Therefore the node density function is $\rho^0(x) \equiv 1$. Note that the weights $w_{N,n}(b, 1 - N, d)$ are not positive for all $n \in \mathbb{Z}_N$ unless further conditions are placed on the remaining real parameters b and d. Insisting that $w_{N,n}(b, 1 - N, d) > 0$ for all $n \in \mathbb{Z}_N$ identifies two disjoint regions in the (b, d)-plane.

One of these regions is delineated by the inequalities $d > 0$ and $b < 2 - N$. In this case, we refer to $w_{N,n}(b, 1 - N, d)$ as the *Hahn weight*, and we call the corresponding polynomials the Hahn polynomials. Let P and Q be positive parameters. Setting $d = P$ and $b = 2 - N - Q$ in the limiting formula for $w_{N,n}(b, 1 - N, d)$, we arrive at a simple formula for the Hahn weights:

$$w_{N,n}(b, 1 - N, d) = w_{N,n}^{\text{Hahn}}(P, Q) := \frac{N^{N-1}}{\Gamma(N)} \cdot \frac{\binom{n + P - 1}{n}\binom{N + Q - 2 - n}{N - 1 - n}}{\binom{N + Q - 2}{Q - 1}}, \quad n \in \mathbb{Z}_N. \qquad (2.71)$$

Note that by taking $P = Q = 1$, the Hahn weights become independent of n, so in this special case the Hahn polynomials are up to a factor the (discrete) Tchebychev polynomials (in the terminology of [AbrS65] and [Sze91]), also known as the Gram polynomials (in the terminology of [DahBA74]).

The other region of the (b, d)-plane for which the weights $w_{N,n}(b, 1 - N, d)$ are positive for all $n \in \mathbb{Z}_N$ is delineated by the inequalities $b > 0$ and $d < 2 - N$. In this case, we refer to $w_{N,n}(b, 1 - N, d)$ as the *associated Hahn weight*, and we call the corresponding polynomials the associated Hahn polynomials. Again, let P and Q be positive parameters. Setting $d = 2 - N - Q$ and $b = P$ in the limiting formula for $w_{N,n}(b, 1 - N, d)$, the associated Hahn weights are

$$w_{N,n}(b, 1 - N, d) = w_{N,n}^{\text{Assoc}}(P, Q) := \frac{N^{N-1}}{\Gamma(N)} \cdot \frac{\Gamma(N)\Gamma(N + Q - 1)\Gamma(P)}{\Gamma(n+1)\Gamma(P+n)\Gamma(N-n)\Gamma(N+Q-1-n)}, \quad n \in \mathbb{Z}_N. \qquad (2.72)$$

Note that

$$w_{N,n}^{\text{Hahn}}(P, Q) w_{N,n}^{\text{Assoc}}(P, Q) \prod_{m \neq n} (x_{N,m} - x_{N,n})^2 = 1,$$

for all $n \in \mathbb{Z}_N$ and all $P > 0$ and $Q > 0$. This means that the associated Hahn polynomials are dual to the Hahn polynomials (compare the general definition (1.46) of dual weights in §1.4.3).

Writing the Hahn weights (2.71) in the form (1.16), we have

$$V_N^{\text{Hahn}}(x_{N,n}; P, Q) = \frac{1}{N} \log\left(\frac{\Gamma(P)\Gamma(N + Q - 1)}{\Gamma(Nx_{N,n} + P - 1/2)\Gamma(N(1 - x_{N,n}) + Q - 1/2)}\right).$$

The interesting case is when P and Q are large. We therefore set $P = NA + 1$ and $Q = NB + 1$ for A and B fixed positive parameters, and from Stirling's formula, we then have

$$V_N^{\text{Hahn}}(x; NA + 1, NB + 1) = V^{\text{Hahn}}(x; A, B) + \frac{\eta^{\text{Hahn}}(x; A, B)}{N},$$

where

$$V^{\text{Hahn}}(x; A, B) := A\log(A) + (B+1)\log(B+1) - (A+x)\log(A+x) - (B+1-x)\log(B+1-x)$$

and

$$\eta^{\text{Hahn}}(x; A, B) := \frac{1}{2}\log\left(\frac{A}{B+1}\right) + O\left(\frac{1}{N}\right).$$

The convergence is uniform for x in compact subsets of $\mathbb{C} \setminus ((-\infty, -A) \cup (B + 1, +\infty))$.

◁ **Remark:** The fact that the leading term in $\eta^{\text{Hahn}}(x; A, B)$ is independent of x can be traced back to the particular choice of the order-1 terms in P and Q that we have made. Other choices consistent with the same leading-order scaling (say, simply taking $P = NA$ and $Q = NB$) introduce genuine analytic x dependence into the leading term of the correction $\eta^{\text{Hahn}}(x; A, B)$. ▷

For the associated Hahn weight (2.72), the case of $P = NA + 1$ and $Q = NB + 1$ is also of interest. By duality,
$$V_N^{\text{Assoc}}(x; AN+1, BN+1) = -V_N^{\text{Hahn}}(x; AN+1, BN+1),$$
and therefore we also have $V^{\text{Assoc}}(x; A, B) = -V^{\text{Hahn}}(z; A, B)$ at the level of the leading term as $N \to \infty$. According to Proposition 2.6, if the equilibrium measure corresponding to the function $V^{\text{Hahn}}(x; A, B)$ and the node density function $\rho^0(z)$ is known for all values of the parameter c, then that corresponding to the function $V^{\text{Assoc}}(x; A, B)$ (and the same node density function) is also known for all values of the parameter c, essentially by means of the involution $c \leftrightarrow 1 - c$.

We have computed the equilibrium measure corresponding to the potential $V^{\text{Hahn}}(x; A, B)$ and $\rho^0(x) \equiv 1$, for $x \in (0,1)$. To describe it, we first define two positive constants c_A and c_B by
$$\begin{aligned} c_A &:= \frac{-(A+B) + \sqrt{(A+B)^2 + 4A}}{2}, \\ c_B &:= \frac{-(A+B) + \sqrt{(A+B)^2 + 4B}}{2}. \end{aligned} \quad (2.73)$$

We can check directly that $0 < c_A, c_B < 1$ for $A, B > 0$, and $c_A < c_B$ if $0 < A < B$. Now let us assume that $A \leq B$ (see the remark below). Then $c_A \leq c_B$, and we consider the three distinct possibilities: $c \in (0, c_A)$, $c \in (c_A, c_B)$, and $c \in (c_B, 1)$. It turns out that in each of these cases, there is one band interval, denoted by $(\alpha, \beta) \subset (0, 1)$, on both sides of which are either saturated regions or voids.

- For $c \in (0, c_A)$, the interval (α, β) is the band, and the intervals $(0, \alpha)$ and $(\beta, 1)$ are voids. We refer to this configuration as *void-band-void*.

- For $c \in (c_A, c_B)$, the interval (α, β) is the band, $(0, \alpha)$ is a saturated region, and $(\beta, 1)$ is a void. We refer to this configuration as *saturated-band-void*.

- For $c \in (c_B, 1)$, the interval (α, β) is the band, and the intervals $(0, \alpha)$ and $(\beta, 1)$ are both saturated regions. We refer to this configuration as *saturated-band-saturated*.

As in the Krawtchouk case, the critical values of $c = c_A$ and $c = c_B$ are somewhat special because either $\alpha = 0$ or $\beta = 1$.

◁ **Remark:** For the case when $A \geq B$, we have $c_B \leq c_A$, and the midregime for c becomes the interval (c_B, c_A). For $c \in (c_B, c_A)$, the interval (α, β) is the band, $(0, \alpha)$ is a void, and $(\beta, 1)$ is a saturated region; we refer to this configuration as *void-band-saturated*. However, there is a symmetry in this problem: if one swaps $A \leftrightarrow B$ and $x \leftrightarrow (1-x)$, then the field $\varphi(x)$ is changed only by a constant that can be absorbed into the Lagrange multiplier ℓ_c. Therefore it is sufficient to consider $A \leq B$. ▷

For all values of c, the band endpoints α and β are the two (real) solutions of the following quadratic equation in X:
$$X^2 - 2\frac{A(A+B) + (A+B)(B-A+2)c + (B-A+2)c^2}{(A+B+2c)^2}X + \left(\frac{c^2 + (A+B)c - A}{(A+B+2c)^2}\right)^2 = 0. \quad (2.74)$$

It is straightforward to check that $\alpha, \beta \in [0, 1]$ for all $0 < c < 1$ and $A, B > 0$, and the formulae for α and β are
$$\alpha := \frac{(B-A+2)c^2 + (A+B)(B-A+2)c + A(A+B) - 2\sqrt{D}}{(A+B+2c)^2} \quad (2.75)$$
and
$$\beta := \frac{(B-A+2)c^2 + (A+B)(B-A+2)c + A(A+B) + 2\sqrt{D}}{(A+B+2c)^2}, \quad (2.76)$$
where the discriminant is given by
$$D := c(1-c)(A+c)(B+c)(A+B+c)(A+B+c+1).$$

Now we describe the density of the equilibrium measure $d\mu^c_{\min}/dx$, assuming (without loss of generality according to the remark above) that $A \leq B$. It is useful to introduce the following notation. Let the positive function $T(x)$ be defined by

$$T(x) := \sqrt{\frac{\beta - x}{x - \alpha}}, \qquad \text{for } \alpha < x < \beta,$$

and define four positive constants by

$$k_1 := \sqrt{\frac{1+B-\alpha}{1+B-\beta}}, \quad k_2 := \sqrt{\frac{1-\alpha}{1-\beta}},$$

$$k_3 := \sqrt{\frac{A+\alpha}{A+\beta}}, \quad k_4 := \sqrt{\frac{\alpha}{\beta}}.$$

Theorem 2.17. *For the functions $V(x) \equiv V^{\text{Hahn}}(x)$ and $\rho^0(x) \equiv 1$ on $x \in [0,1]$, the solution of the variational problem in §2.1 is given by the following formulae when the parameters satisfy $A \leq B$. Let the constants c_A and c_B be given by (2.73) and let α and β be defined by (2.75) and (2.76). If $0 < c < c_A$ (void-band-void), then for $\alpha \leq x \leq \beta$,*

$$\frac{d\mu^c_{\min}}{dx}(x) := \frac{1}{\pi c}\left[\arctan\left(k_2 T(x)\right) + \arctan\left(k_3 T(x)\right) - \arctan\left(k_1 T(x)\right) - \arctan\left(k_4 T(x)\right)\right] \qquad (2.77)$$

and $d\mu^c_{\min}/dx \equiv 0$ if $0 \leq x \leq \alpha$ or $\beta \leq x \leq 1$. The corresponding Lagrange multiplier is given by

$$\ell_c := (\beta - \alpha)\left\{[\log(\beta - \alpha) - 1] K^{(1)}_{\text{VBV}} - 2\log(2) K^{(2)}_{\text{VBV}} + 2 K^{(3)}_{\text{VBV}}\right\} + \varphi(\beta),$$

where

$$K^{(1)}_{\text{VBV}} := \frac{k_1}{1+k_1} - \frac{k_2}{1+k_2} - \frac{k_3}{1+k_3} + \frac{k_4}{1+k_4},$$

$$K^{(2)}_{\text{VBV}} := \frac{k_1}{1-k_1^2} - \frac{k_2}{1-k_2^2} - \frac{k_3}{1-k_3^2} + \frac{k_4}{1-k_4^2},$$

$$K^{(3)}_{\text{VBV}} := \frac{\log(1+k_1)}{1-k_1^2} - \frac{\log(1+k_2)}{1-k_2^2} - \frac{\log(1+k_3)}{1-k_3^2} + \frac{\log(1+k_4)}{1-k_4^2},$$

and $\varphi(\cdot)$ is the external field given in terms of $V(\cdot) = V^{\text{Hahn}}(\cdot; A, B)$ and $\rho^0(\cdot) \equiv 1$ by (2.1). If $c_A < c < c_B$ (saturated-band-void), then for $\alpha \leq x \leq \beta$,

$$\frac{d\mu^c_{\min}}{dx}(x) := \frac{1}{\pi c}\left[\arctan\left(k_2 T(x)\right) + \arctan\left(k_3 T(x)\right) - \arctan\left(k_1 T(x)\right) + \arctan\left(k_4 T(x)\right)\right], \qquad (2.78)$$

and $d\mu^c_{\min}/dx \equiv 1/c$ if $0 \leq x \leq \alpha$ and $d\mu^c_{\min}/dx \equiv 0$ if $\beta \leq x \leq 1$. The corresponding Lagrange multiplier is given by

$$\ell_c := (\beta - \alpha)\left\{[\log(\beta - \alpha) - 1] K^{(1)}_{\text{SBV}} - 2\log(2) K^{(2)}_{\text{SBV}} + 2 K^{(3)}_{\text{SBV}}\right\} + \varphi(\beta) + 2(\beta - \alpha)\log(\beta - \alpha) + 2\alpha - 2\beta\log(\beta),$$

where

$$K^{(1)}_{\text{SBV}} := \frac{k_1}{1+k_1} - \frac{k_2}{1+k_2} - \frac{k_3}{1+k_3} - \frac{k_4}{1+k_4},$$

$$K^{(2)}_{\text{SBV}} := \frac{k_1}{1-k_1^2} - \frac{k_2}{1-k_2^2} - \frac{k_3}{1-k_3^2} - \frac{k_4}{1-k_4^2},$$

$$K^{(3)}_{\text{SBV}} := \frac{\log(1+k_1)}{1-k_1^2} - \frac{\log(1+k_2)}{1-k_2^2} - \frac{\log(1+k_3)}{1-k_3^2} - \frac{\log(1+k_4)}{1-k_4^2}.$$

Finally, if $c_B < c < 1$ (saturated-band-saturated), then for $\alpha \leq x \leq \beta$,

$$\frac{d\mu^c_{\min}}{dx}(x) := \frac{1}{c} + \frac{1}{\pi c}\left[\arctan\left(k_3 T(x)\right) + \arctan\left(k_4 T(x)\right) - \arctan\left(k_1 T(x)\right) - \arctan\left(k_2 T(x)\right)\right], \qquad (2.79)$$

and $d\mu_{\min}^c/dx \equiv 1/c$ if $0 \leq x \leq \alpha$ and $\beta \leq x \leq 1$. The corresponding Lagrange multiplier is given by

$$\ell_c := (\beta - \alpha)\left\{[\log(\beta - \alpha) - 1]\, K_{\text{SBS}}^{(1)} - 2\log(2) K_{\text{SBS}}^{(2)} + 2 K_{\text{SBS}}^{(3)}\right\} + \varphi(\beta) + 2 - 2\beta\log(\beta) - 2(1-\beta)\log(1-\beta),$$

where

$$K_{\text{SBS}}^{(1)} := \frac{k_1}{1+k_1} + \frac{k_2}{1+k_2} - \frac{k_3}{1+k_3} - \frac{k_4}{1+k_4},$$

$$K_{\text{SBS}}^{(2)} := \frac{k_1}{1-k_1^2} + \frac{k_2}{1-k_2^2} - \frac{k_3}{1-k_3^2} - \frac{k_4}{1-k_4^2},$$

$$K_{\text{SBS}}^{(3)} := \frac{\log(1+k_1)}{1-k_1^2} + \frac{\log(1+k_2)}{1-k_2^2} - \frac{\log(1+k_3)}{1-k_3^2} - \frac{\log(1+k_4)}{1-k_4^2}.$$

The shapes of the equilibrium measures for the Hahn weights are illustrated in Figure 2.3, which shows the way the measures change as c is varied for fixed $A < B$. The proof of Theorem 2.17 is simply to check

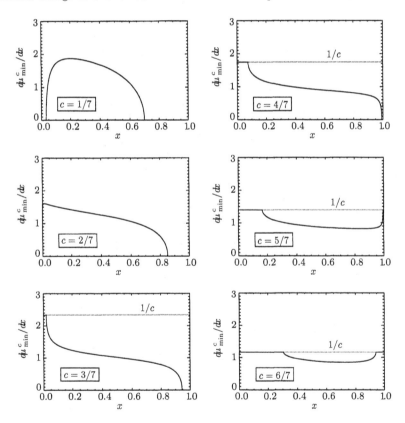

Figure 2.3 *The density of the equilibrium measure for the Hahn polynomials for parameter values $A = 3$ and $B = 7$ and various values of c.*

directly that the following essential conditions are indeed verified:

1. The variational inequality (2.13) holds in all voids, the variational inequality (2.16) holds in all saturated regions, and the equilibrium condition (2.14) holds for $\alpha < x < \beta$.

2. The measure satisfies the normalization constraint (2.5).

3. For $\alpha < x < \beta$, the measure has a density lying strictly between the upper and lower constraints (2.4).

Rather than checking each of these conditions, we will deduce the formulae directly in Appendix B. In general, equilibrium measures may be computed either via integral formulae [KuiV99] relating them to the asymptotics of the corresponding recursion coefficients (which are known for the Hahn polynomials) or by directly solving the variational problem. In Appendix B we adopt the latter approach and follow similar reasoning as in [DeiKM98] to solve the variational problem for the Hahn weight and hence derive the relevant formulae recorded in Theorem 2.17.

◁ **Remark:** The Hahn polynomials have not been studied in the literature to the same extent as the Krawtchouk polynomials. There exists an integral representation of the Hahn polynomials, but it is apparently more difficult to analyze carefully than, for example, the corresponding integral formula for the Krawtchouk polynomials upon which the analysis in [IsmS98] rests. We believe that the formulae for the Hahn equilibrium measure presented in Theorem 2.17 and the corresponding Plancherel-Rotach-type asymptotics formulated in §2.3 are new in the literature. ▷

Chapter Three

Applications

3.1 DISCRETE ORTHOGONAL POLYNOMIAL ENSEMBLES AND THEIR PARTICLE STATISTICS

Recall from §1.1.1 that the discrete orthogonal polynomial ensemble is the statistical ensemble associated with the density (1.1):

$$p^{(N,k)}(x_1,\ldots,x_k) := \mathbb{P}(\text{there are particles at each of the nodes } x_1,\ldots,x_k)$$
$$= \frac{1}{Z_{N,k}} \prod_{1\leq i<j\leq k}(x_i-x_j)^2 \cdot \prod_{j=1}^{k} w(x_j), \qquad (3.1)$$

where $x_1,\ldots,x_k \in X_N$, and X_N is a discrete set of cardinality N as described in §1.2.

Some common properties of discrete orthogonal polynomial ensembles can be read off immediately from formula (3.1). For example, the presence of the Vandermonde factor means that the probability of finding two particles at the same site in X_N is zero. Thus a discrete orthogonal polynomial ensemble always describes an exclusion process. This phenomenon is the discrete analogue of the familiar *level repulsion* phenomenon in random matrix theory. Moreover, because of the discreteness of the underlying space, the particles are separated at least by the distance between consecutive nodes. This strong exclusion due to the discreteness of the space imposes the condition that the particle density has an upper bound, the limiting density of the nodes. This is a new feature in the discrete orthogonal polynomial ensembles that is not present in the orthogonal polynomial ensembles associated with continuous weights (*i.e.*, in random matrix theory). Also, since the weights are associated with nodes, the interpretation is that configurations where particles are concentrated in sets of nodes where the weight is larger are more likely.

Our goal will be to establish asymptotic formulae for various statistics associated with the ensemble (3.1) for a general class of weights in the continuum limit $N \to \infty$ subject to the basic assumptions enumerated in §1.3.1 and the generic simplifying assumptions described in §2.1.2.

Of basic interest is the *m-point correlation function*, defined for $m \leq k$ by

$$R_m^{(N,k)}(x_1,\ldots,x_m) := \mathbb{P}(\text{there are particles at each of the nodes } x_1,\ldots,x_m)$$
$$= \sum_{\substack{x_{m+1}<\cdots<x_k \\ x_j \in X_N}} p^{(N,k)}(x_1,\ldots,x_k). \qquad (3.2)$$

◁ **Remark:** In random matrix theory [Meh91, TraW98] the correlation functions $R_m^{(N,k)}$ are usually introduced with a prefactor of $k!/(k-m)!$ which mediates between a density function for which particles (eigenvalues) are considered to be distinguishable (unordered) and statistics for which order is irrelevant. Since we introduced $p^{(N,k)}(x_1,\ldots,x_k)$ from the start with the interpretation that the particles are indistinguishable, this factor is not present in (3.2). ▷

In particular, the one-point function $R_1^{(N,k)}(x)$ denotes the *density of states*, which is the probability that there is a particle at x. One can also verify the following interpretations: for any set $B \subset X_N$,

$$\sum_{x \in B} R_1^{(N,k)}(x) = \mathbb{E}(\text{number of particles in } B)$$

and
$$\sum_{\substack{x<y \\ x,y \in B}} R_2^{(N,k)}(x,y) = \mathbb{E}(\text{number of pairs of particles in } B),$$
where $\mathbb{E}$ denotes the expected value.

The fundamental calculation in random matrix theory in the case of $\beta = 2$ ensembles, due to Gaudin and Mehta (see, for example, [Meh91] or [TraW98]), shows that the correlation functions may, equivalently, be represented in the form
$$R_m^{(N,k)}(x_1,\ldots,x_m) = \det\left(K_{N,k}(x_i,x_j)\right)_{1 \leq i,j \leq m}, \tag{3.3}$$
where the *reproducing kernel* (Christoffel-Darboux kernel) is defined for nodes x and y by
$$K_{N,k}(x,y) := \sqrt{w(x)w(y)} \sum_{n=0}^{k-1} p_{N,n}(x) p_{N,n}(y). \tag{3.4}$$
With the Christoffel-Darboux formula [Sze91], which holds for all orthogonal polynomials (even in the discrete case) the sum on the right-hand side of (3.4) telescopes. Thus for distinct nodes $x \neq y$,
$$K_{N,k}(x,y) = \sqrt{w(x)w(y)} \frac{\gamma_{N,k-1}}{\gamma_{N,k}} \cdot \frac{p_{N,k}(x) p_{N,k-1}(y) - p_{N,k-1}(x) p_{N,k}(y)}{x - y}$$
$$= \sqrt{w(x)w(y)} \frac{\pi_{N,k}(x) \cdot \gamma_{N,k-1} p_{N,k-1}(y) - \gamma_{N,k-1} p_{N,k-1}(x) \cdot \pi_{N,k}(y)}{x - y}$$
$$= \sqrt{w(x)w(y)} \frac{P_{11}(x;N,k) P_{21}(y;N,k) - P_{21}(x;N,k) P_{11}(y;N,k)}{x - y},$$
where the last line follows from Proposition 1.3. Similarly, for any node x,
$$K_{N,k}(x,x) = w(x) \left(P_{11}'(x;N,k) P_{21}(x;N,k) - P_{21}'(x;N,k) P_{11}(x;N,k) \right).$$
Note that the resulting formulae are expressed in terms of the first column of the solution $\mathbf{P}(x;N,k)$ of Interpolation Problem 1.2 for a single value of k and that, furthermore,
$$P_{11}(x;N,k) P_{21}(y;N,k) - P_{21}(x;N,k) P_{11}(y;N,k) = \left(\mathbf{P}(x;N,k)^{-1} \mathbf{P}(y;N,k)\right)_{21},$$
$$P_{11}'(x;N,k) P_{21}(x;N,k) - P_{21}'(x;N,k) P_{11}(x;N,k) = -\left(\mathbf{P}(x;N,k)^{-1} \mathbf{P}'(x;N,k)\right)_{21}.$$
Therefore the correlation functions are written explicitly in terms of the discrete orthogonal polynomials associated with the nodes X_N and the weights $w_{N,n} = w(x_{N,n})$, and consequently these formulae can be analyzed rigorously in an appropriate continuum limit by using the methods we will present in detail in Chapters 4 and 5.

Consider a set $B \subset X_N$ and an integer m with $0 \leq m \leq \min(\#B, k)$. Another interesting statistic of a discrete orthogonal polynomial ensemble is then
$$A_m^{(N,k)}(B) := \mathbb{P}(\text{there are precisely } m \text{ particles in the set } B), \tag{3.5}$$
which vanishes automatically if $m > \#B$ by exclusion. This statistic is also well known to be expressible by the exact formula
$$A_m^{(N,k)}(B) = \frac{1}{m!} \left(-\frac{d}{dt}\right)^m \bigg|_{t=1} \det\left(1 - tK_{N,k}\big|_B\right), \tag{3.6}$$
where $K_{N,k}$ is the operator (in this case a finite matrix since B is contained in the finite set X_N) acting in $\ell^2(X_N)$ given by the kernel $K_{N,k}(x,y)$ and where $K_{N,k}\big|_B$ denotes the restriction of $K_{N,k}$ to $\ell^2(B)$.

This is by no means an exhaustive list of statistics that can be directly expressed in terms of the orthogonal polynomials associated with the (discrete) weight $w(\cdot)$. For example, one may consider the fluctuations and in particular the variance of the number of particles in an interval $B \subset X_N$. The continuum-limit asymptotics for this statistic were computed in [Joh02] for the Krawtchouk ensemble (see Proposition 2.5 in that paper), with the result that the fluctuations are Gaussian; it would be of some interest to determine whether this is a special property of the Krawtchouk ensemble or a universal property of a large class of ensembles. Also, there are convenient formulae for statistics associated with the spacings between particles; the reader can find such formulae in §5.6 of the book [Dei99].

APPLICATIONS

3.2 DUAL ENSEMBLES AND HOLE STATISTICS

Since the nodes X_N are finite in number, the distribution of the positions $x_1, \ldots, x_k$ of the particles naturally induces a distribution of the positions $y_1, \ldots, y_{\bar{k}}$ of the *holes* (*i.e.*, the nodes not occupied by particles). Here $\bar{k} = N - k$ and $\{x_1, \ldots, x_k\} \cup \{y_1, \ldots, y_{\bar{k}}\} = X_N$. It is interesting to determine the explicit formula of the hole distribution. We will show that when the particle locations x_j are distributed according to the probability density function $p^{(N,k)}(x_1, \ldots, x_k)$ as in (3.1), the density function of the hole locations y_j is always of the same form with only a different choice of weight function.

Let us define the joint hole distribution function as

$$\bar{p}^{(N,\bar{k})}(y_1, \cdots, y_{\bar{k}}) := \mathbb{P}(\text{there are holes at each of the nodes } y_1, \ldots, y_{\bar{k}}). \tag{3.7}$$

Given two complementary sets of nodes $\{x_1, \ldots, x_k\} \cup \{y_1, \ldots, y_{\bar{k}}\} = X_N$, from the definition (3.7) we find

$$\bar{p}^{(N,\bar{k})}(y_1, \ldots, y_{\bar{k}}) = p^{(N,k)}(x_1, \cdots, x_k)$$

$$= \frac{1}{Z_{N,k}} \prod_{1 \le i < j \le k} (x_i - x_j)^2 \cdot \prod_{j=1}^{k} w(x_j).$$

As we may write

$$\prod_{j=1}^{k} w(x_j) = C_N \prod_{j=1}^{\bar{k}} \frac{1}{w(y_j)}, \qquad \text{where } C_N := \prod_{j=0}^{N-1} w_{N,j},$$

we obtain

$$\bar{p}^{(N,\bar{k})}(y_1, \ldots, y_{\bar{k}}) = \frac{C_N}{Z_{N,k}} \prod_{1 \le i < j \le k} (x_i - x_j)^2 \cdot \prod_{j=1}^{\bar{k}} \frac{1}{w(y_j)}$$

$$= \frac{C_N}{Z_{N,k}} \prod_{1 \le i < j \le k} (x_i - x_j)^2 \cdot \prod_{j=1}^{\bar{k}} \prod_{\substack{n=0 \\ y_j \ne x_{N,n}}}^{N-1} (y_j - x_{N,n})^2$$

$$\cdot \prod_{j=1}^{\bar{k}} \left[\frac{1}{w(y_j)} \prod_{\substack{n=0 \\ y_j \ne x_{N,n}}}^{N-1} \frac{1}{(y_j - x_{N,n})^2} \right].$$

A little algebra shows that (compare (9.42) in [Bai99] or Lemma 2.2 in [Joh01])

$$\prod_{1 \le i < j \le k} |x_i - x_j| \cdot \prod_{j=1}^{\bar{k}} \prod_{\substack{n=0 \\ y_j \ne x_{N,n}}}^{N-1} |y_j - x_{N,n}| = \prod_{1 \le i < j \le k} |x_i - x_j| \cdot \prod_{j=1}^{\bar{k}} \prod_{\substack{i=1 \\ i \ne j}}^{\bar{k}} |y_j - y_i| \cdot \prod_{j=1}^{\bar{k}} \prod_{i=1}^{k} |y_j - x_i|$$

$$= \prod_{1 \le i < j \le k} |x_i - x_j| \cdot \prod_{1 \le i < j \le \bar{k}} |y_j - y_i|^2 \cdot \prod_{j=1}^{\bar{k}} \prod_{i=1}^{k} |y_j - x_i|$$

$$= D_N \prod_{1 \le i < j \le \bar{k}} |y_j - y_i|,$$

where D_N is the Vandermonde determinant of the nodes

$$D_N := \prod_{0 \le i < j \le N-1} |x_{N,i} - x_{N,j}|$$

and the identity

$$D_N = \prod_{1 \le i < j \le k} |x_i - x_j| \cdot \prod_{1 \le i < j \le \bar{k}} |y_i - y_j| \cdot \prod_{i=1}^{k} \prod_{j=1}^{\bar{k}} |x_i - y_j|$$

is used in the last line. Therefore the density of the holes is given by

$$\overline{p}^{(N,\bar{k})}(y_1,\ldots,y_{\bar{k}}) = \frac{1}{\overline{Z}_{N,\bar{k}}} \prod_{1 \leq i < j \leq \bar{k}} (y_i - y_j)^2 \cdot \prod_{j=1}^{\bar{k}} \overline{w}(y_j),$$

where the normalization constant is

$$\overline{Z}_{N,\bar{k}} = \frac{Z_{N,k}}{C_N D_N^2} = Z_{N,k} \prod_{j=0}^{N-1} \frac{1}{w_{N,j}} \cdot \prod_{0 \leq i < j \leq N-1} \frac{1}{|x_{N,i} - x_{N,j}|^2}$$

and the weight function is

$$\overline{w}(y_j) = \frac{1}{w(y_j)} \prod_{\substack{n=0 \\ y_j \neq x_{N,n}}}^{N-1} \frac{1}{(y_j - x_{N,n})^2}.$$

Note that this new weight function is precisely the dual weight defined by (1.46) in §1.4.3. Hence when the particles are distributed according to a discrete orthogonal polynomial ensemble, the holes are distributed according to the discrete orthogonal polynomial ensemble corresponding to the dual weights. We will say that the ensembles governed by the density functions $p^{(N,k)}(x_1,\ldots,x_k)$ and $\overline{p}^{(N,\bar{k})}(y_1,\ldots,y_{\bar{k}})$ are dual to each other. Since dual ensembles correspond to weights of similar form, but with the involutions $c \leftrightarrow 1-c$ and $V(x) \leftrightarrow -V(x)$, their statistics are analyzed in exactly the same way. Therefore the universal properties of the particle distribution that we will establish below automatically imply corresponding universal properties of the hole distribution.

3.3 RESULTS ON ASYMPTOTIC UNIVERSALITY FOR GENERAL WEIGHTS

The following theorems all describe the asymptotic behavior as $N \to \infty$ of various statistical quantities connected with the discrete orthogonal polynomial ensemble corresponding to nodes $X_N \subset [a,b]$ characterized by the function $\rho^0(\cdot)$ and weights characterized by the function $V(\cdot)$. These quantities, and the parameter c (asymptotic value of k/N, where k is the number of particles in the ensemble) are presumed to satisfy the same basic assumptions set forth in §1.3.1 and the simplifying assumptions set forth in §2.1.2. The theorems stated in this section will be proved in Chapter 7.

Let ξ_N and η_N be elements of a discrete subset D_N of $\mathbb{R}$ such that $\max D_N - \min D_N$ remains bounded and the distance between neighboring points of D_N converges to a constant as $N \to \infty$. The expression

$$S(\xi_N, \eta_N) := \frac{\sin(\pi(\xi_N - \eta_N))}{\pi(\xi_N - \eta_N)}$$

is called the *discrete sine kernel* ("discrete" reminds us that ξ_N and η_N lie in a discrete set D_N). We extend the definition of the discrete sine kernel to the diagonal by setting

$$S(\xi_N, \xi_N) := 1.$$

Theorem 3.1 (Universality of the discrete sine kernel in bands). *Suppose that $x_1,\ldots,x_l$ and $x_{l+1},\ldots,x_m$ are disjoint sets of nodes in a fixed closed interval F in the interior of any band I and denote by δ_N the distance between the two sets,*

$$\delta_N := \min_{\substack{1 \leq i \leq l \\ l+1 \leq j \leq m}} |x_i - x_j|.$$

Then

$$R_m^{(N,k)}(x_1,\ldots,x_m) = R_l^{(N,k)}(x_1,\ldots,x_l) R_{m-l}^{(N,k)}(x_{l+1},\ldots,x_m) + O\left(\frac{1}{N\delta_N}\right).$$

Fix x in the interior of any band I, let

$$\delta(x) := \left[c \frac{d\mu_{\min}^c}{dx}(x)\right]^{-1}, \tag{3.8}$$

APPLICATIONS 53

and for some integer $n \geq 1$ consider $\xi_N^{(1)}, \ldots, \xi_N^{(n)}$ all to lie in a fixed bounded set $D \subset \mathbb{R}$ such that

$$x_j := x + \xi_N^{(j)} \frac{\delta(x)}{N}, \qquad j = 1, \ldots, n,$$

all satisfy $x_j \in X_N$ and $x_j \to x$ as $N \to \infty$. Then there is a constant $C_{D,n} > 0$ such that for all N sufficiently large,

$$\max_{\xi_N^{(1)}, \ldots, \xi_N^{(n)} \in D} \left| R_n^{(N,k)}(x_1, \ldots, x_n) - \left[\frac{c}{\rho^0(x)} \frac{d\mu_{\min}^c}{dx}(x) \right]^n \det(S(\xi_N^{(i)}, \xi_N^{(j)}))_{1 \leq i,j \leq n} \right| \leq \frac{C_{D,n}}{N}.$$

Thus particles separated by distances large compared to $1/N$ are asymptotically statistically independent, and the asymptotically nontrivial correlations among particles separated by distances comparable to $1/N$ are determined by the discrete sine kernel and the value of the one-point function.

Let the operator $\mathcal{S}(x)$ act on $\ell^2(\mathbb{Z})$ with the kernel (see, for example, [BorOO00]):

$$\mathcal{S}_{ij}(x) := \frac{c}{\rho^0(x)} \frac{d\mu_{\min}^c}{dx}(x) S\left(\frac{c}{\rho^0(x)} \frac{d\mu_{\min}^c}{dx}(x) \cdot i, \frac{c}{\rho^0(x)} \frac{d\mu_{\min}^c}{dx}(x) \cdot j \right)$$

$$= \frac{\sin\left(\frac{c}{\rho^0(x)} \frac{d\mu_{\min}^c}{dx}(x) \cdot \pi(i-j) \right)}{\pi(i-j)},$$

where $i, j \in \mathbb{Z}$.

Theorem 3.2 (Asymptotics of local occupation probabilities in bands). *Let $B_N \subset X_N$ be a set of M nodes of the form*

$$B_N = \{x_{N,j}, x_{N,j+k_1}, x_{N,j+k_2}, \ldots, x_{N,j+k_{M-1}}\},$$

where $\#B_N = M$ is held fixed as $N \to \infty$ and where

$$0 < k_1 < k_2 < \cdots < k_{M-1}$$

are fixed integers. Set $\mathbb{B} := \{0, k_1, k_2, \ldots, k_{M-1}\} \subset \mathbb{Z}$. Suppose also that as $N \to \infty$, $x_{N,j} = \min B_N \to x$, with x lying in a band (and hence the same holds for $x_{N,j+k_{M-1}} = \max B_N$). Then, as $N \to \infty$,

$$A_m^{(N,k)}(B_N) = \frac{1}{m!} \left(-\frac{d}{dt} \right)^m \bigg|_{t=1} \det(1 - t\mathcal{S}(x)|_{\mathbb{B}}) + O\left(\frac{1}{N} \right). \tag{3.9}$$

Theorem 3.3 (Uniform exponential bounds for the correlation functions in voids). *Let F be a fixed closed interval in a void Γ that is bounded away from all bands. Then there is a constant $C_{F,m} > 0$ such that for all N sufficiently large,*

$$\max_{x_1, \ldots, x_m \in X_N \cap F} \left| R_m^{(N,k)}(x_1, \ldots, x_m) \right| \leq C_{F,m} \frac{e^{-mK_F N}}{N^m},$$

where the constant K_F is defined by

$$K_F := \min_{z \in F} \left[\frac{\delta E_c}{\delta \mu}(z) - \ell_c \right]. \tag{3.10}$$

Note that $K_F > 0$ because F is closed and disjoint from the support of the equilibrium measure $\mu_{\min}^c$.

To explain our next result, we introduce the following notation. For any $x \in (a,b)$, any $H > 0$, and any $N > 0$, let

$$E_{\text{int}}([A,B]; x, H, N) := \mathbb{E}\left(\text{number of particles at nodes } z \text{ of the form } z = x + \frac{\xi_N}{H\sqrt{N}}, \text{ with } A \leq \xi_N \leq B \right).$$

Theorem 3.4 (Normal particle number distribution near interior local minima of $\delta E_c/\delta\mu$ in voids). *There is a finite set Q such that for each point $x \notin Q$ lying in the interior of a void Γ where*

$$\frac{\delta E_c}{\delta \mu}(z) - \ell_c = W + H^2 \cdot (z-x)^2 + O\left((z-x)^3\right) \tag{3.11}$$

holds with $\mu = \mu_{\min}^c$ for some $H > 0$ as $z \to x$, there is a subsequence of integers N tending to infinity for which we have

$$\frac{E_{\text{int}}([C,D] \subset [A,B]; x, H, N)}{E_{\text{int}}([A,B]; x, H, N)} = \frac{\int_C^D e^{-\xi^2} d\xi}{\int_A^B e^{-\xi^2} d\xi} + O\left(\frac{1}{\sqrt{N}}\right). \tag{3.12}$$

That is, the expected number of particles in a certain interval of size $1/\sqrt{N}$ near x is given by a normal distribution.

◁ **Remark:** Whether in the interior of a given void Γ there may exist a local minimum of $\delta E_c/\delta\mu - \ell_c$ at x depends on the parameter c and the nature of the functions $V(x)$ and $\rho^0(x)$ characterizing the equilibrium measure. ▷

The higher (multipoint) correlation functions for particles in a neighborhood of size $1/\sqrt{N}$ of the interior local minimum at x are smaller in magnitude by a factor proportional to $1/\sqrt{N}$ than the one-point function. This implies that although the one-point function is Gaussian, the statistics of distinct particles near x are far from independent.

◁ **Remark:** Another interesting possibility would be a local minimum of $\delta E_c/\delta\mu - \ell_c$ occurring at either endpoint a or b or the interval of accumulation of nodes if this endpoint lies in a void. But a direct calculation gives, for x in a void Γ,

$$\frac{d}{dx}\left[\frac{\delta E_c}{\delta \mu}(x) - \ell_c\right] = \text{P.V.} \int_a^b \frac{\rho^0(y)\, dy}{x-y} - 2c \int_a^b \frac{\mu'(y)\, dy}{x-y} + V'(x)\,.$$

Here $\mu = \mu_{\min}^c$. The second integral is nonsingular because x lies outside the support of the equilibrium measure. As x tends to an endpoint of $[a,b]$ in a void Γ (necessarily an exterior gap), the latter two terms remain finite and the first term tends to $-\infty$ as $x \downarrow a$ and to $+\infty$ as $x \uparrow b$ (under our assumptions about $V(x)$ and $\rho^0(x)$). Thus neither endpoint can correspond to a local minimum. ▷

The analogue of Theorem 3.3 for saturated regions is the following.

Theorem 3.5 (Uniform exponential bounds for the correlation functions in saturated regions). *Let F be a fixed closed interval in a saturated region Γ that is bounded away from all bands. Then there is a constant $C_{F,m} > 0$ such that for all N sufficiently large,*

$$\max_{x_1,\ldots,x_m \in X_N \cap F} \left| R_m^{(N,k)}(x_1,\ldots,x_m) - 1 \right| \leq C_{F,m} \frac{e^{-L_F N}}{N}\,,$$

where the constant L_F is defined by

$$L_F := -\max_{z \in F}\left[\frac{\delta E_c}{\delta \mu}(z) - \ell_c\right]. \tag{3.13}$$

Note that $L_F > 0$ because F is a closed subinterval of an interval in which the the variational inequality (2.16) holds.

To explain our next result, we introduce notation for the total number of nodes near x. For $x \in (a,b)$, any $H > 0$, and any $N > 0$, let

$$M_{\text{int}}([A,B]; x, H, N) := \#\left\{\text{nodes } z \text{ of the form } z = x + \frac{\xi_N}{H\sqrt{N}}, \text{ with } A < \xi_N < B\right\},$$

which is asymptotically proportional to $\sqrt{N}$ for fixed H and fixed $A < B$. Then the analogue of Theorem 3.4 for saturated regions is the following.

APPLICATIONS 55

Theorem 3.6 (Normal particle number deviations near interior local maxima of $\delta E_c/\delta\mu$ in saturated regions). *There is a finite set Q such that for each point $x \notin Q$ lying in the interior of a saturated region Γ where*

$$\frac{\delta E_c}{\delta \mu}(z) - \ell_c = -W - H^2 \cdot (z-x)^2 + O\left((z-x)^3\right)$$

holds with $\mu = \mu_{\min}^c$ for some $H > 0$ as $z \to x$, there is a subsequence of integers N tending to infinity for which we have

$$\frac{M_{\text{int}}([C,D] \subset [A,B]; x, H, N) - E_{\text{int}}([C,D] \subset [A,B]; x, H, N)}{M_{\text{int}}([A,B]; x, H, N) - E_{\text{int}}([A,B]; x, H, N)} = \frac{\int_C^D e^{-\xi^2}\, d\xi}{\int_A^B e^{-\xi^2}\, d\xi} + O\left(\frac{1}{\sqrt{N}}\right).$$

That is, the deviation of the expected number of particles from the number of available nodes in a certain interval of size $1/\sqrt{N}$ near x is given by a normal distribution.

◁ **Remark:** It is not possible for a local maximum to occur at an endpoint of $[a,b]$ lying in a saturated region, since for x in a saturated region Γ,

$$\frac{d}{dx}\left[\frac{\delta E_c}{\delta \mu}(x) - \ell_c\right] = -c\,\text{P.V.}\int_a^b \frac{\mu'(y)\, dy}{x-y} + \int_a^b \frac{\rho^0(y) - c\mu'(y)}{x-y}\, dy + V'(x),$$

where $\mu = \mu_{\min}^c$ and the second term is nonsingular because the upper constraint is satisfied by the equilibrium measure in saturated regions. The latter two terms remain finite as x tends to an endpoint of $[a,b]$, but the first term tends to $+\infty$ as $x \downarrow a$ and to $-\infty$ as $x \uparrow b$. This shows that a local maximum may not occur at either endpoint in saturated regions. ▷

The expression

$$A(\xi_N, \eta_N) := \frac{\text{Ai}(\xi_N)\text{Ai}'(\eta_N) - \text{Ai}'(\xi_N)\text{Ai}(\eta_N)}{\xi_N - \eta_N} \tag{3.14}$$

is called the *Airy kernel*.

Theorem 3.7 (Universality of the Airy kernel near band edges adjacent to voids). *For each fixed $M > 0$, each left band edge α separating the band from a void, and each positive integer m, there is a constant $G_\alpha^m(M) > 0$ such that for sufficiently large N,*

$$\max_{\substack{x_1,\ldots,x_m \in X_N \\ \alpha - MN^{-1/2} < x_j < \alpha + MN^{-2/3}, \forall j}} \left| R_m^{(N,k)}(x_1, \ldots, x_m) - \left[\frac{(\pi c B_\alpha^L)^{2/3}}{N^{1/3}\rho^0(\alpha)}\right]^m \det\left(A(\xi_N^{(i)}, \xi_N^{(j)})\right)_{1 \le i,j \le m} \right| \le \frac{G_\alpha^m(M)}{N^{(m+1)/3}},$$

where

$$B_\alpha^L := \lim_{x \downarrow \alpha} \frac{1}{\sqrt{x-\alpha}} \frac{d\mu_{\min}^c}{dx}(x) > 0 \tag{3.15}$$

and $\xi_N^{(j)} = -\left(N\pi c B_\alpha^L\right)^{2/3}(x_j - \alpha)$. Similarly, for each fixed $M > 0$, each right band edge β separating the band from a void, and each positive integer m, there is a constant $G_\beta^m(M) > 0$ such that for sufficiently large N,

$$\max_{\substack{x_1,\ldots,x_m \in X_N \\ \beta - MN^{-2/3} < x_j < \beta + MN^{-1/2}, \forall j}} \left| R_m^{(N,k)}(x_1, \ldots, x_m) - \left[\frac{(\pi c B_\beta^R)^{2/3}}{N^{1/3}\rho^0(\beta)}\right]^m \det\left(A(\xi_N^{(i)}, \xi_N^{(j)})\right)_{1 \le i,j \le m} \right| \le \frac{G_\beta^m(M)}{N^{(m+1)/3}},$$

where

$$B_\beta^R := \lim_{x \uparrow \beta} \frac{1}{\sqrt{\beta - x}} \frac{d\mu_{\min}^c}{dx}(x) > 0 \tag{3.16}$$

and $\xi_N^{(j)} = \left(N\pi c B_\beta^R\right)^{2/3}(x_j - \beta)$.

Theorem 3.8 (Universality of the Airy kernel near band edges adjacent to saturated regions). *For each fixed $M > 0$, each left band edge α separating the band from a saturated region, and each positive integer m, there is a constant $H_\alpha^m(M) > 0$ such that for sufficiently large N,*

$$\max_{\substack{x_1,\ldots,x_m \in X_N \\ \alpha - MN^{-1/2} < x_j < \alpha + MN^{-2/3}, \forall j}} \left| R_m^{(N,k)}(x_1,\ldots,x_m) - 1 + \frac{(\pi \bar{c} \bar{B}_\alpha^L)^{2/3}}{N^{1/3} \rho^0(\alpha)} \sum_{j=1}^m A(\xi_N^{(j)}, \xi_N^{(j)}) \right| \leq \frac{H_\alpha^m(M)}{N^{2/3}},$$

where $\bar{c} := 1 - c$,

$$\bar{B}_\alpha^L := \lim_{x \downarrow \alpha} \frac{1}{\sqrt{x - \alpha}} \frac{c}{\bar{c}} \left[\frac{1}{c} \rho^0(x) - \frac{d\mu_{\min}^c}{dx}(x) \right] > 0, \quad (3.17)$$

and $\xi_N^{(j)} = -(N \pi \bar{c} \bar{B}_\alpha^L)^{2/3} (x_j - \alpha)$. Similarly, for each fixed $M > 0$, each right band edge β separating the band from a saturated region, and each positive integer m, there is a constant $H_\beta^m(M) > 0$ such that for sufficiently large N,

$$\max_{\substack{x_1,\ldots,x_m \in X_N \\ \beta - MN^{-2/3} < x_j < \beta + MN^{-1/2}, \forall j}} \left| R_m^{(N,k)}(x_1,\ldots,x_m) - 1 + \frac{(\pi \bar{c} \bar{B}_\beta^R)^{2/3}}{N^{1/3} \rho^0(\beta)} \sum_{j=1}^m A(\xi_N^{(j)}, \xi_N^{(j)}) \right| \leq \frac{H_\beta^m(M)}{N^{2/3}},$$

where again $\bar{c} = 1 - c$,

$$\bar{B}_\beta^R := \lim_{x \uparrow \beta} \frac{1}{\sqrt{\beta - x}} \frac{c}{\bar{c}} \left[\frac{1}{c} \rho^0(x) - \frac{d\mu_{\min}^c}{dx}(x) \right] > 0, \quad (3.18)$$

and $\xi_N^{(j)} = (N \pi \bar{c} \bar{B}_\beta^R)^{2/3} (x_j - \beta)$.

A statistic more interesting than the correlation functions near a band edge is the limiting distribution of the location of the leftmost or rightmost particle or hole. It is well known that the distribution of the largest eigenvalue of a random matrix from the Gaussian unitary ensemble converges, after proper centering and scaling, to a certain one-parameter family of Fredholm determinants constructed from the Airy kernel. The dependence of the determinant on the parameter can also be expressed in terms of a particular solution to the Painlevé II equation [TraW94]. This universal distribution function is known as the *Tracy-Widom distribution*. We claim that the distribution of the location of the leftmost or rightmost particle or hole has the same limit for general discrete orthogonal polynomial ensembles of the type corresponding to the assumptions about the nodes, weights, and equilibrium measures described in §1.3.1 and §2.1.2.

Let $x_{\min} \in X_N$ and $x_{\max} \in X_N$ be the nodes occupied by the leftmost and rightmost particles, respectively. Also denote by $\mathcal{A}|_{[s,\infty)}$ the (trace class) integral operator acting on $L^2[s,\infty)$ with the Airy kernel (3.14). Recall the generic assumption that the equilibrium measure of the k-particle ensemble has either a void or a saturated region adjacent to each endpoint of the interval $[a,b]$ in which the nodes accumulate (the exterior gaps). Then we have the following result.

Theorem 3.9 (Tracy-Widom distribution for the leftmost and rightmost particles). *If the left endpoint a is adjacent to a void (exterior gap) (a, α), then for each fixed $s \in \mathbb{R}$,*

$$\lim_{N \to \infty} \mathbb{P}\left((x_{\min} - \alpha) \cdot (\pi N c B_\alpha^L)^{2/3} \geq -s \right) = \det(1 - \mathcal{A}|_{[s,\infty)}), \quad (3.19)$$

where B_α^L is defined by (3.15). If the right endpoint b is adjacent to a void (exterior gap) (β, b), then for each fixed $s \in \mathbb{R}$,

$$\lim_{N \to \infty} \mathbb{P}\left((x_{\max} - \beta) \cdot (\pi N c B_\beta^R)^{2/3} \leq s \right) = \det(1 - \mathcal{A}|_{[s,\infty)}), \quad (3.20)$$

where B_β^R is defined by (3.16).

We also obtain a similar result for the leftmost and the rightmost holes. Let $y_{\min}$ and $y_{\max}$ be the nodes occupied by the leftmost and rightmost holes respectively.

APPLICATIONS

Theorem 3.10 (Tracy-Widom distribution for the leftmost and rightmost holes). *If the left endpoint a is adjacent to a saturated region (a, α), then for each fixed $s \in \mathbb{R}$,*

$$\lim_{N \to \infty} \mathbb{P}\left((y_{\min} - \alpha) \cdot (\pi N \bar{c} \bar{B}_\alpha^L)^{2/3} \geq -s\right) = \det(1 - \mathcal{A}|_{[s,\infty)}),$$

where $\bar{B}_\alpha^L$ is defined by (3.17) and $\bar{c} = 1 - c$. If the right endpoint b is adjacent to a saturated region (β, b), then for each fixed $s \in \mathbb{R}$,

$$\lim_{N \to \infty} \mathbb{P}\left((y_{\max} - \beta) \cdot (\pi N \bar{c} \bar{B}_\beta^R)^{2/3} \leq s\right) = \det(1 - \mathcal{A}|_{[s,\infty)}),$$

where $\bar{B}_\beta^R$ is defined by (3.18) and $\bar{c} = 1 - c$.

3.4 RANDOM RHOMBUS TILINGS OF A HEXAGON

3.4.1 Relation to the Hahn and associated Hahn ensembles

We state the result in [Joh00] providing expressions for probability density functions related to rhombus tilings of the $\mathfrak{abc}$-hexagon in terms of discrete orthogonal polynomial ensembles, as mentioned in §1.1.1.

Consider tiling the $\mathfrak{abc}$-hexagon as shown in Figure 1.1 with rhombi (see Figure 1.2) having sides of unit length. See Figure 1.3 for an example. Assume without loss of generality that $\mathfrak{a} \geq \mathfrak{b}$ (by the symmetry of the hexagon, the case when $\mathfrak{a} < \mathfrak{b}$ is completely analogous). Consider the mth vertical line of the lattice $\mathcal{L}$ counted from the left. We denote by $\mathcal{L}_m$ the intersection of this line and the lattice $\mathcal{L}$. The number of points in $\mathcal{L}_m$ is

$$N = N(\mathfrak{a}, \mathfrak{b}, \mathfrak{c}, m) := \mathfrak{c} + \frac{\mathfrak{a} - \mathfrak{a}_m}{2} + \frac{\mathfrak{b} - \mathfrak{b}_m}{2},$$

where

$$\mathfrak{a}_m := |m - \mathfrak{a}| \quad \text{and} \quad \mathfrak{b}_m := |m - \mathfrak{b}|.$$

In a given tiling, the N points in $\mathcal{L}_m$ correspond to the positions (in the sense defined earlier; see Figure 1.2) of a number of rhombi of types I, II, and III. We call the positions of horizontal rhombi (types I and II) the *particles*, and the positions of vertical rhombi (type III) the *holes*. See Figure 3.1 for an example of $\mathcal{L}_m$, when $m = 3$, illustrating the corresponding particles and holes.

Let Q_m be the lowest point in the sublattice $\mathcal{L}_m$. On the sublattice $\mathcal{L}_m$, there are always exactly $\mathfrak{c}$ particles, and $L_m := N - \mathfrak{c}$ holes. Now let $x_1 < \cdots < x_\mathfrak{c}$, where $x_j \in \{0, 1, 2, \ldots, N-1\}$, denote the (ordered) distances of the particles in $\mathcal{L}_m$ from Q_m and let $\xi_1 < \cdots < \xi_{L_m}$, where $\xi_j \in \{0, 1, 2, \ldots, N-1\}$, denote the distances of the holes in $\mathcal{L}_m$ from Q_m. In particular, we then have $\{x_1, \ldots, x_\mathfrak{c}\} \cup \{\xi_1, \ldots, \xi_{L_m}\} = \{0, 1, 2, \ldots, N-1\}$. The uniform probability distribution on the ensemble of tilings induces probability distributions for finding particles and holes at particular locations in the one-dimensional finite lattice $\mathcal{L}_m$. Let $\tilde{P}_m(x_1, \ldots, x_\mathfrak{c})$ denote the probability of finding the particle configuration $x_1, \ldots, x_\mathfrak{c}$ and let $P_m(\xi_1, \ldots, \xi_{L_m})$ denote the probability of finding the hole configuration $\xi_1, \ldots, \xi_{L_m}$.

Proposition 3.11 (Theorem 4.1 in [Joh00]). *Let $\mathfrak{a}, \mathfrak{b}, \mathfrak{c} \geq 1$ be given integers with $\mathfrak{a} \geq \mathfrak{b}$. Then*

$$\tilde{P}_m(x_1, \ldots, x_\mathfrak{c}) = \frac{1}{\tilde{Z}_m} \prod_{1 \leq j < k \leq \mathfrak{c}} (x_j - x_k)^2 \prod_{j=1}^{\mathfrak{c}} \tilde{w}(x_j),$$

where $\tilde{Z}_m$ is the normalization constant (partition function) and the weight function is the associated Hahn weight (see (2.72))

$$\tilde{w}(x) := w_{N,x}^{\text{Assoc}}(\mathfrak{a}_m + 1, \mathfrak{b}_m + 1) = \frac{\tilde{C}}{x!(\mathfrak{a}_m + x)!(N - x - 1)!(N - x - 1 + \mathfrak{b}_m)!},$$

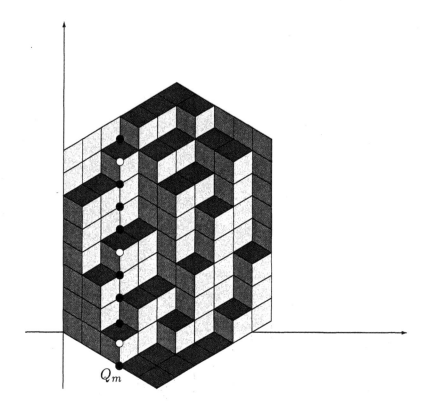

Figure 3.1 *A rhombus tiling of the* $\mathfrak{abc}$*-hexagon, and the lattice* $\mathcal{L}_m$, *when* $m = 3$; *holes are represented by white dots, and particles are represented by black dots.*

for a certain constant $\tilde{C}$. Also,

$$P_m(\xi_1, \ldots, \xi_{L_m}) = \frac{1}{Z_m} \prod_{1 \leq j < k \leq L_m} (\xi_j - \xi_k)^2 \prod_{j=1}^{L_m} w(\xi_j),$$

where Z_m is the normalization constant and the weight function is the Hahn weight (see (2.71))

$$w(\xi) := w_{N,\xi}^{\text{Hahn}}(\mathfrak{a}_m + 1, \mathfrak{b}_m + 1) = C \frac{(\xi + \mathfrak{a}_m)!(N - \xi - 1 + \mathfrak{b}_m)!}{\xi!(N - \xi - 1)!},$$

for a certain constant C.

Together with the scaling (1.2), we set

$$m = \tau n,$$

for some fixed $\tau > 0$. The mean density of particles in $\mathcal{L}_m$ is then

$$\bar{c} := \frac{\mathfrak{c}}{N} = \frac{2\mathfrak{C}}{2\mathfrak{C} + \mathfrak{A} + \mathfrak{B} - |\tau - \mathfrak{A}| - |\tau - \mathfrak{B}|}, \tag{3.21}$$

and the mean density of holes in $\mathcal{L}_m$ is

$$c := \frac{N - \mathfrak{c}}{N} = \frac{\mathfrak{A} + \mathfrak{B} - |\tau - \mathfrak{A}| - |\tau - \mathfrak{B}|}{2\mathfrak{C} + \mathfrak{A} + \mathfrak{B} - |\tau - \mathfrak{A}| - |\tau - \mathfrak{B}|}. \tag{3.22}$$

APPLICATIONS

3.4.2 Statistical asymptotics

The general asymptotic results stated in §3.3 combined with the specific calculations of the equilibrium measure for the Hahn weight in §2.4.2 imply several facts in the random tiling of the $\mathfrak{abc}$-hexagon. First, Theorems 3.1, 3.3, and 3.5 predict the asymptotic behavior of the one-point correlation function, implying that as $n \to \infty$, the one-dimensional lattice $\mathcal{L}_m$, when rescaled to a finite size independent of n, consists of three disjoint intervals: one band surrounded by two gaps (either saturated regions or voids, depending on the parameters α, β, γ, and τ). The saturated regions and voids correspond to the polar zones, while the central band is a section of the temperate zone. Hence in particular, the endpoints of the band (see equations (2.74)–(2.76), where $A := \mathfrak{a}_m/N$ and $B := \mathfrak{b}_m/N$ are functions of $\mathfrak{A}$, $\mathfrak{B}$, $\mathfrak{C}$, and τ only), when considered as functions of τ for fixed $\mathfrak{A}$, $\mathfrak{B}$, and $\mathfrak{C}$, determine the typical shape of the boundary between the polar and temperate zones of the rescaled $\mathfrak{abc}$-hexagon. It may be checked that this curve, as calculated directly from the quadratic equation (2.74), coincides with the inscribed ellipse first shown to be the expected shape of the boundary by Cohn, Larsen, and Propp [CohLP98].

Moreover, we find that the one-point functions for particles and holes converge uniformly to the equilibrium measures, respectively, for the associated Hahn weight corresponding to the value of $\bar{c}$ given in (3.21) and for the Hahn weight corresponding to the value of c given in (3.22), and we obtain a precise error bound. This result thus improves upon those obtained in [CohLP98] and [Joh00]. We expect that with additional analysis of the same formulae it should be possible to show that the error is locally uniform with respect to τ, in which case the same bounds should hold for more general regions $U \in \mathbb{R}^2$. We state our result in this direction as follows.

Theorem 3.12 (Strong asymptotics of the one-point function in the $\mathfrak{abc}$-hexagon). *Consider holes in the line $\mathcal{L}_m$ of length N, where $m = \tau n$ and τ is fixed as $n \to \infty$. The corresponding one-point function for holes, $R_1^{(N,cN)}(\xi)$, satisfies*

$$R_1^{(N,cN)}(\xi) \to c \frac{d\mu_{\min}^c}{dx}(x), \qquad \text{where } x = \frac{\xi}{N},$$

as $n \to \infty$, with $\mathfrak{a} = \mathfrak{A}n$, $\mathfrak{b} = \mathfrak{B}n$, $\mathfrak{c} = \mathfrak{C}n$, and $\mathfrak{A}$, $\mathfrak{B}$, and $\mathfrak{C}$ are held fixed. Here the equilibrium measure is that corresponding to the Hahn weight with parameters $A = \mathfrak{a}_m/N$ and $B = \mathfrak{b}_m/N$ (see (2.77)–(2.79) in §2.4.2). The convergence is uniform for $\xi = 0, 1, 2, \ldots, N-1$. Note that the limit function $cd\mu_{\min}^c/dx(x)$ is identically equal to 1 in the polar zones near the vertices P_2 and P_5 and is identically equal to 0 in the polar zones near the vertices P_1, P_3, P_4, and P_6. The rate of convergence is uniformly exponentially fast (the error is of the order $O(e^{-Kn})$ for some $K > 0$) for ξ in any polar zone such that $x = \xi/N$ is uniformly bounded away from the temperate zone as $n \to \infty$. For ξ in the temperate zone such that $x = \xi/N$ is uniformly bounded away from all polar zones as $n \to \infty$, the rate of convergence is such that the error is uniformly of the order $O(1/n)$.

In the temperate zone, in addition to the one-point function, we can control all the multipoint correlation functions under proper scaling (see Theorem 3.1). One consequence of this is the following theorem concerning the scaling limit for the locations of the holes (see Theorem 3.2) in the line $\mathcal{L}_m$.

Theorem 3.13 (Local occupation probabilities in the temperate zone of the $\mathfrak{abc}$-hexagon). *Consider a vertical line $\mathcal{L}_m$ of length N in the $\mathfrak{abc}$-hexagon with $\mathfrak{a} = \mathfrak{A}n$, $\mathfrak{b} = \mathfrak{B}n$, $\mathfrak{c} = \mathfrak{C}n$, and $m = \tau n$ for fixed positive $\mathfrak{A}$, $\mathfrak{B}$, $\mathfrak{C}$, and τ. Let $x > 0$ be fixed such that $Nx \in \mathbb{Z}_N$ and such that the location $\xi = Nx$ units above Q_m in $\mathcal{L}_m$ lies in the temperate zone bounded away from the expected boundary between the polar and temperate zones by a distance proportional to n. Let $B = \{Nx, Nx + j_1, Nx + j_2, \ldots, Nx + j_M\}$, where $\mathbb{B} = \{0, j_1, j_2, \ldots, j_M\} \subset \mathbb{Z}_N$ is a fixed set of integers. Then*

$$\lim_{n \to \infty} \mathbb{P}(\text{there are precisely } p \text{ holes in the set } B) = \frac{1}{p!}\left(-\frac{d}{dt}\right)^p\bigg|_{t=1} \det(1 - t\mathcal{S}(x)|_{\mathbb{B}}),$$

where $\mathcal{S}(x)$ acts on $\ell^2(\mathbb{Z})$ with the kernel

$$\mathcal{S}_{ij}(x) = \frac{\sin(\pi q(x)(i-j))}{\pi(i-j)}, \qquad \text{for } i, j \in \mathbb{Z},$$

where $q(x) = c\, d\mu_{\min}^c/dx(x)$ is the limiting one-point function (i.e., the density of states) at x.

Finally, we obtain the limiting distribution of the fluctuation of the boundary separating the polar and temperate zones. From Theorem 3.9, we have the following result, which was conjectured in [Joh00]. Recall that the Fredholm determinant $\det(1 - \mathcal{A}|_{[x,\infty)})$ (see (3.23) below) has an alternative expression in terms of a particular solution of the Painlevé II equation in the independent variable x, which is referred to in random matrix theory as the Tracy-Widom law.

Theorem 3.14 (Tracy-Widom distribution for extreme particles and holes in the $\mathfrak{abc}$-hexagon). *Consider a vertical line $\mathcal{L}_m$ of length N in the $\mathfrak{abc}$-hexagon with $\mathfrak{a} = \mathfrak{A}n$, $\mathfrak{b} = \mathfrak{B}n$, $\mathfrak{c} = \mathfrak{C}n$, and $m = \tau n$ for fixed positive $\mathfrak{A}, \mathfrak{B}, \mathfrak{C}$, and τ. Suppose further that τ is sufficiently small or sufficiently large that the polar zone at the top of $\mathcal{L}_m$ is a void for holes (equivalently, is saturated with particles). Denote by ξ_* the height above the point Q_m of the topmost hole in $\mathcal{L}_m$ and recall that for β defined by (2.76) in §2.4.2 with $A = \mathfrak{a}_m/N$ and $B = \mathfrak{b}_m/N$, the limiting expected height above Q_m of the boundary between the temperate and polar zones is $N\beta$. Then, for some constant $t > 0$,*

$$\lim_{n \to \infty} \mathbb{P}\left(\frac{\xi_* - N\beta}{(tn)^{1/3}} \leq x\right) = \det(1 - \mathcal{A}|_{[x,\infty)}), \tag{3.23}$$

for each $x \in \mathbb{R}$, where $\mathcal{A}|_{[x,\infty)}$ is the Airy operator acting on $L^2[x, \infty)$ with the Airy kernel (3.14).

The above result applies to the boundary between the polar zones near the vertices P_4 and P_6 and the temperate zone. The analogous results hold for the boundary near P_1 and P_3 with the use of the other endpoint α (see (2.75) in §2.4.2) in place of β, a change of sign in the inequality, the interpretation of ξ_* as the location of the bottommost hole in $\mathcal{L}_m$, and a proper adjustment of the constant t. Similarly, for the boundary near P_2 and P_5 where the polar zones are voids for particles (or packed with holes), the analogous results hold with the interpretation of ξ_* as the height above Q_m of the bottommost or topmost particle.

◁ **Remark:** Similar results for domino tilings of the Aztec diamond are obtained in [Joh01]. In [OkoR03], a q-version or grand canonical ensemble version of the uniform probability measure on the set of rhombus tilings of the $\mathfrak{abc}$-hexagon is considered; thus the size of the hexagon also becomes a random variable. These authors computed the correlation functions of holes in the temperate region that do not necessarily lie along the same line, in a proper limit that corresponds to the limit $N \to \infty$. The result of this calculation is a kernel built from the incomplete beta function, referred to as the "discrete incomplete beta kernel." This kernel reduces to the discrete sine kernel when the holes all lie along the same line. We expect that the same kernel should appear in the Hahn ensemble if one computes the asymptotic correlation function for holes lying in a two-dimensional region. The Airy limit of the boundary of the polar zones for this model was obtained by Ferrari and Spohn [FerS03]. ▷

3.5 THE CONTINUUM LIMIT OF THE TODA LATTICE

3.5.1 Solution procedure

The fundamental assumption in force in studying the continuum limit of the finite Toda lattice is that there are smooth functions $A(c)$ and $B(c)$ defined for $c \in [0, 1]$, with $B(0) = B(1) = 0$, such that for each N, the initial data for the Toda lattice is

$$a_{N,k}(0) = A\left(\frac{k}{N-1}\right), \quad \text{for } k = 0, \ldots, N-1, \tag{3.24}$$

and

$$b_{N,k}(0) = B\left(\frac{2k+1}{2N-2}\right), \quad \text{for } k = 0, \ldots, N-2. \tag{3.25}$$

For convenience, we assume that the combination $\alpha(c) := A(c) - 2B(c)$ is strictly concave up, while the combination $\beta(c) := A(c) + 2B(c)$ is strictly concave down, as illustrated in Figure 3.2. For each N, these initial data determine, by the three-term recurrence relations (see (1.11)–(1.13)), a sequence of polynomials

APPLICATIONS

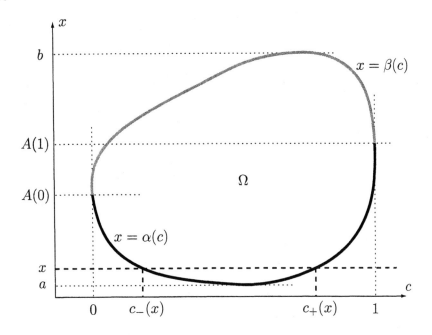

Figure 3.2 *The functions $\alpha(c)$ and $\beta(c)$ and the definition of the turning points $c_\pm(x)$.*

guaranteed to be orthogonal with respect to a discrete measure that is the spectral measure of the corresponding Jacobi matrix. The nodes X_N of orthogonality for these polynomials are the eigenvalues of this Jacobi matrix, and the weights are the squares of the first components of the corresponding $\ell^2(X_N)$ normalized eigenvectors. See [DeiM98] or [Dei99] for more details about these correspondences.

In [DeiM98] the eigenvalue problem for the Jacobi matrix corresponding to sampled data of the form (3.24) and (3.25) is carefully analyzed in the limit $N \to \infty$ by a WKB approach in which the functions $c_\pm(x)$ indicated in Figure 3.2 play the role of turning points. This analysis shows that the nodes X_N become asymptotically distributed as $N \to \infty$ in the interval $[a,b]$ (at whose endpoints the turning points coalesce) according to the density function $\rho^0(x)$ defined by

$$\rho^0(x) := \frac{1}{\pi} \int_{c_-(x)}^{c_+(x)} \frac{dc}{\sqrt{(2B(c))^2 - (x - A(c))^2}}, \qquad (3.26)$$

and that the weights have Nth-root asymptotics corresponding to the external field $\varphi(x)$ whose derivative is given by

$$\varphi'(x) := 2\,\mathrm{sgn}\,(x - A(0)) \int_0^{c_-(x)} \frac{dc}{\sqrt{(x - A(c))^2 - (2B(c))^2}}.$$

Note that (3.26) is indeed the density of a probability measure on $[a,b]$ since if Ω is the region between the curves $x = \alpha(c)$ and $x = \beta(c)$ in Figure 3.2, then

$$\int_a^b \rho^0(x)\,dx = \frac{1}{\pi} \iint_\Omega \frac{dc\,dx}{\sqrt{(2B(c))^2 - (x - A(c))^2}} = \int_0^1 dc = 1\,.$$

The Toda flow in the spectral domain leaves the nodes X_N fixed and modifies the weights $w_{N,n}$ by an exponential factor according to (1.32) for $p = 1$. This in turn implies that the potential function $V(x)$ evolves on the rescaled time scale $T = t/N$ according to

$$V(x;T) = V(x) - 2Tx\,.$$

It then follows from (2.1) that the external field appearing in the equilibrium energy problem becomes T-dependent as $\varphi(x;T) = \varphi(x) - 2Tx$. Given $\rho^0(x)$ and $\varphi(x;T)$, the inverse problem for the continuum limit of the Toda lattice is to determine the asymptotic behavior as $N \to \infty$ of the recurrence coefficients $a_{N,k}(t)$ and $b_{N,k}(t)$, where $t = NT$ and T is approaching a fixed value.

This inverse problem is solved in detail by Theorem 2.8, which characterizes the recurrence coefficients in terms of the deformed equilibrium measure $\mu_{\min}^{c,T}$ solving the equilibrium energy problem with input data $\rho^0(x)$ and $\varphi(x;T)$. The asymptotic description of the $\{a_{N,k}(t)\}$ and $\{b_{N,k}(t)\}$ afforded by Theorem 2.8 involves slowly varying behavior on the macroscopic spatial scale c and time scale T through the band endpoints $\alpha_j = \alpha_j(c,T)$ and $\beta_j = \beta_j(c,T)$, for $j = 0,\ldots,G$. It is proved in [DeiM98] directly from the equilibrium energy problem that these endpoints necessarily satisfy a hyperbolic nonlinear system of partial differential equations in Riemann invariant (diagonal) form:

$$\frac{\partial \alpha_j}{\partial T} + C_{\alpha,j}(\alpha_0,\ldots,\alpha_G,\beta_0,\ldots,\beta_G)\frac{\partial \alpha_j}{\partial c} = 0,$$
$$\frac{\partial \beta_j}{\partial T} + C_{\beta,j}(\alpha_0,\ldots,\alpha_G,\beta_0,\ldots,\beta_G)\frac{\partial \beta_j}{\partial c} = 0. \tag{3.27}$$

These equations are called the *Whitham equations* or *modulation equations*. In the special case of $G = 0$, the Whitham equations are simply

$$\frac{\partial \alpha_0}{\partial T} + \frac{\beta_0 - \alpha_0}{4} \cdot \frac{\partial \alpha_0}{\partial c} = 0 \quad \text{and} \quad \frac{\partial \beta_0}{\partial T} - \frac{\beta_0 - \alpha_0}{4} \cdot \frac{\partial \beta_0}{\partial c} = 0, \tag{3.28}$$

which is the formal continuum-limit system for the Toda lattice (1.6) written in the variables (Riemann invariants) $\alpha_0 = A - 2B$ and $\beta_0 = A + 2B$. In this case, Theorem 2.8 proves the validity of the Toda lattice equations as a numerical scheme for solving the partial differential equations (1.6). As pointed out before, the existence of strong limits of $a_{N,k}(t)$ and $b_{N,k}(t)$ as $N \to \infty$ with $k/N \to c$ and $t/N \to T$ in the case that the equilibrium measure configuration involves only one band (α_0, β_0) (in particular this holds for T small enough) has been proved before; see [DeiM98] or [BloGPU03].

3.5.2 Strong asymptotics beyond shock formation

When singularities form from smooth initial data in solutions of the hyperbolic system (1.6) or (3.28) near some point (c,T), one notices a coincident birth of a new band in the solution of the equilibrium energy problem. For such (c,T) where the equilibrium measure has more than one band of unconstrained support, there is no longer a strong limit for the recurrence coefficients due to the rapid oscillations (on the microscopic spatial scale k and time scale t) contributed by the Riemann theta functions in the asymptotic formulae predicted by Theorem 2.8. The vector $\mathbf{r}$ is proportional to t under the Toda flow, and on the microscopic level the formulae in Theorem 2.8 show that the recurrence coefficients oscillate rapidly in a multiphase fashion with G real phases (the components of $\mathbf{r} - \kappa\mathbf{\Omega}$) that are linear in t and Ω. The macroscopic properties of the oscillations (weak limits) may be obtained by multiphase averaging of these oscillations, and thus one arrives in a different way at the Whitham equations (3.27) governing the weak limit as proved in [DeiM98]. However, the formulae in Theorem 2.8 provide more information, namely, strong asymptotics in the oscillatory regions like those evident in Figures 1.4 and 1.5.

Not all smooth initial data for the hyperbolic system (1.6) or (3.28) leads to shock formation in finite time. Indeed, a particular family of initial data leading to a global in time solution is that corresponding to equally spaced eigenvalues of the Jacobi matrix with Krawtchouk weights. Recall from the discussion in §2.4.1 that the family of Krawtchouk weights is invariant under the Toda flow, with only a change of the parameter $p = p(t)$ given explicitly by (2.68). Since the paper of Dragnev and Saff [DraS00] proves that for all p and q the equilibrium measure for the Krawtchouk weights has only one band of unconstrained support, we have $G = 0$ for all T. The formulae in [DraS00] for the motion of the band endpoints $(\alpha_0(c,T), \beta_0(c,T))$ induced by the deformation (2.68) provide an explicit global solution of the hyperbolic system (3.28).

3.5.3 Defect motion in saturated regions

Another interesting aspect of the Toda flow (or for that matter, any continuous one-parameter deformation of the weights $w_{N,n}$) is the effect it can have on defects in the pattern of zeros in saturated regions for

APPLICATIONS

the equilibrium measure as explained in Theorem 2.12 and the subsequent discussion. To demonstrate the dynamics of defects with and without spurious zeros, it is enough to find a density function $\rho^0(x)$ and a potential function $V(x)$ such that the solution of the equilibrium energy problem involves, for some $c \in (0,1)$, a saturated region surrounded by two bands (*i.e.*, one that is an interior gap). Thus to see defects, we need to have $G > 0$, but it is irrelevant for the current discussion whether $V(x)$ arose by the Toda flow from initial data with $G = 0$. One chooses a large value of N and a degree k close to Nc, constructs numerically the orthogonal polynomial $\pi_{N,k}(z;t)$ for several values of t corresponding to the Toda deformation of the weights (1.32) for $p = 1$, and computes the zeros in the saturated region. For example, with a nonconstant node density $\rho^0(x)$ on $[a,b] = [0,1]$ proportional to $1 + 10x$ and weights corresponding to a potential $V(x) = \cos(2\pi x)/2$, we set $N = 40$ and constructed the Toda-deformed weights and then the orthogonal polynomial of degree $k = 25$ for several values of the rapid time scale t in the range $1 < t < 6$. Snapshots of a subset of the saturated region showing the zeros therein as circles superimposed on the number line, with tick marks indicating the nodes, are shown in Figures 3.3 and 3.4. In these figures, each of the circles representing a zero contains a slash whose slope indicates whether it lies just to the right of the nearest node (slope of -1) or just to the left (slope of 1). A zero moving continuously from one node to the next will therefore appear as a rotating circle with the slash rotating smoothly from one diagonal slope to its opposite, passing through the vertical in between. On the other hand, a zero passing continuously through a node will appear as a circle with a slash whose slope suddenly changes sign.

Figure 3.3 illustrates the motion of a defect carrying a spurious zero. As discussed earlier in §2.3, the defect moves because the spurious zero approaches a node from the left and then "exchanges identity" with the zero just on the other side of the node, with that zero becoming spurious and free to move away from the node to which it had previously been exponentially confined. This can be seen in the figure as, more or less one at a time, the circle representing each zero rolls from one node to the next.

A short time after the defect leaves the saturated region for the band to the right it reappears, now traveling from right to left through the saturated region, this time not carrying a spurious zero. This is illustrated in Figure 3.4. The defect propagates to the left as the Hurwitz zeros move through the nodes to the right one at a time. The slope of the slash changes suddenly when the corresponding zero crosses a node, and not much else interesting happens in between the snapshots. Thus, unlike the more-or-less continuous motion of a defect illustrated by the snapshots in Figure 3.3, the motion of a defect not carrying a spurious zero appears to take place in discrete jumps. Also note that the zeros are moving in the direction *opposite* the motion of the defect. These numerics illustrate concretely the detailed phenomena predicted by Theorem 2.12.

3.5.4 Asymptotics of the linearized problem

In §1.4.2 we introduced explicit formulae in terms of the matrix elements of $\mathbf{P}(z;N,k,t)$ for solutions of the linearized Toda lattice equations when the weights $w_{N,n}$ are deformed according to the Toda flow (see (1.32) with $p = 1$). The available freedom in these formulae is the choice of the traceless matrix $\mathbf{C}$, which may be written in the Pauli basis of $\mathfrak{sl}(2)$ as

$$\mathbf{C} = c_1\sigma_1 + c_2\sigma_2 + c_3\sigma_3$$

(here σ_2 is the Pauli matrix

$$\sigma_2 := \begin{pmatrix} 0 & -i \\ i & 0 \end{pmatrix}$$

and the other Pauli matrices have been defined previously) and the arbitrary complex parameter $z \in \mathbb{C}\setminus X_N$. This is more than enough freedom to span the $2N - 1$ dimensional solution space of the linearized Toda lattice equations.

If c_1 and c_2 are not both zero, then the solutions apparently involve exponential factors in t that arise because the matrix $e^{zt\sigma_3}$ relating $\mathbf{P}(z;N,k,t)$ and $\mathbf{M}(z;N,k,t)$ does not commute with σ_1 or σ_2. However, in the large-N limit there is also exponential behavior in the matrix elements of $\mathbf{P}(z;N,k,t)$ as described, for example, in Theorem 2.7. It is necessarily the case that, at least while the solution of the hyperbolic system

(1.6) is classical, this exponential behavior cancels any present because of the factor e^{zt} if the parameters c_1, c_2, c_3, and z are chosen so that the solution is also uniformly bounded in k. This follows from the Lax Equivalence Theorem applied to the Toda lattice equations viewed as a numerical scheme for (1.6), taking account of the strong convergence of this scheme before shock time for general initial data $A(c)$ and $B(c)$.

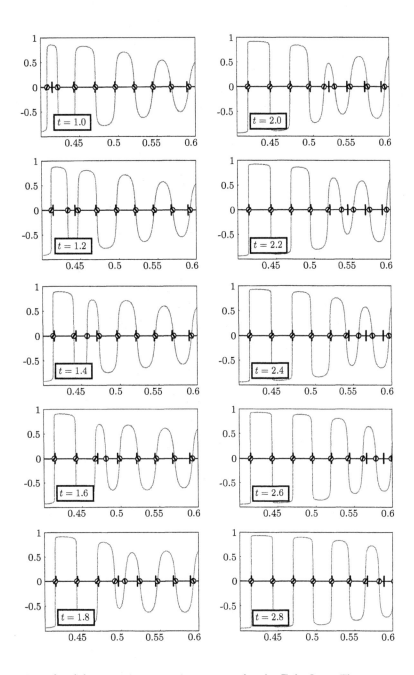

Figure 3.3 *The motion of a defect carrying a spurious zero under the Toda flow. The gray curve is the graph of* $2\tan^{-1}(\sinh^{-1}(C\pi_{40,25}(z)))/\pi$ *for an appropriate constant* C.

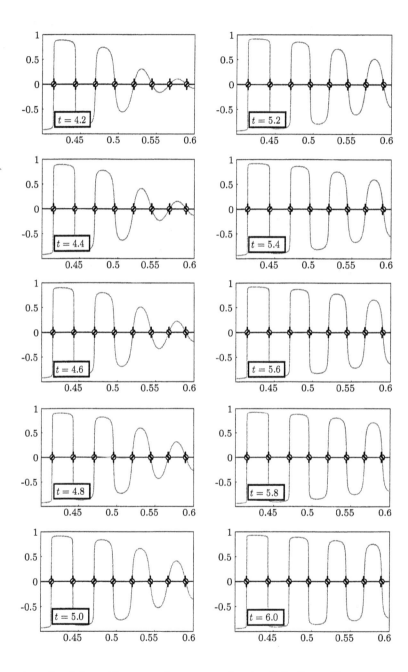

Figure 3.4 *The motion of a defect with no spurious zero under the Toda flow. The gray curve is the graph of* $2\tan^{-1}(\sinh^{-1}(C\pi_{40,25}(z)))/\pi$ *for an appropriate constant* C.

Chapter Four

An Equivalent Riemann-Hilbert Problem

In this chapter we introduce a sequence of exact transformations relating the matrix $\mathbf{P}(z; N, k)$ to a matrix $\mathbf{X}(z)$ satisfying an equivalent Riemann-Hilbert problem. The Riemann-Hilbert problem characterizing the matrix $\mathbf{X}(z)$ will be amenable to asymptotic analysis in the joint limit of large degree k and large parameter N. This asymptotic analysis will be carried out in Chapter 5.

4.1 CHOICE OF Δ: THE TRANSFORMATION FROM $\mathbf{P}(z; N, k)$ TO $\mathbf{Q}(z; N, k)$

It turns out that Interpolation Problem 1.2 will be amenable to analysis without any modification of the triangularity of some of the residue matrices only if the equilibrium measure never realizes its upper constraint. This is because the variational inequality (2.16) associated with this constraint leads to exponential growth as $N \to \infty$ in each situation that we wish to exploit the inequality (2.13) to obtain exponential decay. This difficulty was recognized, for example, in [BorO05], where for a specific weight it was circumvented using representations of the corresponding polynomials in terms of hypergeometric functions. We need to handle the problem of the upper constraint in full generality, and we will do so by using an explicit transformation of the form (1.43) to reverse the triangularity of the residue matrices near only those poles where the upper constraint is active and leaving the triangularity of the remaining residues unchanged. The result of the change of variables (1.43) is a matrix $\mathbf{Q}(z; N, k)$ that depends on the choice of a subset $\Delta \subset \mathbb{Z}_N$. Our immediate goal is to describe how the set Δ must be chosen to prepare for the subsequent asymptotic analysis to be described in Chapter 5 in the limit $N \to \infty$.

The continuity of $d\mu_{\min}^c/dx$ (which follows from our basic assumptions outlined in §1.3.1; see also §2.1) along with the assumption (1.14) implies that voids and saturated regions cannot be adjacent to each other but must always be separated by bands. A band that lies between a void and a saturated region (rather than between two voids or between two saturated regions) will be called a *transition band*. In each transition band, we select arbitrarily a fixed point y_k. There are a finite number, say M, of transition bands, and we label the points we select, one from each, in increasing order: $y_1, \ldots, y_M$.

With each y_k we associate a sequence $\{y_{k,N}\}_{N=0}^{\infty}$ that converges to y_k as $N \to \infty$. Each element of the sequence is defined by the quantization rule:

$$N \int_a^{y_{k,N}} \rho^0(x)\, dx = \left\lceil N \int_a^{y_k} \rho^0(x)\, dx \right\rceil, \tag{4.1}$$

where $\lceil u \rceil$ denotes the least integer greater than or equal to u. We call the points $y_{k,N}$ *transition points* and use the notation Y_N for the set $\{y_{k,N}\}_{k=1}^{M}$ and Y_∞ for the set $\{y_k\}_{k=1}^{M}$. Since $\rho^0(x)$ is analytic and nonzero in (a, b), we have $y_{k,N} = y_k + O(1/N)$ as $N \to \infty$. Also, comparing with the condition (1.15) that defines the nodes X_N, we see that each of the transition points $y_{k,N}$ asymptotically lies halfway between two adjacent nodes. Note that if only one constraint is active in $[a, b]$, then there are no transition bands at all and therefore no transition points; so $Y_N = \emptyset$ in this case. For all sufficiently large fixed N, the transition points $y_{k,N}$ are ordered in the same way as the points y_k. For each N, we take the transition points in Y_N to be the common endpoints of two complementary systems Σ_0^∇ and Σ_0^Δ of open subintervals of (a, b):

Definition 4.1 (The systems of subintervals Σ_0^∇ and Σ_0^Δ). *The set Σ_0^∇ is the union of those among the open subintervals $(y_{k,N}, y_{k+1,N})$ for $k = 1, \ldots, M$, $(a, y_{1,N})$, and $(y_{M,N}, b)$ (or (a, b) if there are no transition points) that contain no saturated regions. The set Σ_0^Δ is the union of those among the open subintervals*

$(y_{k,N}, y_{k+1,N})$ for $k = 1, \ldots, M$, $(a, y_{1,N})$, and $(y_{M,N}, b)$ (or (a, b) if there are no transition points) that contain no voids.

See Figure 4.1. The sets Σ_0^∇ and Σ_0^Δ depend on N in a very mild way, but they depend more crucially on the fixed parameter c and on the analytic functions $V(x)$ and $\rho^0(x)$.

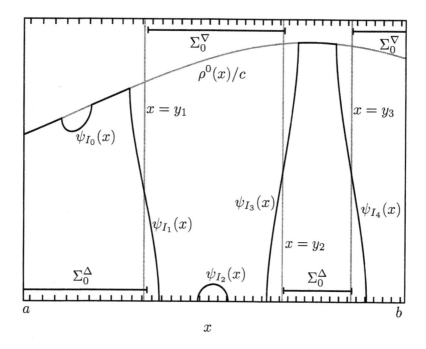

Figure 4.1 *A diagram showing the relation of a hypothetical equilibrium measure $\mu_{\min}^c$ to the interval systems Σ_0^∇ and Σ_0^Δ. The nodes X_N are indicated at the top and bottom of the figure with tick marks; their density is proportional to the upper constraint (gray curve). The endpoints of subintervals of Σ_0^∇ and Σ_0^Δ are the transition points Y_N that converge as $N \to \infty$ to the fixed points $x = y_k$ whose positions within each transition band are indicated with gray vertical lines. The analytic unconstrained components $\psi_I(x)$ of the density $d\mu_{\min}^c/dx(x)$ are also indicated.*

With this notation, we now describe how we will choose the set Δ involved in the change of variables (1.43) from $\mathbf{P}(z; N, k)$ to $\mathbf{Q}(z; N, k)$. The set Δ will be taken to contain precisely those indices n corresponding to nodes $x_{N,n}$ contained in Σ_0^Δ:

$$\Delta := \{n \in \mathbb{Z}_N \text{ such that } x_{N,n} \in \Sigma_0^\Delta\}. \tag{4.2}$$

In particular, this choice has the effect of reversing the triangularity of the residue matrices at those nodes $x_{N,n}$ where the upper constraint is active. Note that $\#\Delta$ is roughly proportional to N; we will define a rational constant d_N by writing

$$d_N := \frac{\#\Delta}{N}.$$

Note that d_N has a limiting value d as $N \to \infty$; for technical reasons (see (4.5) below) we will assume without loss of generality (because we have considerable freedom in choosing the points in Y_∞) that $d \neq c$.

4.2 REMOVAL OF POLES IN FAVOR OF DISCONTINUITIES ALONG CONTOURS: THE TRANSFORMATION FROM $\mathbf{Q}(z;N,k)$ TO $\mathbf{R}(z)$

The transformation in this section is based on an idea that was first used in [KamMM03]. There, an analytic function was employed to simultaneously interpolate the residues of many poles at the pole locations. A generalization of this procedure involving two distinct analytic interpolants was introduced in [Mil02]. The approach we take in this section will also use two interpolants.

Note that by definition of the nodes $x_{N,j} \in X_N$ (see §1.3.1) and using (2.50), we have the *interpolation identity*

$$ie^{-iN\theta^0(x_{N,n})/2} = -ie^{iN\theta^0(x_{N,n})/2} = (-1)^{N-1-n}, \quad \text{for } N \in \mathbb{N} \text{ and } n \in \mathbb{Z}_N. \tag{4.3}$$

Let $\epsilon > 0$ be a fixed parameter (independent of N) and consider the contour Σ illustrated in Figure 4.2. The figure is drawn to correspond to the hypothetical equilibrium measure illustrated in Figure 4.1. The

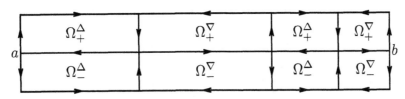

Figure 4.2 *The oriented contour Σ and regions $\Omega_\pm^\nabla$ and $\Omega_\pm^\Delta$. The vertical segments of Σ lie, from left to right, along $\Re(z) = a$, $\Re(z) = y_{1,N}$, $\Re(z) = y_{2,N}$, $\Re(z) = y_{3,N}$, and $\Re(z) = b$. The horizontal segments of Σ lie, from bottom to top, along $\Im(z) = -\epsilon$, $\Im(z) = 0$, and $\Im(z) = \epsilon$.*

contour Σ consists of the subintervals Σ_0^∇ and Σ_0^Δ and additional horizontal segments with $|\Im(z)| = \epsilon$ and vertical line segments with $\Re(z) = a$, $\Re(z) = b$, and $\Re(z) \in Y_N$. We take the parameter ϵ to be sufficiently small so that the contour Σ lies entirely in the region of analyticity of $V(x)$ and $\rho^0(x)$. Further restrictions will be placed on ϵ later on.

From the solution of Interpolation Problem 1.2 transformed into the matrix $\mathbf{Q}(z;N,k)$ via (1.43) using the choice of Δ given in (4.2), we define a new matrix $\mathbf{R}(z)$ as follows. Set

$$\mathbf{R}(z) := \mathbf{Q}(z;N,k) \begin{pmatrix} 1 & \mp i e^{\mp i N\theta^0(z)/2} e^{-NV_N(z)} \dfrac{\prod_{n\in\Delta}(z-x_{N,n})}{\prod_{n\in\nabla}(z-x_{N,n})} \\ 0 & 1 \end{pmatrix}, \quad \text{for } z \in \Omega_\pm^\nabla, \tag{4.4}$$

$$\mathbf{R}(z) := \mathbf{Q}(z;N,k) \begin{pmatrix} 1 & 0 \\ \mp i e^{\mp i N\theta^0(z)/2} e^{NV_N(z)} \dfrac{\prod_{n\in\nabla}(z-x_{N,n})}{\prod_{n\in\Delta}(z-x_{N,n})} & 1 \end{pmatrix}, \quad \text{for } z \in \Omega_\pm^\Delta,$$

and $\mathbf{R}(z) := \mathbf{Q}(z;N,k)$, for all other $z \in \mathbb{C} \setminus \Sigma$.

The significance of this explicit change of variables is that all poles have completely disappeared from the problem. Using the residue conditions (1.44) and (1.45) in conjunction with the interpolation identity (4.3), it is easy to check that $\mathbf{R}(z)$ is an analytic function for $z \in \mathbb{C} \setminus \Sigma$ that takes continuous, and in fact analytic, boundary values on Σ. In fact, $\mathbf{R}(z)$ can easily be seen to be the solution of a Riemann-Hilbert problem relative to the contour Σ. This problem is sufficiently similar to that introduced in [FokIK91] for the continuous-weights case that it may, in principle, be analyzed by methods like those used in [DeiKMVZ99a, DeiKMVZ99b]. We now proceed to describe the steps required for the corresponding analysis in the discrete case.

4.3 USE OF THE EQUILIBRIUM MEASURE: THE TRANSFORMATION FROM R(z) TO S(z)

4.3.1 The complex potential $g(z)$ and the matrix $\mathbf{S}(z)$

The parameter c and the analytic functions $V(z)$ and $\rho^0(z)$ all influence the large-N behavior of the orthogonal polynomials. Thus we recall the equilibrium measure $\mu_{\min}^c$ obtained in terms of these quantities in §2.1, and for $x \in \Sigma_0^\nabla \cup \Sigma_0^\Delta \subset [a,b]$ we define the piecewise real-analytic density function as follows:

$$\rho(x) := \begin{cases} \dfrac{c}{c-d_N} \dfrac{d\mu_{\min}^c}{dx}(x), & x \in \Sigma_0^\nabla \\ \dfrac{c}{c-d_N} \left(\dfrac{d\mu_{\min}^c}{dx}(x) - \dfrac{1}{c}\rho^0(x) \right), & x \in \Sigma_0^\Delta. \end{cases} \quad (4.5)$$

We extend the domain of ρ to the whole interval $[a,b]$, say, by defining the function at its jump discontinuities to be the average of its left and right limits. Noting the denominators in (4.5), we recall that we have assumed without any loss of generality that $\lim_{N\to\infty} d_N \neq c$. Since $\mu_{\min}^c$ is a probability measure and since we may equivalently express d_N in the form

$$d_N = \int_{\Sigma_0^\Delta} \rho^0(x)\, dx,$$

we see that

$$\int_a^b \rho(x)\, dx = 1. \quad (4.6)$$

We also introduce the associated complex logarithmic potential

$$g(z) := \int_a^b \log(z-x)\rho(x)\, dx. \quad (4.7)$$

The logarithm in (4.7) is the principal branch; thus this function is analytic for $z \in \mathbb{C} \setminus (-\infty, b]$. As a consequence of (4.6), we have $g(z) \sim \log(z)$ as $z \to \infty$. The function $g(z)$ takes boundary values on $(-\infty, b]$ that are Hölder-continuous with any exponent $\alpha < 1$.

Recall the constant γ defined in (2.30). This constant is bounded in the limit $N \to \infty$. Consider the transformation

$$\mathbf{S}(z) := e^{(N\ell_c+\gamma)\sigma_3/2} \mathbf{R}(z) e^{(\#\Delta-k)g(z)\sigma_3} e^{-(N\ell_c+\gamma)\sigma_3/2}. \quad (4.8)$$

Now the identity (4.6) implies that the exponential $e^{(\#\Delta-k)g(z)}$ is analytic for $z \in \mathbb{C} \setminus [a,b]$ (in fact, since we are assuming a constraint to be active at both ends of the interval, the support of $\rho(x)$ is a closed subinterval of (a,b), and we may replace $[a,b]$ by $\mathrm{supp}(\rho(x))$ in this statement). Thus, like $\mathbf{R}(z)$, the matrix $\mathbf{S}(z)$ is also analytic for $z \in \mathbb{C} \setminus \Sigma$, and the boundary values taken on Σ are continuous. However, since $\mathbf{R}(z)z^{(\#\Delta-k)\sigma_3} \to \mathbb{I}$ as $z \to \infty$, we see that $\mathbf{S}(z)$ satisfies the normalization condition

$$\mathbf{S}(z) = \mathbb{I} + O\left(\frac{1}{z}\right) \quad \text{as } z \to \infty.$$

4.3.2 The jump of $\mathbf{S}(z)$ on the real axis

The point of introducing the equilibrium measure in this way is that the matrix $\mathbf{S}(z)$ satisfies jump conditions across the voids, bands, and saturated regions of $[a,b]$ that are analytically tractable as a consequence of the variational inequalities that $\mu_{\min}^c$ imposes on $\delta E_c/\delta\mu$ in the gaps. To describe these jump conditions, we first introduce, for $z \in [a,b]$, the functions

$$\theta(z) := 2\pi(d_N - c)\int_z^b \rho(s)\, ds \quad \text{and} \quad \phi(z) := -2\pi\kappa \int_z^b \rho(s)\, ds. \quad (4.9)$$

AN EQUIVALENT RIEMANN-HILBERT PROBLEM

Recalling the upper and lower constraints on the equilibrium measure, we see that the definition (4.5) implies that the function $\theta(z)$ is real and nondecreasing for $z \in \Sigma_0^\nabla$ and real and nonincreasing for $z \in \Sigma_0^\Delta$. Next, for $z \in \Sigma_0^\nabla$, we define the function

$$T_\nabla(z) := 2\cos\left(\frac{N\theta^0(z)}{2}\right) \frac{\prod_{n \in \Delta}(z - x_{N,n})}{\prod_{n \in \nabla}(z - x_{N,n})} \exp\left(N\left[\int_{\Sigma_0^\nabla} \log|z-x|\rho^0(x)\,dx - \int_{\Sigma_0^\Delta} \log|z-x|\rho^0(x)\,dx\right]\right), \tag{4.10}$$

and for $z \in \Sigma_0^\Delta$, we define the function

$$T_\Delta(z) := 2\cos\left(\frac{N\theta^0(z)}{2}\right) \frac{\prod_{n \in \nabla}(z - x_{N,n})}{\prod_{n \in \Delta}(z - x_{N,n})} \exp\left(-N\left[\int_{\Sigma_0^\nabla} \log|z-x|\rho^0(x)\,dx - \int_{\Sigma_0^\Delta} \log|z-x|\rho^0(x)\,dx\right]\right). \tag{4.11}$$

Note that both $T_\nabla(z)$ and $T_\Delta(z)$ are positive real-analytic functions throughout their respective intervals of definition (the cosine function cancels the poles contributed by the denominator in each case).

Now, denoting the boundary value taken by $\mathbf{S}(z)$ on Σ from the left by $\mathbf{S}_+(z)$ and that taken from the right by $\mathbf{S}_-(z)$, we can easily derive the relation

$$\mathbf{S}_+(z) = \mathbf{S}_-(z) \begin{pmatrix} e^{iN\theta(z)}e^{i\phi(z)} & -iT_\nabla(z)e^{\gamma - \eta(z) + \kappa(g_+(z)+g_-(z))}\exp\left(N\left[\ell_c - \frac{\delta E_c}{\delta\mu}(z)\right]\right) \\ 0 & e^{-iN\theta(z)}e^{-i\phi(z)} \end{pmatrix} \tag{4.12}$$

holding for z in any subinterval of Σ_0^∇. Similarly, if z is in any subinterval of Σ_0^Δ, then

$$\mathbf{S}_+(z) = \mathbf{S}_-(z) \begin{pmatrix} e^{-iN\theta(z)}e^{-i\phi(z)} & 0 \\ iT_\Delta(z)e^{\eta(z) - \gamma - \kappa(g_+(z)+g_-(z))}\exp\left(N\left[\frac{\delta E_c}{\delta\mu}(z) - \ell_c\right]\right) & e^{iN\theta(z)}e^{i\phi(z)} \end{pmatrix}. \tag{4.13}$$

Here $g_+(z) + g_-(z)$ is the sum of the upper and lower boundary values taken by the complex potential $g(z)$ on the real axis, and the variational derivative is evaluated on the equilibrium measure $\mu_{\min}^c$.

As z varies within a gap Γ, the definition (4.5) implies that the functions $\theta(z)$ and $\phi(z)$ remain constant. In particular, to each gap Γ we may assign a constant

$$\phi_\Gamma := \phi(z), \qquad \text{for } z \in \Gamma. \tag{4.14}$$

The constant values of $\theta(z)$ in the gaps have essentially already been defined. Recalling the definitions (2.21)–(2.23), depending on whether Γ is (respectively) a void between two bands, a saturated region between two bands, or one of the intervals (a, α_0) or (β_G, b), we see from (4.1) that

$$\theta(z) \equiv \theta_\Gamma \quad \left(\text{mod } \frac{2\pi}{N}\right),$$

for z in any gap Γ. Note that the constants θ_Γ are by definition independent of the transition points Y_N. Note also that

$$e^{\pm iN\theta_\Gamma}e^{\pm i\phi_\Gamma} = 1, \qquad \text{when } \Gamma = (a, \alpha_0) \text{ or } \Gamma = (\beta_G, b),$$

because $\#\Delta$ and k are both integers.

Now, for z in a void Γ, the strict variational inequality (2.13) holds. Subject to the claim that $T_\nabla(z)$ remains bounded as $N \to \infty$ (this claim is established in Proposition 4.3 below), we therefore see that the jump matrix relating the boundary values in (4.12) is exponentially close to the constant matrix $e^{iN\theta_\Gamma\sigma_3}e^{i\phi_\Gamma\sigma_3}$ as $N \to \infty$. Similarly, for z in a saturated region Γ, the strict variational inequality (2.16) holds, which shows that the jump matrix relating the boundary values in (4.13) is exponentially close to the constant matrix $e^{-iN\theta_\Gamma\sigma_3}e^{-i\phi_\Gamma\sigma_3}$ in the limit $N \to \infty$.

A band interval I can be contained in Σ_0^∇, in Σ_0^Δ, or (if it is a transition band) partly in Σ_0^∇ and partly in Σ_0^Δ. Throughout I, the equilibrium condition (2.14) holds identically. Thus, for $z \in I \cap \Sigma_0^\nabla$, we have a factorization of the jump condition:

$$\mathbf{S}_+(z) = \mathbf{S}_-(z) \begin{pmatrix} e^{iN\theta(z)}e^{i\phi(z)} & -iT_\nabla(z)e^{\gamma-\eta(z)+\kappa(g_+(z)+g_-(z))} \\ 0 & e^{-iN\theta(z)}e^{-i\phi(z)} \end{pmatrix} = \mathbf{S}_-(z)\mathbf{L}_-(z)\mathbf{J}(z)\mathbf{L}_+(z), \qquad (4.15)$$

where, for $z \in \Sigma_0^\nabla$,

$$\mathbf{L}_\pm(z) := \begin{pmatrix} T_\nabla(z)^{\mp 1/2} & 0 \\ iT_\nabla(z)^{-1/2}e^{\eta(z)-\gamma-2\kappa g_\pm(z)}e^{\pm iN\theta(z)} & T_\nabla(z)^{\pm 1/2} \end{pmatrix}$$

and

$$\mathbf{J}(z) := \begin{pmatrix} 0 & -ie^{\gamma-\eta(z)+\kappa(g_+(z)+g_-(z))} \\ -ie^{\eta(z)-\gamma-\kappa(g_+(z)+g_-(z))} & 0 \end{pmatrix}. \qquad (4.16)$$

As noted earlier, the function $T_\nabla(z)$ is a strictly positive analytic function throughout $I \cap \Sigma_0^\nabla$, and we take $T_\nabla(z)^{\pm 1/2}$ to be positive also. Similarly, for $z \in I \cap \Sigma_0^\Delta$, (4.13) becomes

$$\mathbf{S}_+(z) = \mathbf{S}_-(z) \begin{pmatrix} e^{-iN\theta(z)}e^{-i\phi(z)} & 0 \\ iT_\Delta(z)e^{\eta(z)-\gamma-\kappa(g_+(z)+g_-(z))} & e^{iN\theta(z)}e^{i\phi(z)} \end{pmatrix} = \mathbf{S}_-(z)\mathbf{U}_-(z)\mathbf{J}(z)^{-1}\mathbf{U}_+(z), \qquad (4.17)$$

where, for $z \in \Sigma_0^\Delta$,

$$\mathbf{U}_\pm(z) := \begin{pmatrix} T_\Delta(z)^{\pm 1/2} & -iT_\Delta(z)^{-1/2}e^{\gamma-\eta(z)+2\kappa g_\pm(z)}e^{\pm iN\theta(z)} \\ 0 & T_\Delta(z)^{\mp 1/2} \end{pmatrix}$$

and $\mathbf{J}(z)$ is defined as in (4.16). Note that since $T_\Delta(z)$ is strictly positive for $z \in I \subset \Sigma_0^\Delta$, we are choosing the square roots $T_\Delta(z)^{\pm 1/2}$ to be positive also.

4.3.3 Important properties of the functions $T_\nabla(z)$ and $T_\Delta(z)$

Here we establish for later use several properties of $T_\nabla(z)$ and $T_\Delta(z)$. We first introduce the related function $Y(z)$ defined by

$$Y(z) := \frac{\prod_{n \in \Delta}(z - x_{N,n})}{\prod_{n \in \nabla}(z - x_{N,n})} \exp\left(N\left[\int_{\Sigma_0^\nabla} \log(z-x)\rho^0(x)\,dx - \int_{\Sigma_0^\Delta} \log(z-x)\rho^0(x)\,dx\right]\right), \qquad (4.18)$$

for all z in the domain of analyticity of $\rho^0(z)$ with $\Im(z) \neq 0$.

We begin by explicitly relating $T_\nabla(z)$, $T_\Delta(z)$, and $Y(z)$.

Proposition 4.2 (Analytic Properties of $T_\nabla(z)$, $T_\Delta(z)$, and $Y(z)$). *There exists an open complex neighborhood G of the closed interval $[a,b]$ such that the following statements are true.*

1. *$T_\nabla(z)$ admits analytic continuation to the domain*

$$D_\nabla := (\mathbb{C} \setminus (\Sigma_0^\Delta \cup (-\infty, a] \cup [b, +\infty))) \cap G\,.$$

2. *$T_\Delta(z)$ admits analytic continuation to the domain*

$$D_\Delta := (\mathbb{C} \setminus (\Sigma_0^\nabla \cup (-\infty, a] \cup [b, +\infty))) \cap G\,.$$

3. *$Y(z)$ admits analytic continuation to the domain $G \setminus [a,b]$.*

4. *The function $T_\nabla(z)$ is real and positive for $z \in \Sigma_0^\nabla \subset D_\nabla$, the function $T_\Delta(z)$ is real and positive for $z \in \Sigma_0^\Delta \subset D_\Delta$, and the continuations of $T_\nabla(z)$ and $T_\Delta(z)$ map the open domains D_∇ and D_Δ, respectively, into the cut plane $\mathbb{C} \setminus (-\infty, 0]$.*

AN EQUIVALENT RIEMANN-HILBERT PROBLEM

5. The square roots $T_\nabla(z)^{1/2}$ and $T_\Delta(z)^{1/2}$ exist as analytic functions defined in the open domains D_∇ and D_Δ, respectively, that are real and positive for $z \in \Sigma_0^\nabla \subset D_\nabla$ and $z \in \Sigma_0^\Delta \subset D_\Delta$, respectively.

6. We have the identities
$$T_\nabla(z)^{1/2} T_\Delta(z)^{1/2} = T_\nabla(z) Y(z)^{-1} = T_\Delta(z) Y(z) = 1 + e^{-iN\theta^0(z)}, \qquad \text{for } z \in G \text{ with } \Im(z) > 0, \tag{4.19}$$
and
$$T_\nabla(z)^{1/2} T_\Delta(z)^{1/2} = T_\nabla(z) Y(z)^{-1} = T_\Delta(z) Y(z) = 1 + e^{iN\theta^0(z)}, \qquad \text{for } z \in G \text{ with } \Im(z) < 0. \tag{4.20}$$

These formulae also hold on the real axis in the sense of boundary values taken from the upper and lower half-planes.

Proof. We take the domain G to be contained in the domain of analyticity of $\rho^0(z)$. Let $z \in \Sigma_0^\nabla$. We then have
$$\int_{\Sigma_0^\nabla} \log|z-x| \rho^0(x)\, dx - \int_{\Sigma_0^\Delta} \log|z-x| \rho^0(x)\, dx$$
$$= \lim_{\epsilon \downarrow 0} \left[\int_{\Sigma_0^\nabla} \log(z \pm i\epsilon - x) \rho^0(x)\, dx - \int_{\Sigma_0^\Delta} \log(z \pm i\epsilon - x) \rho^0(x)\, dx \right] \mp \frac{i\theta^0(z)}{2} \pm 2\pi i M,$$
where
$$M = \int_{z < x \in \Sigma_0^\Delta} \rho^0(x)\, dx.$$

The integral M is a constant since $z \in \Sigma_0^\nabla$, and by virtue of the quantization condition (4.1) it is an integer. This proves that $T_\nabla(z)$ may be analytically continued from any subinterval of Σ_0^∇ to all of the open domain D_∇ and that the continuation does not depend on the particular subinterval of Σ_0^∇ from which the continuation is performed. The analytic continuation of $T_\Delta(z)$ to the open domain D_Δ is obtained in a similar way. The function $Y(z)$ clearly admits analytic continuation to $z > b$, and for $z < a$, we have
$$\lim_{\epsilon \downarrow 0} \frac{Y(z+i\epsilon)}{Y(z-i\epsilon)} = \exp\left(2\pi i N \left[\int_{\Sigma_0^\nabla} \rho^0(x)\, dx - \int_{\Sigma_0^\Delta} \rho^0(x)\, dx \right] \right) = 1,$$
where the last equality follows from the quantization condition (4.1) that determines the endpoints of the subintervals of Σ_0^∇ and Σ_0^Δ. This proves statements 1, 2, and 3.

This line of argument immediately establishes some of the identities claimed in statement 6, namely, that $T_\nabla(z) Y(z)^{-1} = T_\Delta(z) Y(z) = 1 + e^{-iN\theta^0(z)}$ holds for $\Im(z) > 0$ and that $T_\nabla(z) Y(z)^{-1} = T_\Delta(z) Y(z) = 1 + e^{iN\theta^0(z)}$ holds for $\Im(z) < 0$. Combining these, one easily obtains the identities
$$T_\nabla(z) T_\Delta(z) = (1 + e^{-iN\theta^0(z)})^2, \qquad \text{for } \Im(z) > 0, \tag{4.21}$$
and
$$T_\nabla(z) T_\Delta(z) = (1 + e^{iN\theta^0(z)})^2, \qquad \text{for } \Im(z) < 0. \tag{4.22}$$

Let G_+ and G_- denote the intersections of the neighborhood G with the upper and lower open half-planes, respectively. By choosing G to be sufficiently small but independent of N, we may ensure (because the analytic function $\rho^0(z)$ is strictly positive for $z \in [a,b]$) that for all $N > 0$ the function $w = 1 + e^{\mp iN\theta^0(z)}$ maps the open set $G_\pm$ into the open disc $|w-1| < 1$. It follows that the image of $G_\pm$ under the map $(1 + e^{\mp iN\theta^0(z)})^2$ is an open set disjoint from the negative real axis. In particular, from (4.21) we see that the analytic functions $T_\nabla(z)$ and $T_\Delta(z)$ have no zeros in the open set G_+, and similarly from (4.22) we see that neither function has any zeros in the open set G_-. Now, the strict positivity of $T_\nabla(z)$ for $z \in \Sigma_0^\nabla$ is a simple consequence of the definition (4.10), and that of $T_\Delta(z)$ for $z \in \Sigma_0^\Delta$ is a simple consequence of the definition (4.11). So while $T_\nabla(z)$ has no zeros in G away from the real axis or in Σ_0^∇, (4.21) and (4.22) show that the boundary values taken by $T_\nabla(z)$ on any subinterval of Σ_0^Δ from above or below have many double zeros. Similarly, the boundary values taken by $T_\Delta(z)$ on any subinterval of Σ_0^∇ have many double zeros. However, it is clear from the preceding statements and from (4.21) that if C is a contour homotopic to a subinterval of Σ_0^Δ that lies (with the exception of its endpoints) in the open upper half-plane, and if C is close enough to the real axis, then $T_\nabla(z)$ maps C into the cut plane $\mathbb{C} \setminus (-\infty, 0]$. If instead C lies in the lower half-plane, then (4.22) shows that it is again mapped by $T_\nabla(z)$ into the cut plane $\mathbb{C} \setminus (-\infty, 0]$ if it lies close enough to the real axis. Similar arguments show that contours in D_Δ homotopic to subintervals of Σ_0^∇ and close enough to the real axis are mapped by $T_\Delta(z)$ into the cut plane $\mathbb{C} \setminus (-\infty, 0]$. This is sufficient to establish statement 4.

Statement 5 follows from statement 4 with an appropriate choice of the square root. The remaining identities in statement 6 are then obtained by taking the square root of (4.21) and (4.22) and choosing the sign to be consistent with taking the limit $z \to y_{k,N}$, in which the left-hand side is positive. $\square$

In a suitable precise sense, the functions $T_\nabla(z)$, $T_\Delta(z)$, and $Y(z)$ may all be regarded as being approximately equal to 1 when N is large. This is the content of the following proposition.

Proposition 4.3 (Asymptotic Properties of $T_\nabla(z)$, $T_\Delta(z)$, and $Y(z)$). *There exists an open complex neighborhood G of the closed interval $[a, b]$ such that the following statements are true.*

1. *(Asymptotics away from the boundary.) For any fixed compact subset $K \subset D_\nabla$, there exists a constant $C_K^\nabla > 0$ for which the estimate*

$$\sup_{z \in K} |T_\nabla(z) - 1| \leq \frac{C_K^\nabla}{N} \tag{4.23}$$

holds for all sufficiently large N. Similarly, for any fixed compact subset $K \subset D_\Delta$, there exists a constant $C_K^\Delta > 0$ for which the estimate

$$\sup_{z \in K} |T_\Delta(z) - 1| \leq \frac{C_K^\Delta}{N} \tag{4.24}$$

holds for all sufficiently large N. Finally, for any fixed compact subset $K \subset G \setminus [a, b]$, there is a constant $C_K > 0$ for which the estimate

$$\sup_{z \in K} |Y(z) - 1| \leq \frac{C_K}{N}, \tag{4.25}$$

holds for all sufficiently large N.

2. *(Asymptotics near $z = a$ and $z = b$.) If $K \subset G$ is a compact neighborhood of $z = a$ and Σ_0^Δ is bounded away from K, then there is a constant $C_K^{\nabla,a} > 0$, and for each $\delta > 0$ there is a constant $C_{K,\delta}^{\nabla,a}$ such that for sufficiently large N,*

$$\sup_{z \in K, |\arg(z-a)| < \pi} \left| T_\nabla(z) - \frac{\sqrt{2\pi} e^{-\zeta_a} \zeta_a^{\zeta_a}}{\Gamma(\zeta_a + 1/2)} \right| \leq \frac{C_K^{\nabla,a}}{N},$$

$$\sup_{z \in K, \delta \leq |\arg(z-a)| \leq \pi} \left| Y(z) - \frac{\Gamma(1/2 - \zeta_a)}{\sqrt{2\pi} e^{\zeta_a} (-\zeta_a)^{-\zeta_a}} \right| \leq \frac{C_{K,\delta}^{\nabla,a}}{N}, \tag{4.26}$$

where

$$\zeta_a := N \int_a^z \rho^0(s)\, ds\,.$$

If instead it is Σ_0^∇ that is bounded away from K, then there is a constant $C_K^{\Delta,a} > 0$, and for each $\delta > 0$ there is a constant $C_{K,\delta}^{\Delta,a} > 0$ such that for sufficiently large N,

$$\sup_{z \in K, |\arg(z-a)| < \pi} \left| T_\Delta(z) - \frac{\sqrt{2\pi} e^{-\zeta_a} \zeta_a^{\zeta_a}}{\Gamma(\zeta_a + 1/2)} \right| \leq \frac{C_K^{\Delta,a}}{N},$$

$$\sup_{z \in K, \delta \leq |\arg(z-a)| \leq \pi} \left| Y(z)^{-1} - \frac{\Gamma(1/2 - \zeta_a)}{\sqrt{2\pi} e^{\zeta_a} (-\zeta_a)^{-\zeta_a}} \right| \leq \frac{C_{K,\delta}^{\Delta,a}}{N}. \tag{4.27}$$

AN EQUIVALENT RIEMANN-HILBERT PROBLEM

Similarly, if $K \subset G$ is a compact neighborhood of $z = b$ and Σ_0^Δ is bounded away from K, then there is a constant $C_K^{\nabla,b} > 0$, and for each $\delta > 0$ there is a constant $C_{K,\delta}^{\nabla,b} > 0$ such that for sufficiently large N,

$$\sup_{z \in K, |\arg(b-z)| < \pi} \left| T_\nabla(z) - \frac{\sqrt{2\pi} e^{-\zeta_b} \zeta_b^{\zeta_b}}{\Gamma(\zeta_b + 1/2)} \right| \leq \frac{C_K^{\nabla,b}}{N},$$

$$\sup_{z \in K, \delta \leq |\arg(b-z)| \leq \pi} \left| Y(z) - \frac{\Gamma(1/2 - \zeta_b)}{\sqrt{2\pi} e^{\zeta_b} (-\zeta_b)^{-\zeta_b}} \right| \leq \frac{C_{K,\delta}^{\nabla,b}}{N}, \quad (4.28)$$

where

$$\zeta_b := N \int_z^b \rho^0(s)\,ds.$$

If instead it is Σ_0^∇ that is bounded away from K, then there is a constant $C_K^{\Delta,b} > 0$, and for each $\delta > 0$ there is a constant $C_{K,\delta}^{\Delta,b} > 0$ such that for sufficiently large N,

$$\sup_{z \in K, |\arg(b-z)| < \pi} \left| T_\Delta(z) - \frac{\sqrt{2\pi} e^{-\zeta_b} \zeta_b^{\zeta_b}}{\Gamma(\zeta_b + 1/2)} \right| \leq \frac{C_K^{\Delta,b}}{N},$$

$$\sup_{z \in K, \delta \leq |\arg(b-z)| \leq \pi} \left| Y(z)^{-1} - \frac{\Gamma(1/2 - \zeta_b)}{\sqrt{2\pi} e^{\zeta_b} (-\zeta_b)^{-\zeta_b}} \right| \leq \frac{C_{K,\delta}^{\Delta,b}}{N}. \quad (4.29)$$

Proof. For $z, x \in G$, let

$$D(z,x) := \frac{1}{z-x} \int_x^z \rho^0(s)\,ds.$$

This function is analytic in both variables, and since $D(z,x)$ is strictly positive for both z and x in $[a,b]$, we may choose G to be a sufficiently small neighborhood of $[a,b]$ to ensure that $\Re(D(z,x))$ is strictly positive for all z and x in G. In particular, $D(z,x)$ is nonzero. It follows that $\log(D(z,x))$ is well defined as an analytic function for z and x in G. Next we define an analytic function for $x \in G$ by the integral

$$m(x) := \int_a^x \rho^0(s)\,ds.$$

Since $\rho^0(s)$ is strictly positive in $[a,b]$, there is a unique analytic inverse function, which we denote by $x(m)$, that is defined for $m \in m(G)$, where $m(G)$ is an open complex neighborhood of $[0,1]$. It follows that

$$\frac{\partial^2}{\partial m^2} \log(D(z, x(m))) \text{ is uniformly bounded for } z \in G \text{ and } m \in m(G). \quad (4.30)$$

Using this fact, we see that there are constants $C^\nabla > 0$ and $C^\Delta > 0$ such that for sufficiently large N,

$$\sup_{z \in G} \left| \int_{m(\Sigma_0^\nabla)} \log(D(z, x(s)))\,ds - \sum_{n \in \nabla} \log(D(z, x(s_{N,n}))) \frac{1}{N} \right| \leq \frac{C^\nabla}{N^2} \quad (4.31)$$

and

$$\sup_{z \in G} \left| \int_{m(\Sigma_0^\Delta)} \log(D(z, x(s)))\,ds - \sum_{n \in \Delta} \log(D(z, x(s_{N,n}))) \frac{1}{N} \right| \leq \frac{C^\Delta}{N^2}, \quad (4.32)$$

where

$$s_{N,n} := \frac{2n+1}{2N}.$$

Indeed, these estimates follow from (4.30) because the sums are Riemann sum estimates of the corresponding integrals with the midpoints $s_{N,n}$ of the subintervals $(n/N, (n+1)/N)$ chosen as sample points. The midpoint rule is second-order accurate if the second derivative of the integrand is uniformly bounded. The constants C^∇ and C^Δ depend on the bound implied by (4.30), which depends in turn on G.

For $z \in \Sigma_0^\nabla$, we define

$$\tilde{T}_\nabla(z) := 2\cos(\pi N - \pi N m(z)) \frac{\prod_{n \in \Delta}(m(z) - s_{N,n})}{\prod_{n \in \nabla}(m(z) - s_{N,n})}$$
$$\cdot \exp\left(N\left[\int_{m(\Sigma_0^\nabla)} \log|m(z) - s|\,ds - \int_{m(\Sigma_0^\Delta)} \log|m(z) - s|\,ds\right]\right),$$

which is extended by analytic continuation to $z \in D_\nabla$. The estimates (4.31) and (4.32) imply that uniformly for all $z \in D_\nabla$,

$$T_\nabla(z) = \tilde{T}_\nabla(z)\left(1 + O\left(\frac{1}{N}\right)\right) \quad \text{as } N \to \infty. \tag{4.33}$$

Indeed, some straightforward calculations show that

$$\log\left(\frac{T_\nabla(z)}{\tilde{T}_\nabla(z)}\right) = \left(\sum_{n \in \nabla} \log(D(z, x_{N,n})) - N\int_{m(\Sigma_0^\nabla)} \log(D(z, x(s)))\,ds\right)$$
$$- \left(\sum_{n \in \Delta} \log(D(z, x_{N,n})) - N\int_{m(\Sigma_0^\Delta)} \log(D(z, x(s)))\,ds\right),$$

from which (4.33) follows. Similarly, for $z \in \Sigma_0^\Delta$, we define

$$\tilde{T}_\Delta(z) := 2\cos(\pi N - \pi N m(z)) \frac{\prod_{n \in \nabla}(m(z) - s_{N,n})}{\prod_{n \in \Delta}(m(z) - s_{N,n})}$$
$$\cdot \exp\left(N\left[\int_{m(\Sigma_0^\Delta)} \log|m(z) - s|\,ds - \int_{m(\Sigma_0^\nabla)} \log|m(z) - s|\,ds\right]\right),$$

which is extended to $z \in D_\Delta$ by analytic continuation, and we have

$$T_\Delta(z) = \tilde{T}_\Delta(z)\left(1 + O\left(\frac{1}{N}\right)\right) \quad \text{as } N \to \infty \tag{4.34}$$

holding uniformly for $z \in D_\Delta$. The uniform asymptotic relations (4.33) and (4.34) effectively reduce the asymptotic analysis of the functions $T_\nabla(z)$ and $T_\Delta(z)$ to that of the functions $\tilde{T}_\nabla(z)$ and $\tilde{T}_\Delta(z)$. This is advantageous because the discrete points $s_{N,n}$ are equally spaced, while the nodes $x_{N,n}$ are not necessarily so.

Thus it remains to study $\tilde{T}_\nabla(z)$ and $\tilde{T}_\Delta(z)$. We will consider $\tilde{T}_\nabla(z)$ since the analysis of $\tilde{T}_\Delta(z)$ is similar. Assume that K is a compact subset of the open set D_∇. Let

$$\delta_K := \frac{1}{2} \inf_{z \in K \cap \Sigma_0^\nabla, w \in \Sigma_0^\Delta \cup \{a,b\}} |z - w| > 0$$

be half the minimum distance of $K \cap \Sigma_0^\nabla$ from the boundary of Σ_0^∇. Also, define the open covering U by

$$U := \bigcup_{z \in K \cap \Sigma_0^\nabla} (z - \delta_K, z + \delta_K)$$

and let $F = \overline{U}$ be the closure. Finally, set

$$\epsilon_K := \inf_{z \in K, \Re(z) \notin F} |\Im(z)| > 0.$$

This is strictly positive because K is compact and can touch the real axis only in the interior of subintervals of Σ_0^∇. Thus each $z \in K$ satisfies either $|\Im(z)| \geq \epsilon_K > 0$ (because $\Re(z) \notin F$) or
$$\inf_{w \in \Sigma_0^\Delta \cup \{a,b\}} |w - \Re(z)| \geq \delta_K > 0$$
(because $\Re(z) \in F$).

We may extend $\tilde{T}_\nabla(z)$ into the complex plane from Σ_0^∇ by the following formula:

$$\tilde{T}_\nabla(z) = \left(1 + e^{2\pi i \operatorname{sgn}(\Im(z)) Nm(z)}\right)$$
$$\cdot \exp\left(N\left[\int_{m(\Sigma_0^\nabla)} \log(m(z) - s)\,ds - \sum_{n \in \nabla} \log(m(z) - s_{N,n}) \frac{1}{N}\right]\right)$$
$$\cdot \exp\left(N\left[\sum_{n \in \Delta} \log(m(z) - s_{N,n}) \frac{1}{N} - \int_{m(\Sigma_0^\Delta)} \log(m(z) - s)\,ds\right]\right). \quad (4.35)$$

Suppose that for some $\epsilon > 0$, we have $|\Im(z)| \geq \epsilon$, a condition that also bounds $\Im(m(z))$ away from zero. Therefore $\log(m(z) - s)$ has a second derivative with respect to s that is uniformly bounded for all $s \in [a,b]$. The bound on the second derivative will depend on ϵ and the function $\rho^0(s)$ used to define the function $m(z)$. In any case, an argument involving midpoint-rule Riemann sums shows that the second and third lines of (4.35) are each uniformly of the form $1 + O(1/N)$ as $N \to \infty$ for $|\Im(z)| \geq \epsilon$. Furthermore, a Cauchy-Riemann argument shows that the first line of (4.35) is exponentially close to 1 as $N \to \infty$ for $z \in G$ with $|\Im(z)| \geq \epsilon$. Thus we have shown that there is a constant $C_\epsilon > 0$ such that for sufficiently large N,
$$\sup_{z \in G, |\Im(z)| \geq \epsilon > 0} |\tilde{T}_\nabla(z) - 1| \leq \frac{C_\epsilon}{N}.$$

Next suppose that $\Re(z) \in \Sigma_0^\nabla$, bounded away from $\Sigma_0^\Delta \cup \{a,b\}$ by a distance $\delta > 0$. Let J denote the maximal component interval of Σ_0^∇ that contains $\Re(z)$ and suppose that the corresponding index subset of ∇ consists of the contiguous list of integers $A, A+1, \ldots, B-1, B$. Then from the representation (4.35), one sees once again by a midpoint-rule Riemann sum argument that the factor on the third line of (4.35) is of the form $1 + O(1/N)$ as $N \to \infty$ with a constant on the $O(1/N)$ term that depends on δ. A similar argument applies to the factor on the second line of (4.35), with the exception of the contribution of the integral over J and the corresponding discrete sum. Thus, uniformly for $\Re(z)$ as above, we have
$$\tilde{T}_\nabla(z) = \tilde{T}_\nabla^J(z)\left(1 + O\left(\frac{1}{N}\right)\right) \quad \text{as } N \to \infty, \quad (4.36)$$
where
$$\tilde{T}_\nabla^J(z) := \frac{2N^{B+1-A} e^{i\pi \operatorname{sgn}(\Im(z)) Nm(z)} \cos(\pi Nm(z))}{\displaystyle\prod_{n=A}^{B}\left(Nm(z) - n - \frac{1}{2}\right)} \exp\left(N \int_{\frac{A}{N}}^{\frac{B+1}{N}} \log(m(z) - s)\,ds\right). \quad (4.37)$$

When we evaluate the integral exactly and rewrite the product in terms of the Euler gamma function, we have
$$\tilde{T}_\nabla^J(z) = 2(-1)^{B+1} e^{-(B+1-A)} \frac{\Gamma\left(Nm(z) - B - \frac{1}{2}\right)}{\Gamma\left(Nm(z) - A + \frac{1}{2}\right)} \cos(\pi Nm(z))$$
$$\cdot e^{(Nm(z)-A)\log(Nm(z)-A)} e^{(B+1-Nm(z))\log(B+1-Nm(z))},$$
and with the use of the reflection identity $\Gamma(1/2 + z)\Gamma(1/2 - z) = \pi \sec(\pi z)$, we get
$$\tilde{T}_\nabla^J(z) = \frac{2\pi e^{-(B+1-A)} e^{(Nm(z)-A)\log(Nm(z)-A)} e^{(B+1-Nm(z))\log(B+1-Nm(z))}}{\Gamma\left(Nm(z) - A + \frac{1}{2}\right)\Gamma\left(B + 1 - Nm(z) + \frac{1}{2}\right)}.$$

Now the condition that $\Re(z)$ be bounded away from the endpoints of J by at least $\delta > 0$ fixed implies that $Nm(z) - A + 1/2$ and $B + 1 - Nm(z) + 1/2$ are both quantities in the right half-plane that scale like N; an application of Stirling's formula then gives, uniformly for such z,
$$\tilde{T}_\nabla^J(z) = 1 + O\left(\frac{1}{N}\right) \quad \text{as } N \to \infty.$$

Taking $\delta = \delta_K$ and $\epsilon = \epsilon_K$ then completes the proof of (4.23), that $T_\nabla(z) - 1$ is uniformly of order $1/N$ as $N \to \infty$ for $z \in K$, where K is bounded away from $(-\infty, a) \cup \Sigma_0^\Delta \cup (b, +\infty)$. Analogous arguments establish the corresponding result (4.24) for $T_\Delta(z)$. Using (4.19) and (4.20) then proves (4.25). Thus statement 1 is established.

If K is a compact set containing the left endpoint $z = a$ and bounded away from Σ_0^Δ (so that the lower constraint is active at the left endpoint), then one may follow nearly identical arguments to arrive at the asymptotic relation (4.36) now holding uniformly for $z \in K$, where J is the leftmost subinterval of Σ_0^∇ and $\tilde{T}_\nabla^J(z)$ is defined by (4.37). In this case we have $A = 0$, so we expand only the gamma function involving B. This proves the first line of (4.26); the second line follows upon using (4.19) and (4.20). On the other hand, if K contains $z = b$ where the lower constraint is active, then again we have (4.36) holding uniformly for $z \in K$ where now J is the rightmost subinterval of Σ_0^∇. Thus $B = N - 1$, and we expand only the gamma function involving A to prove (4.28). The analogous statements (4.27) and (4.29) are proved similarly. Thus statement 2 is established. □

4.4 STEEPEST DESCENT: THE TRANSFORMATION FROM $\mathbf{S}(z)$ TO $\mathbf{X}(z)$

Now from any band interval $I \cap \Sigma_0^\nabla$, the matrix $\mathbf{L}_+(z)$ admits an analytic continuation into the upper half-plane and the matrix $\mathbf{L}_-(z)$ admits an analytic continuation into the lower half-plane. Since the function $\theta(z)$ is real and increasing in $I \cap \Sigma_0^\nabla$, its analytic continuation from I, which we denote by $\theta_I^\nabla(z)$, will have a positive imaginary part near the real axis in the upper half-plane and a negative imaginary part near the real axis in the lower half-plane, as a simple Cauchy-Riemann argument shows. Thus the factors $e^{\pm iN\theta_I^\nabla(z)}$ present in $\mathbf{L}_\pm(z)$ continued into their respective half-planes become exponentially small as $N \to \infty$. Subject to the claim that the analytic function $T_\nabla(z) - 1$ remains uniformly small upon analytic continuation, we see that the analytic continuations of $\mathbf{L}_+(z)$ and $\mathbf{L}_-(z)$ into the upper and lower half-planes, respectively, become small perturbations of the identity matrix.

Similarly, from a band interval $I \cap \Sigma_0^\Delta$, the analytic continuation of the matrix $\mathbf{U}_-(z)$ into the upper half-plane and that of $\mathbf{U}_+(z)$ into the lower half-plane will be small perturbations of the identity matrix in the limit $N \to \infty$ because the real function $\theta(z)$ is strictly decreasing. This implies that the analytic continuation of $\theta(z)$, which in this case we refer to as $\theta_I^\Delta(z)$, has an imaginary part that is positive in the lower half-plane and negative in the upper half-plane.

Therefore, if the factors $\mathbf{U}_+(z)$ and $\mathbf{L}_+(z)$ can be deformed into the upper half-plane, and at the same time if the factors $\mathbf{U}_-(z)$ and $\mathbf{L}_-(z)$ can be deformed into the lower half-plane, then the rapidly oscillatory jump matrix for $\mathbf{S}(z)$ in the bands will be resolved into near-identity factors and a central slowly varying factor. This idea is the essence of the steepest-descent method for matrix Riemann-Hilbert problems developed by Deift and Zhou.

To carry out the deformation, it will be convenient to introduce some explicit formulae for the analytic continuations $\theta_I^\nabla(z)$ and $\theta_I^\Delta(z)$. If $I \subset \Sigma_0^\nabla$ is a band containing a point (or endpoint) x, then we have
$$\theta_I^\nabla(z) := \theta(x) + 2\pi c \int_x^z \psi_I(s)\, ds. \tag{4.38}$$

If $I \subset \Sigma_0^\Delta$ is a band containing a point (or endpoint) x, then we have
$$\theta_I^\Delta(z) := \theta(x) - 2\pi c \int_x^z \overline{\psi}_I(s)\, ds. \tag{4.39}$$

If I is a transition band, then it is divided into two halves, $I \cap \Sigma_0^\nabla$ and $I \cap \Sigma_0^\Delta$, by the transition point $y_{k,N}$ therein. From $I \cap \Sigma_0^\nabla$ we obtain a continuation $\theta_I^\nabla(z)$ of $\theta(z)$ using the formula (4.38) for $x \in I \cap \Sigma_0^\nabla$, and from $I \cap \Sigma_0^\Delta$ we obtain a continuation $\theta_I^\Delta(z)$ of $\theta(z)$ using the formula (4.39) for $x \in I \cap \Sigma_0^\Delta$.

AN EQUIVALENT RIEMANN-HILBERT PROBLEM

Based on the factorizations (4.15) and (4.17), we now carry out the steepest-descent deformation, introducing a final change of variables defining a new unknown $\mathbf{X}(z)$ in terms of $\mathbf{S}(z)$ with the aim of obtaining a jump condition for $\mathbf{X}(z)$ in the bands involving only the matrix $\mathbf{J}(z)$. Let Σ_{SD} be the oriented contour illustrated in Figure 4.3. For each band interval $I \subset (a, b)$, we make the following definitions. If z

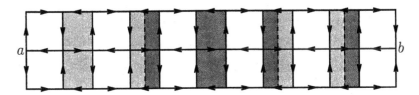

Figure 4.3 *The oriented contour Σ_{SD} consists of the interval $[a, b]$, corresponding horizontal segments $\Im(z) = \pm\epsilon$, and vertical segments aligned at the edges of all band intervals. The dashed vertical lines separating the lighter- and darker-shaded regions are not part of Σ_{SD}.*

lies in the open rectangle $I \cap \Sigma_0^\nabla + i(0, \epsilon)$ (one of the darker-shaded rectangles lying in the upper half-plane in Figure 4.3), we set

$$\mathbf{X}(z) := \mathbf{S}(z) \begin{pmatrix} T_\nabla(z)^{1/2} & 0 \\ -iT_\nabla(z)^{-1/2} e^{\eta(z)-\gamma-2\kappa g(z)} e^{iN\theta_I^\nabla(z)} & T_\nabla(z)^{-1/2} \end{pmatrix}. \quad (4.40)$$

If z lies in the open rectangle $I \cap \Sigma_0^\nabla - i(0, \epsilon)$ (one of the darker-shaded rectangles in the lower half-plane), we set

$$\mathbf{X}(z) := \mathbf{S}(z) \begin{pmatrix} T_\nabla(z)^{1/2} & 0 \\ iT_\nabla(z)^{-1/2} e^{\eta(z)-\gamma-2\kappa g(z)} e^{-iN\theta_I^\nabla(z)} & T_\nabla(z)^{-1/2} \end{pmatrix}. \quad (4.41)$$

Next, if z lies in the open rectangle $I \cap \Sigma_0^\Delta + i(0, \epsilon)$ (one of the lighter-shaded rectangles in the upper half-plane), we set

$$\mathbf{X}(z) := \mathbf{S}(z) \begin{pmatrix} T_\Delta(z)^{-1/2} & -iT_\Delta(z)^{-1/2} e^{\gamma-\eta(z)+2\kappa g(z)} e^{-iN\theta_I^\Delta(z)} \\ 0 & T_\Delta(z)^{1/2} \end{pmatrix}. \quad (4.42)$$

And if z lies in the open rectangle $I \cap \Sigma_0^\Delta - i(0, \epsilon)$ (one of the lighter-shaded rectangles in the lower half-plane), we set

$$\mathbf{X}(z) := \mathbf{S}(z) \begin{pmatrix} T_\Delta(z)^{-1/2} & iT_\Delta(z)^{-1/2} e^{\gamma-\eta(z)+2\kappa g(z)} e^{iN\theta_I^\Delta(z)} \\ 0 & T_\Delta(z)^{1/2} \end{pmatrix}. \quad (4.43)$$

Finally, for all remaining $z \in \mathbb{C} \setminus \Sigma_{\mathrm{SD}}$, we set

$$\mathbf{X}(z) := \mathbf{S}(z). \quad (4.44)$$

4.5 PROPERTIES OF $\mathbf{X}(z)$

This change of variables is the last of a sequence of exact and explicit transformations relating $\mathbf{P}(z; N, k)$ to $\mathbf{Q}(z; N, k)$, $\mathbf{Q}(z; N, k)$ to $\mathbf{R}(z)$, $\mathbf{R}(z)$ to $\mathbf{S}(z)$, and finally $\mathbf{S}(z)$ to $\mathbf{X}(z)$. For future reference it will be useful to summarize this sequence of transformations by presenting the explicit formulae directly giving $\mathbf{X}(z)$ in terms of $\mathbf{P}(z; N, k)$, the solution of Interpolation Problem 1.2. In general, the transformation may be written as

$$\mathbf{X}(z) := e^{(N\ell_c+\gamma)\sigma_3/2} \mathbf{P}(z; N, k) \mathbf{D}(z) e^{(N(d_N-c)-\kappa)g(z)\sigma_3} e^{-(N\ell_c+\gamma)\sigma_3/2}, \quad (4.45)$$

where the matrix $\mathbf{D}(z)$ takes different forms in different regions of the complex plane as follows. For z in the unbounded component of $\mathbb{C} \setminus \Sigma_{\mathrm{SD}}$, we have

$$\mathbf{D}(z) := \begin{pmatrix} \prod_{n \in \Delta}(z - x_{N,n})^{-1} & 0 \\ 0 & \prod_{n \in \Delta}(z - x_{N,n}) \end{pmatrix}. \quad (4.46)$$

For z in the regions $\Omega_\pm^\nabla$ such that $\Re(z)$ lies in a void of $[a,b]$, we have

$$\mathbf{D}(z) := \begin{pmatrix} \displaystyle\prod_{n\in\Delta}(z-x_{N,n})^{-1} & \mp i e^{\mp iN\theta^0(z)/2} e^{-NV_N(z)} \displaystyle\prod_{n\in\nabla}(z-x_{N,n})^{-1} \\ 0 & \displaystyle\prod_{n\in\Delta}(z-x_{N,n}) \end{pmatrix}. \tag{4.47}$$

For z in the regions $\Omega_\pm^\Delta$ such that $\Re(z)$ lies in a saturated region of $[a,b]$, we have

$$\mathbf{D}(z) := \begin{pmatrix} \displaystyle\prod_{n\in\Delta}(z-x_{N,n})^{-1} & 0 \\ \mp i e^{\mp iN\theta^0(z)/2} e^{NV_N(z)} \displaystyle\prod_{n\in\nabla}(z-x_{N,n}) & \displaystyle\prod_{n\in\Delta}(z-x_{N,n}) \end{pmatrix}. \tag{4.48}$$

For z in the regions Ω_+^∇ such that $\Re(z)$ lies in a band I of $[a,b]$ (*i.e.*, in one of the darker-shaded regions in the upper half-plane in Figure 4.3), we have

$$\begin{aligned}
D_{11}(z) &:= T_\nabla(z)^{1/2} \prod_{n\in\Delta}(z-x_{N,n})^{-1} \\
&\qquad - T_\nabla(z)^{-1/2} e^{N(\ell_c - 2(d_N-c)g(z) - V(z) + i\theta_I^\nabla(z) - i\theta^0(z)/2)} \prod_{n\in\nabla}(z-x_{N,n})^{-1}, \\
D_{12}(z) &:= -i T_\nabla(z)^{-1/2} e^{-\eta(z)} e^{-N(V(z) + i\theta^0(z)/2)} \prod_{n\in\nabla}(z-x_{N,n})^{-1}, \\
D_{21}(z) &:= -i T_\nabla(z)^{-1/2} e^{\eta(z)} e^{N(\ell_c - 2(d_N-c)g(z) + i\theta_I^\nabla(z))} \prod_{n\in\Delta}(z-x_{N,n}), \\
D_{22}(z) &:= T_\nabla(z)^{-1/2} \prod_{n\in\Delta}(z-x_{N,n}).
\end{aligned} \tag{4.49}$$

For z in the regions Ω_-^∇ such that $\Re(z)$ lies in a band I of $[a,b]$ (*i.e.*, in one of the darker-shaded regions in the lower half-plane in Figure 4.3), we have

$$\begin{aligned}
D_{11}(z) &:= T_\nabla(z)^{1/2} \prod_{n\in\Delta}(z-x_{N,n})^{-1} \\
&\qquad - T_\nabla(z)^{-1/2} e^{N(\ell_c - 2(d_N-c)g(z) - V(z) - i\theta_I^\nabla(z) + i\theta^0(z)/2)} \prod_{n\in\nabla}(z-x_{N,n})^{-1}, \\
D_{12}(z) &:= i T_\nabla(z)^{-1/2} e^{-\eta(z)} e^{-N(V(z) - i\theta^0(z)/2)} \prod_{n\in\nabla}(z-x_{N,n})^{-1}, \\
D_{21}(z) &:= i T_\nabla(z)^{-1/2} e^{\eta(z)} e^{N(\ell_c - 2(d_N-c)g(z) - i\theta_I^\nabla(z))} \prod_{n\in\Delta}(z-x_{N,n}), \\
D_{22}(z) &:= T_\nabla(z)^{-1/2} \prod_{n\in\Delta}(z-x_{N,n}).
\end{aligned} \tag{4.50}$$

For z in the regions Ω_+^Δ such that $\Re(z)$ lies in a band I of $[a,b]$ (*i.e.*, in one of the lighter-shaded regions in the upper half-plane in Figure 4.3), we have

$$\begin{aligned}
D_{11}(z) &:= T_\Delta(z)^{-1/2} \prod_{n\in\Delta}(z-x_{N,n})^{-1}, \\
D_{12}(z) &:= -i T_\Delta(z)^{-1/2} e^{-\eta(z)} e^{-N(\ell_c - 2(d_N-c)g(z) + i\theta_I^\Delta(z))} \prod_{n\in\Delta}(z-x_{N,n})^{-1}, \\
D_{21}(z) &:= -i T_\Delta(z)^{-1/2} e^{\eta(z)} e^{N(V(z) - i\theta^0(z)/2)} \prod_{n\in\nabla}(z-x_{N,n}), \\
D_{22}(z) &:= T_\Delta(z)^{1/2} \prod_{n\in\Delta}(z-x_{N,n}) \\
&\qquad - T_\Delta(z)^{-1/2} e^{-N(\ell_c - 2(d_N-c)g(z) - V(z) + i\theta_I^\Delta(z) + i\theta^0(z)/2)} \prod_{n\in\nabla}(z-x_{N,n}).
\end{aligned} \tag{4.51}$$

AN EQUIVALENT RIEMANN-HILBERT PROBLEM

Finally, for z in the regions Ω_-^Δ such that $\Re(z)$ lies in a band I of $[a,b]$ (*i.e.*, in one of the lighter-shaded regions in the lower half-plane in Figure 4.3), we have

$$D_{11}(z) := T_\Delta(z)^{-1/2} \prod_{n \in \Delta} (z - x_{N,n})^{-1},$$

$$D_{12}(z) := iT_\Delta(z)^{-1/2} e^{-\eta(z)} e^{-N(\ell_c - 2(d_N - c)g(z) - i\theta_I^\Delta(z))} \prod_{n \in \Delta} (z - x_{N,n})^{-1},$$

$$D_{21}(z) := iT_\Delta(z)^{-1/2} e^{\eta(z)} e^{N(V(z) + i\theta^0(z)/2)} \prod_{n \in \nabla} (z - x_{N,n}), \qquad (4.52)$$

$$D_{22}(z) := T_\Delta(z)^{1/2} \prod_{n \in \Delta} (z - x_{N,n})$$
$$\quad - T_\Delta(z)^{-1/2} e^{-N(\ell_c - 2(d_N - c)g(z) - V(z) - i\theta_I^\Delta(z) - i\theta^0(z)/2)} \prod_{n \in \nabla} (z - x_{N,n}).$$

Unlike the contour Σ, the new contour Σ_{SD} does not contain the vertical segments $Y_N \pm i(0, \epsilon)$ that form the common boundary of the lighter- and darker-shaded rectangles and that are illustrated with dashed lines in Figure 4.3. Since the matrix $\mathbf{X}(z)$ is defined by different formulae in the lighter- and darker-shaded regions, one should suspect that $\mathbf{X}(z)$ cannot be defined on the common boundary so as to make $\mathbf{X}(z)$ continuous there. In other words, it might seem that there should be a jump discontinuity of $\mathbf{X}(z)$ on these vertical segments. On the contrary, we have the following result.

Proposition 4.4. *The matrix $\mathbf{X}(z)$ defined from (4.40)–(4.44) extends to a function analytic in $\mathbb{C} \setminus \Sigma_{\text{SD}}$. In particular, $\mathbf{X}(z)$ is continuous and analytic on the vertical segments $Y_N \pm i(0, \epsilon)$. Moreover, on each subset of Σ_{SD} that contains no self-intersection points, the matrix-valued ratio of boundary values taken by $\mathbf{X}(z)$ is an analytic function of z.*

Proof. Let $\mathbf{X}^{\nabla,+}(z)$ denote the matrix $\mathbf{X}(z)$ defined by (4.45), with $\mathbf{D}(z)$ given by (4.49), and let $\mathbf{X}^{\Delta,+}(z)$ denote the matrix $\mathbf{X}(z)$ defined by (4.45), with $\mathbf{D}(z)$ given by (4.51). We will show that $\mathbf{X}^{\nabla,+}(z)$ and $\mathbf{X}^{\Delta,+}(z)$ are the same analytic function in the common region $0 < \Im(z) < \epsilon$ and $\Re(z) \in I$, where I is a transition band. By direct calculation, we obtain

$$e^{(N(d_N-c)-\kappa)g(z)\sigma_3} e^{-(N\ell_c+\gamma)\sigma_3/2} \mathbf{X}^{\nabla,+}(z)^{-1} \mathbf{X}^{\Delta,+}(z) e^{(N\ell_c+\gamma)\sigma_3/2} e^{-(N(d_N-c)-\kappa)g(z)\sigma_3} = \mathbf{A}^+(z),$$

where

$$A_{11}^+(z) := \frac{1 + e^{-iN\theta^0(z)}}{T_\nabla(z)^{1/2} T_\Delta(z)^{1/2}},$$

$$A_{12}^+(z) := ie^{-\eta(z)} e^{-N(\ell_c - 2(d_N-c)g(z) + i\theta_I^\Delta(z))} \left[F_\Delta^+(z)^{-1} - \frac{1 + e^{-iN\theta^0(z)}}{T_\nabla(z)^{1/2} T_\Delta(z)^{1/2}} \right],$$

$$A_{21}^+(z) := ie^{\eta(z)} e^{N(\ell_c - 2(d_N-c)g(z) + i\theta_I^\nabla(z))} \left[\frac{1 + e^{-iN\theta^0(z)}}{T_\nabla(z)^{1/2} T_\Delta(z)^{1/2}} - F_\nabla^+(z)^{-1} \right], \qquad (4.53)$$

$$A_{22}^+(z) := \frac{1 + e^{-iN\theta^0(z)}}{T_\nabla(z)^{1/2} T_\Delta(z)^{1/2}} e^{iN(\theta_I^\nabla(z) - \theta_I^\Delta(z))} + T_\nabla(z)^{1/2} T_\Delta(z)^{1/2} - e^{-iN\theta^0(z)} \left[F_\nabla^+(z) + F_\Delta^+(z) \right]$$

and where

$$F_\nabla^+(z) := \frac{T_\Delta(z)^{1/2} \prod_{n \in \Delta} (z - x_{N,n})}{T_\nabla(z)^{1/2} \prod_{n \in \nabla} (z - x_{N,n})} e^{N(\ell_c - 2(d_N-c)g(z) + i\theta_I^\nabla(z) - V(z) + i\theta^0(z)/2)},$$

$$F_\Delta^+(z) := \frac{T_\nabla(z)^{1/2} \prod_{n \in \nabla} (z - x_{N,n})}{T_\Delta(z)^{1/2} \prod_{n \in \Delta} (z - x_{N,n})} e^{-N(\ell_c - 2(d_N-c)g(z) + i\theta_I^\Delta(z) - V(z) - i\theta^0(z)/2)}. \qquad (4.54)$$

Now taking the base points x in the formulae (4.38) and (4.39) to both coincide with the transition point $y_{k,N}$ in the transition band I, then recalling (2.15) and using the quantization condition (4.1), and finally comparing with the definition (2.50) of $\theta^0(z)$, we obtain the identity

$$e^{iN(\theta_I^\nabla(z)-\theta_I^\Delta(z))} = e^{-iN\theta^0(z)}, \qquad \text{for } |\Im(z)| < \epsilon,\, \Re(z) \in I,\text{ and } N \in \mathbb{Z}. \tag{4.55}$$

Taking this identity into account, along with the identity (4.19) from Proposition 4.2 valid for $\Im(z) > 0$, we therefore see that the matrix elements (4.53) simplify:

$$\begin{aligned}
A_{11}^+(z) &= 1, \\
A_{12}^+(z) &= ie^{-\eta(z)}e^{-N(\ell_c-2(d_N-c)g(z)+i\theta_I^\Delta(z))}\left[F_\Delta^+(z)^{-1}-1\right], \\
A_{21}^+(z) &= ie^{\eta(z)}e^{N(\ell_c-2(d_N-c)g(z)+i\theta_I^\nabla(z))}\left[1-F_\nabla^+(z)^{-1}\right], \\
A_{22}^+(z) &= 1+e^{-iN\theta^0(z)}\left[2-F_\nabla^+(z)-F_\Delta^+(z)\right].
\end{aligned}$$

Thus to show that $\mathbf{X}^{\nabla,+}(z) \equiv \mathbf{X}^{\Delta,+}(z)$, it suffices to show that $F_\nabla^+(z) \equiv 1$ and $F_\Delta^+(z) \equiv 1$.

Let us calculate the boundary value taken by the function $F_\nabla^+(z)$ on the real interval $I \cap \Sigma_0^\nabla$ from the upper half-plane. For such z, we have three facts at our disposal, namely, the identity $\theta_I^\nabla(z) \equiv \theta(z)$, the identity (4.19) from Proposition 4.2, and the formula (4.10). Applying these, and in particular first using the latter to eliminate the ratio of products in the definition (4.54) of $F_\nabla^+(z)$, we obtain simply

$$F_\nabla^+(z) = \exp\left(N\left[\int_{\Sigma_0^\Delta} \log|z-x|\rho^0(x)\,dx - \int_{\Sigma_0^\nabla} \log|z-x|\rho^0(x)\,dx + \ell_c - 2(d_N-c)g_+(z) + i\theta(z) - V(z)\right]\right),$$

where $g_+(z)$ indicates a boundary value taken from the upper half-plane. Using (4.5), (4.7), and (4.9), we see that for real $z \in [a,b]$,

$$2(d_N-c)g_+(z) - i\theta(z) = -2c\int_a^b \log|z-x|\,d\mu_{\min}^c(x) + 2\int_{\Sigma_0^\Delta} \log|z-x|\rho^0(x)\,dx.$$

Therefore, recalling (2.1) and (2.9), we have simply

$$F_\nabla^+(z) = \exp\left(N\left[\ell_c - \frac{\delta E_c}{\delta \mu}(z)\right]\right),$$

where the variational derivative is evaluated on the equilibrium measure $\mu_{\min}^c$. It follows that $F_\nabla^+(z) \equiv 1$ as a consequence of (2.14) since z is in a band I. By analytic continuation this identity holds in the whole region $0 < \Im(z) < \epsilon$ with $\Re(z) \in I$.

We may also compute a boundary value of the function $F_\Delta^+(z)$, letting z tend toward the real interval $I \cap \Sigma_0^\Delta$ from the upper half-plane. In this case, instead of (4.10), we use the identity (4.11) to eliminate the ratio of products, and we may write $\theta_I^\Delta(z) \equiv \theta(z)$. The rest of the argument is exactly the same, and we thus deduce that the identity $F_\Delta^+(z) \equiv 1$ holds for $z \in I \cap \Sigma_0^\Delta$ in the sense of a boundary value taken from the upper half-plane. But by analytic continuation it also holds in the whole region of interest: $0 < \Im(z) < \epsilon$ and $\Re(z) \in I$. This completes the proof that $\mathbf{X}(z)$ has no jump discontinuity along the vertical segments between the lighter- and darker-shaded regions illustrated in the upper half-plane in Figure 4.3.

Now let $\mathbf{X}^{\nabla,-}(z)$ denote the matrix $\mathbf{X}(z)$ defined by (4.45), with $\mathbf{D}(z)$ given by (4.50), and let $\mathbf{X}^{\Delta,-}(z)$ denote the matrix $\mathbf{X}(z)$ defined by (4.45), with $\mathbf{D}(z)$ given by (4.52). We will now show that $\mathbf{X}^{\nabla,-}(z)$ and $\mathbf{X}^{\Delta,-}(z)$ are the same analytic function in the common region $-\epsilon < \Im(z) < 0$ and $\Re(z) \in I$, where I is a transition band. As before, by direct calculation we have

$$e^{(N(d_N-c)-\kappa)g(z)\sigma_3}e^{-(N\ell_c+\gamma)\sigma_3/2}\mathbf{X}^{\nabla,-}(z)^{-1}\mathbf{X}^{\Delta,-}(z)e^{(N\ell_c+\gamma)\sigma_3/2}e^{-(N(d_N-c)-\kappa)g(z)\sigma_3} = \mathbf{A}^-(z),$$

where

$$A_{11}^-(z) := \frac{1+e^{iN\theta^0(z)}}{T_\nabla(z)^{1/2}T_\Delta(z)^{1/2}},$$

$$A_{12}^-(z) := ie^{-\eta(z)}e^{-N(\ell_c-2(d_N-c)g(z)-i\theta_I^\Delta(z))}\left[\frac{1+e^{iN\theta^0(z)}}{T_\nabla(z)^{1/2}T_\Delta(z)^{1/2}} - F_\Delta^-(z)^{-1}\right],$$

$$A_{21}^-(z) := ie^{\eta(z)}e^{N(\ell_c-2(d_N-c)g(z)-i\theta_I^\nabla(z))}\left[F_\nabla^-(z)^{-1} - \frac{1+e^{iN\theta^0(z)}}{T_\nabla(z)^{1/2}T_\Delta(z)^{1/2}}\right],$$

$$A_{22}^-(z) := \frac{1+e^{iN\theta^0(z)}}{T_\nabla(z)^{1/2}T_\Delta(z)^{1/2}}e^{iN(\theta_I^\Delta(z)-\theta_I^\nabla(z))} + T_\nabla(z)^{1/2}T_\Delta(z)^{1/2} - e^{iN\theta(z)}\left[F_\nabla^-(z) + F_\Delta^-(z)\right]$$

and where

$$F_\nabla^-(z) := \frac{T_\Delta(z)^{1/2}\displaystyle\prod_{n\in\Delta}(z-x_{N,n})}{T_\nabla(z)^{1/2}\displaystyle\prod_{n\in\nabla}(z-x_{N,n})}e^{N(\ell_c-2(d_N-c)g(z)-i\theta_I^\nabla(z)-V(z)-i\theta^0(z)/2)},$$

$$F_\Delta^-(z) := \frac{T_\nabla(z)^{1/2}\displaystyle\prod_{n\in\nabla}(z-x_{N,n})}{T_\Delta(z)^{1/2}\displaystyle\prod_{n\in\Delta}(z-x_{N,n})}e^{-N(\ell_c-2(d_N-c)g(z)-i\theta_I^\Delta(z)-V(z)+i\theta^0(z)/2)}.$$

Taking from Proposition 4.2 the identity (4.20), valid for $\Im(z) < 0$, and using the identity (4.55), we see that these formulae simplify:

$$A_{11}^-(z) = 1,$$
$$A_{12}^-(z) = ie^{-\eta(z)}e^{-N(\ell_c-2(d_N-c)g(z)-i\theta_I^\Delta(z))}\left[1 - F_\Delta^-(z)^{-1}\right],$$
$$A_{21}^-(z) = ie^{\eta(z)}e^{N(\ell_c-2(d_N-c)g(z)-i\theta_I^\nabla(z))}\left[F_\nabla^-(z)^{-1} - 1\right],$$
$$A_{22}^-(z) = 1 + e^{iN\theta^0(z)}\left[2 - F_\nabla^-(z) - F_\Delta^-(z)\right].$$

Therefore the problem again reduces to showing that $F_\nabla^-(z) \equiv 1$ and $F_\Delta^-(z) \equiv 1$.

Taking the boundary value of the function $F_\nabla^-(z)$ from the lower half-plane on the real interval $I \cap \Sigma_0^\nabla$, we may substitute for the ratio of products from (4.10) and use the identity (4.20) from Proposition 4.2 along with $\theta_I^\nabla(z) \equiv \theta(z)$. Since (4.5), (4.7), and (4.9) imply that that for all real $z \in [a,b]$,

$$2(d_N - c)g_-(z) + i\theta(z) = -2c\int_a^b \log|z-x|\,d\mu_{\min}^c(x) + 2\int_{\Sigma_0^\Delta}\log|z-x|\rho^0(x)\,dx,$$

where $g_-(z)$ indicates a boundary value taken from the lower half-plane, the definitions (2.1) and (2.9), along with the equilibrium condition (2.14), show that $F_\nabla^-(z) \equiv 1$ in the sense of a boundary value taken from the lower half-plane on $I \cap \Sigma_0^\nabla$. But by analytic continuation, this identity also holds throughout the region $-\epsilon < \Im(z) < 0$ and $\Re(z) \in I$.

To show that $F_\Delta^-(z) \equiv 1$ in the region $-\epsilon < \Im(z) < 0$ and $\Re(z) \in I$, we repeat the above arguments but take the boundary value from the lower-half plane in the interval $I \cap \Sigma_0^\Delta$, where the identity (4.11) may be used to eliminate the ratio of products and where the identity $\theta_I^\Delta(z) \equiv \theta(z)$ holds. This completes the proof that $\mathbf{X}(z)$ has no jump discontinuity along the vertical segments between the lighter- and darker-shaded regions illustrated in the lower half-plane in Figure 4.3. □

◁ **Remark:** Part of the significance of Proposition 4.4 is that all essential dependence on the set Y_∞, the choice of which was somewhat arbitrary, has disappeared. In particular, when we approximate $\mathbf{X}(z)$ in the limit of large N, we will be able to obtain error estimates that are of the same magnitude regardless of the number of transition points, or indeed regardless of whether there are any transition points at all. This is an improvement over the bounds stated in our announcement [BaiKMM03], which identified different estimates in two cases (there called Case I and Case II) depending on whether any transition points are present. ▷

Having defined the matrix $\mathbf{X}(z)$ explicitly in terms of the solution $\mathbf{P}(z; N, k)$ of Interpolation Problem 1.2 by the formula (4.45), with $\mathbf{D}(z)$ given by (4.46)–(4.52) allows us to replace that problem with an equivalent problem for the new unknown $\mathbf{X}(z)$. This is advantageous because the problem whose solution is $\mathbf{X}(z)$ is more amenable to analysis. In order to correctly pose the problem, we must introduce some additional notation for particular segments of Σ_{SD}, as illustrated in Figure 4.4. The labelling of the figure relates to

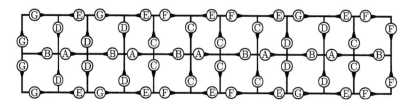

Figure 4.4 *Components of the oriented contour Σ_{SD}. See the text for the key.*

this new notation as follows.

- Bands I of (a, b) in Σ_{SD} are labelled A.

- Gaps Γ of (a, b) in Σ_{SD} are labelled B.

- Vertical segments of Σ_{SD} that are connected to band endpoints will be denoted by $\Sigma_{0\pm}^{\nabla}$ (labelled C) or $\Sigma_{0\pm}^{\Delta}$ (labelled D) depending on whether the endpoint lies in Σ_0^{∇} or Σ_0^{Δ}; the additional subscript indicates whether the segment lies in the upper $(+)$ or lower $(-)$ half-plane.

- Horizontal segments lying above (below) bands will be denoted by Σ_{I+} (Σ_{I-}), and both are labelled E.

- Horizontal segments lying above (below) voids will be denoted by $\Sigma_{\Gamma+}^{\nabla}$ ($\Sigma_{\Gamma-}^{\nabla}$), and both are labelled F.

- Horizontal segments lying above (below) saturated regions will be denoted by $\Sigma_{\Gamma+}^{\Delta}$ ($\Sigma_{\Gamma-}^{\Delta}$), and both are labelled G.

We also denote each vertical segment passing through an endpoint a or b by the same symbol as the component of Σ_{SD} to which it is joined at $|\Im(z)| = \epsilon$.

The problem equivalent to Interpolation Problem 1.2 is the subject of the following proposition.

Proposition 4.5. *The matrix $\mathbf{X}(z)$ defined by (4.45) and (4.46)–(4.52) is the unique solution of the following Riemann-Hilbert problem.*

Riemann-Hilbert Problem 4.6. *Find a 2×2 matrix $\mathbf{X}(z)$ with the following properties:*

1. **Analyticity**: *$\mathbf{X}(z)$ is an analytic function of z for $z \in \mathbb{C} \setminus \Sigma_{\text{SD}}$.*

2. **Normalization**: *As $z \to \infty$,*
$$\mathbf{X}(z) = \mathbb{I} + O\left(\frac{1}{z}\right). \tag{4.56}$$

3. **Jump Conditions**: *$\mathbf{X}(z)$ takes uniformly continuous boundary values on Σ_{SD} from each connected component of $\mathbb{C} \setminus \Sigma_{\text{SD}}$. For each non-self-intersection point $z \in \Sigma_{\text{SD}}$, we denote by $\mathbf{X}_+(z)$ ($\mathbf{X}_-(z)$) the limit of $\mathbf{X}(w)$ as $w \to z$ from the left (right). If we let $g_+(z) + g_-(z)$, for real z, denote the sum*

AN EQUIVALENT RIEMANN-HILBERT PROBLEM

of boundary values taken by $g(z)$ from the upper and lower half-planes, the boundary values taken on Σ_{SD} by $\mathbf{X}(z)$ satisfy the following conditions. For z in a void $\Gamma \subset \Sigma_0^\nabla$,

$$\mathbf{X}_+(z) = \mathbf{X}_-(z) \begin{pmatrix} e^{-iN\theta_\Gamma}e^{-i\phi_\Gamma} & iT_\nabla(z)e^{\gamma-\eta(z)+\kappa(g_+(z)+g_-(z))}e^{-N\xi_\Gamma(z)} \\ 0 & e^{iN\theta_\Gamma}e^{i\phi_\Gamma} \end{pmatrix}.$$

For z in a saturated region $\Gamma \subset \Sigma_0^\Delta$,

$$\mathbf{X}_+(z) = \mathbf{X}_-(z) \begin{pmatrix} e^{-iN\theta_\Gamma}e^{-i\phi_\Gamma} & 0 \\ iT_\Delta(z)e^{\eta(z)-\gamma-\kappa(g_+(z)+g_-(z))}e^{-N\xi_\Gamma(z)} & e^{iN\theta_\Gamma}e^{i\phi_\Gamma} \end{pmatrix}.$$

For z in any band I,

$$\mathbf{X}_+(z) = \mathbf{X}_-(z) \begin{pmatrix} 0 & -ie^{\gamma-\eta(z)+\kappa(g_+(z)+g_-(z))} \\ -ie^{\eta(z)-\gamma-\kappa(g_+(z)+g_-(z))} & 0 \end{pmatrix}.$$

For z in any vertical segment $\Sigma_{0\pm}^\nabla$ meeting the real axis at an endpoint z_0 of a band I,

$$\mathbf{X}_+(z) = \mathbf{X}_-(z) \begin{pmatrix} T_\nabla(z)^{\pm 1/2} & 0 \\ -iT_\nabla(z)^{-1/2}e^{\eta(z)-\gamma-2\kappa g(z)}e^{\pm iN\theta(z_0)}\exp\left(\pm 2\pi iNc \int_{z_0}^z \psi_I(s)\,ds\right) & T_\nabla(z)^{\mp 1/2} \end{pmatrix}.$$

For z in any vertical segment $\Sigma_{0\pm}^\Delta$ meeting the real axis at an endpoint z_0 of a band I,

$$\mathbf{X}_+(z) = \mathbf{X}_-(z) \begin{pmatrix} T_\Delta(z)^{\mp 1/2} & -iT_\Delta(z)^{-1/2}e^{\gamma-\eta(z)+2\kappa g(z)}e^{\mp iN\theta(z_0)}\exp\left(\pm 2\pi iNc \int_{z_0}^z \overline{\psi}_I(z)\,ds\right) \\ 0 & T_\Delta(z)^{\pm 1/2} \end{pmatrix}.$$

For z in any segment $\Sigma_{\Gamma\pm}^\nabla$ parallel to a void $\Gamma \subset \Sigma_0^\nabla$ or with $\Re(z) = a$ or $\Re(z) = b$,

$$\mathbf{X}_+(z) = \mathbf{X}_-(z) \begin{pmatrix} 1 & iY(z)e^{\gamma-\eta(z)+2\kappa g(z)}e^{\mp iN\theta_\Gamma}e^{\mp iN\theta^0(z)}e^{-N\xi_\Gamma(z)} \\ 0 & 1 \end{pmatrix}.$$

For z in any segment $\Sigma_{\Gamma\pm}^\Delta$ parallel to a saturated region $\Gamma \subset \Sigma_0^\Delta$ or with $\Re(z) = a$ or $\Re(z) = b$,

$$\mathbf{X}_+(z) = \mathbf{X}_-(z) \begin{pmatrix} 1 & 0 \\ iY(z)^{-1}e^{\eta(z)-\gamma-2\kappa g(z)}e^{\pm iN\theta_\Gamma}e^{\mp iN\theta^0(z)}e^{-N\xi_\Gamma(z)} & 1 \end{pmatrix}.$$

To express as concisely as possible the relationship between the boundary values taken by $\mathbf{X}(z)$ on segments $\Sigma_{I\pm}$ parallel to a band I, it is convenient to choose some fixed $y \in I$ and then define y_N for each $N \in \mathbb{N}$ by the rule

$$N \int_a^{y_N} \rho^0(x)\,dx = \left\lceil N \int_a^y \rho^0(x)\,dx \right\rceil,$$

which may be compared with (4.1). Thus if I is a transition band, we may take y_N to be the transition point $y_{k,N} \in Y_N$ contained therein. Otherwise we may think of y_N as a "virtual transition point." With the sequence $\{y_N\}_{N=0}^\infty$ so determined, we have that for z in any segment $\Sigma_{I\pm}$ parallel to any band I,

$$\mathbf{X}_+(z) = \mathbf{X}_-(z) \begin{pmatrix} T_\Delta(z)^{-1/2} & v_{12}^\pm(z) \\ v_{21}^\pm(z) & T_\nabla(z)^{-1/2} \end{pmatrix}^{\pm 1},$$

where

$$v_{12}^\pm(z) := \mp iT_\Delta(z)^{-1/2}e^{\gamma-\eta(z)+2\kappa g(z)}e^{\mp iN\theta(y_N)}\exp\left(\pm 2\pi iNc \int_{y_N}^z \overline{\psi}_I(s)\,ds\right),$$

$$v_{21}^\pm(z) := \mp iT_\nabla(z)^{-1/2}e^{\eta(z)-\gamma-2\kappa g(z)}e^{\pm iN\theta(y_N)}\exp\left(\pm 2\pi iNc \int_{y_N}^z \psi_I(s)\,ds\right).$$

Proof. The domain of analyticity of $\mathbf{X}(z)$ is clear from the nature of the definition (4.45) with (4.46)–(4.52), and from Proposition 4.4. The normalization condition follows from the corresponding normalization of $\mathbf{P}(z; N, k)$ and from (4.6). The continuity of the boundary values is obvious everywhere except on the real axis, but here the poles in $\mathbf{P}(z; N, k)$ are cancelled by corresponding zeros in the boundary values of $T_\nabla(z)^{1/2}$ and $T_\Delta(z)^{1/2}$. Finally, the jump conditions are a direct consequence of the continuity of $\mathbf{P}(z; N, k)$ and the known discontinuities of $\mathbf{D}(z)$.

This shows that $\mathbf{X}(z)$ defined by (4.45) with (4.46)–(4.52) indeed satisfies all of the conditions of Riemann-Hilbert Problem 4.6. The uniqueness of the solution follows from Liouville's Theorem because the matrix ratio of any two solutions is necessarily an entire function of z that tends to the identity matrix as $z \to \infty$. □

Chapter Five

Asymptotic Analysis

In this chapter we provide all the tools for a complete asymptotic analysis of discrete orthogonal polynomials with a large class of (generally nonclassical) weights, in the joint limit of large degree and a large number of nodes. These results will then be used in Chapter 6 to establish precise convergence theorems about the discrete orthogonal polynomials, and in Chapter 7 to prove a number of universality results concerning statistics of related discrete orthogonal polynomial ensembles.

5.1 CONSTRUCTION OF A GLOBAL PARAMETRIX FOR $\mathbf{X}(z)$

5.1.1 Outer asymptotics

Our immediate goal is to use the deformations we have carried out to construct a model for the matrix $\mathbf{X}(z)$ that we expect to be asymptotically accurate pointwise in z as $N \to \infty$. The proof of validity will be given in §5.2.

The basic observation at this point, which we will justify more precisely in §5.2, is that the jump matrix relating $\mathbf{X}_+(z)$ and $\mathbf{X}_-(z)$ in Riemann-Hilbert Problem 4.6 is closely approximated by the identity matrix in the limit $N \to \infty$ for $z \in \Sigma_{\mathrm{SD}} \setminus [a,b]$. Moreover, the jump matrix in any gap $\Gamma \subset [a,b]$ is closely approximated in the same limit by a constant matrix $e^{-iN\theta_\Gamma \sigma_3} e^{-i\phi_\Gamma \sigma_3}$. Neglecting the errors on an ad hoc basis leads to a model Riemann-Hilbert problem.

Riemann-Hilbert Problem 5.1. *Let $\{\Gamma_j = (\beta_{j-1}, \alpha_j), \text{ for } j = 1, \ldots, G\}$ denote the set of interior gaps in (a,b) and let the bands be denoted by $\{I_j = (\alpha_j, \beta_j), \text{ for } j = 0, \ldots, G\}$. Let Σ_{model} denote the interval $[\alpha_0, \beta_G]$, oriented from left to right. Find a 2×2 matrix $\dot{\mathbf{X}}(z)$ with the following properties:*

1. **Analyticity**: *$\dot{\mathbf{X}}(z)$ is an analytic function of z for $z \in \mathbb{C} \setminus \Sigma_{\mathrm{model}}$.*

2. **Normalization**: *As $z \to \infty$,*
$$\dot{\mathbf{X}}(z) = \mathbb{I} + O\left(\frac{1}{z}\right).$$

3. **Jump Conditions**: *$\dot{\mathbf{X}}(z)$ takes continuous boundary values on Σ_{model} except at the endpoints of the bands, where inverse fourth-root singularities are admitted. For $z \in \Sigma_{\mathrm{model}}$, let $\dot{\mathbf{X}}_+(z)$ ($\dot{\mathbf{X}}_-(z)$) denote the boundary value taken by $\dot{\mathbf{X}}(z)$ on the left (right) of Σ_{model} according to its orientation. For z in the gap Γ_j, the boundary values satisfy*
$$\dot{\mathbf{X}}_+(z) = \dot{\mathbf{X}}_-(z) \begin{pmatrix} e^{iN\theta_{\Gamma_j}} e^{i\phi_{\Gamma_j}} & 0 \\ 0 & e^{-iN\theta_{\Gamma_j}} e^{-i\phi_{\Gamma_j}} \end{pmatrix}, \tag{5.1}$$

where the constant θ_{Γ_j} is defined by (2.21) or (2.22) depending on whether Γ_j is a void or a saturated region, and ϕ_{Γ_j} is defined by (4.14). For z in any band I_j, the boundary values satisfy
$$\dot{\mathbf{X}}_+(z) = \dot{\mathbf{X}}_-(z) \begin{pmatrix} 0 & -ie^{\gamma - \eta(z) + \kappa(g_+(z) + g_-(z))} \\ -ie^{\eta(z) - \gamma - \kappa(g_+(z) + g_-(z))} & 0 \end{pmatrix}.$$

Here the expression $g_+(z) + g_-(z)$ refers to the sum of the boundary values taken for $z \in I_j \subset \mathbb{R}$ from the upper and lower half-planes.

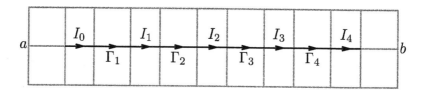

Figure 5.1 *The contour Σ_{model} corresponding to the hypothetical equilibrium measure illustrated in Figure 4.1 shown against the gray background of Σ_{SD}. Note that by contrast with Σ_{SD}, the gap intervals Γ_j are now oriented from left to right. Thus the boundary value $\dot{\mathbf{X}}_+(z)$ ($\dot{\mathbf{X}}_-(z)$) refers to a limit from the upper (lower) half-plane.*

The contour Σ_{model} corresponding to the hypothetical situation first illustrated in Figure 4.1 is shown in Figure 5.1. Problems of this sort are solved in terms of Riemann theta functions of genus G, where $G + 1$ is the number of bands $I_0, \ldots, I_G$ (see, for example, [DeiKMVZ99b]). Our subsequent analysis and error estimates will not rely heavily on the specific formulae for the solution, although as is clear from §2.3 these details do emerge in the leading-order asymptotics justified by our analysis. For completeness, the solution of Riemann-Hilbert Problem 5.1 is explained in Appendix A.

The essential facts we will require later are the following.

Proposition 5.2. *Riemann-Hilbert Problem 5.1 has a unique solution $\dot{\mathbf{X}}(z)$ that is uniformly bounded with bound independent of N in any neighborhood that does not contain any of the endpoints of the bands $I_0, \ldots, I_G$. Although the numbers ϕ_{Γ_j} depend on the choice of transition points in the set Y_N, the combination $\dot{\mathbf{X}}(z)e^{\kappa g(z)\sigma_3}$ is independent of any particular choice of transition points. Also, $\det(\dot{\mathbf{X}}(z)) = 1$.*

Proof. A solution is developed in detail in Appendix A, and uniqueness can be established by an argument based on Liouville's Theorem. A similar argument proves that $\det(\dot{\mathbf{X}}(z)) = 1$. The uniform boundedness of $\dot{\mathbf{X}}(z)$ away from the band endpoints and the invariance of the combination $\dot{\mathbf{X}}(z)e^{\kappa g(z)\sigma_3}$ are consequences of the solution formulae given in Appendix A; a discussion of these features can be found there. □

The boundary values taken by the solution of Riemann-Hilbert Problem 5.1 have the following useful properties.

Proposition 5.3. *For z in any interior gap (void or saturated region) $\Gamma_j = (\beta_{j-1}, \alpha_j) \subset \Sigma_{\text{model}}$, we have the identity*

$$\dot{\mathbf{X}}_+(z)e^{\kappa g_+(z)\sigma_3} = \left(\dot{\mathbf{X}}_+(z)e^{\kappa g_+(z)\sigma_3}\right)^* \begin{pmatrix} e^{iN\theta_{\Gamma_j}} & 0 \\ 0 & e^{-iN\theta_{\Gamma_j}} \end{pmatrix},$$

where the star denotes componentwise complex conjugation. Similarly, for real $z < \alpha_0$, we have

$$\dot{\mathbf{X}}(z)e^{\kappa g_+(z)\sigma_3} = \left(\dot{\mathbf{X}}(z)e^{\kappa g_+(z)\sigma_3}\right)^* \begin{pmatrix} e^{-2\pi iNc} & 0 \\ 0 & e^{2\pi iNc} \end{pmatrix},$$

and for real $z > \beta_G$, we have

$$\dot{\mathbf{X}}(z)e^{\kappa g(z)\sigma_3} = \left(\dot{\mathbf{X}}(z)e^{\kappa g(z)\sigma_3}\right)^*.$$

Moreover, the product $p(z) := \dot{X}_{11}(z)\dot{X}_{12}(z)$ extends to $\mathbb{C} \setminus ([\alpha_0, \beta_0] \cup \ldots \cup [\alpha_G, \beta_G])$ as a real-analytic function satisfying $p(z) < 0$ for all real $z < \alpha_0$ and $p(z) > 0$ for all real $z > \beta_G$. For all $j = 1, \ldots, G$, there is a real number $z_j \in [\beta_{j-1}, \alpha_j]$ such that $p(z) > 0$ for $\beta_{j-1} < z < z_j$ and $p(z) < 0$ for $z_j < z < \alpha_j$. If in fact $z_j \in (\beta_{j-1}, \alpha_j)$, then z_j is a simple zero of $p(z)$. The zeros z_j depend on the parameter κ in a quasiperiodic fashion, with G frequencies that depend on the parameters $c \in (0, 1)$ and N, the function $\eta(z)$, and the equilibrium measure. Generically, $z_j \in (\beta_{j-1}, \alpha_j)$, and the situation in which $z_j = \beta_{j-1}$ or $z_j = \alpha_j$, for some j, should be regarded as exceptional. In the generic case, the boundary values $\dot{X}_{11+}(z)$ and $\dot{X}_{12+}(z)$

ASYMPTOTIC ANALYSIS

are analytic at $z = z_j$, and thus either $\dot{X}_{11+}(z)$ has a simple zero only at $z = z_j$ and $\dot{X}_{12+}(z)$ is bounded away from zero in Γ_j, or $\dot{X}_{12+}(z)$ has a simple zero only at $z = z_j$ and $\dot{X}_{11+}(z)$ is bounded away from zero in Γ_j.

For z in any band I_j, the identity

$$\dot{\mathbf{X}}_+(z)e^{\kappa g_+(z)\sigma_3} = \left(\dot{\mathbf{X}}_+(z)e^{\kappa g_+(z)\sigma_3}\right)^* \begin{pmatrix} 0 & -ie^{\gamma-\eta(z)} \\ -ie^{\eta(z)-\gamma} & 0 \end{pmatrix} \tag{5.2}$$

holds. Furthermore, for $z \in I_j$ the elements of $\dot{\mathbf{X}}_+(z)$ are strictly nonzero.

Proof. The matrix $\mathbf{M}(z) := \dot{\mathbf{X}}(z)e^{\kappa g(z)\sigma_3}$ and the corresponding matrix $\mathbf{N}(z) := \mathbf{M}(z^*)^*$ are both analytic for $z \in \mathbb{C} \setminus (-\infty, \beta_G]$, where β_G is the rightmost band endpoint. As $z \to \infty$, we have $\mathbf{M}(z)e^{-\kappa \log(z)\sigma_3} = \mathbb{I} + O(1/z)$ and also $\mathbf{N}(z)e^{-\kappa \log(z)\sigma_3} = \mathbb{I} + O(1/z)$. Furthermore, it is easily checked that at each point $z \in (-\infty, \beta_G]$, we have $\mathbf{M}_-(z)^{-1}\mathbf{M}_+(z) = \mathbf{N}_-(z)^{-1}\mathbf{N}_+(z)$. This means that both matrices satisfy the same Riemann-Hilbert problem. Uniqueness of solutions for this problem follows as usual from Liouville's Theorem. Thus $\mathbf{M}(z) = \mathbf{N}(z) = \mathbf{M}(z^*)^*$. The claimed relations follow from the jump relations for $\dot{\mathbf{X}}(z)$ since for each real z, $\mathbf{M}_-(z) = \mathbf{M}_+(z)^*$.

Suppose that at some point z in a band I_j we have $M_{11+}(z) = 0$. Then it follows from (5.2) that $M_{12+}(z) = 0$ also. But this implies that $\det(\mathbf{M}(z)) = 0$, which contradicts the fact that (see Proposition 5.2) $\det(\mathbf{M}(z)) = 1$. In a similar way, one sees that any other matrix element of $\mathbf{M}_\pm(z)$ having a zero in I_j leads to a contradiction.

The fact that the product $p(z) = \dot{X}_{11}(z)\dot{X}_{12}(z)$ extends to the complement of the bands I_j as an analytic function follows from the jump condition (5.1) and the analyticity of $\dot{\mathbf{X}}(z)$ for $z \in \mathbb{C} \setminus \Sigma_{\text{model}}$. The sign of $p(z)$ is discussed in detail in Appendix A. $\square$

By using the explicit formulae given in Appendix A, one can obtain the identities $W(z) \equiv \dot{X}_{11}(z)e^{\kappa g(z)}$ and $Z(z) \equiv \dot{X}_{12}(z)e^{-\kappa g(z)}$, where $W(z)$ and $Z(z)$ are the functions defined in (2.43) and (2.44) (or for $G = 0$, (2.45) and (2.46)), respectively.

5.1.2 Inner asymptotics near band edges

In any neighborhood of a point in the interior of either Σ_0^∇ or $\Sigma_0^\triangle$ that marks the boundary between a band and a gap, the pointwise asymptotics used to arrive at the jump conditions for the matrix $\dot{\mathbf{X}}(z)$ starting from those for the matrix $\mathbf{X}(z)$ are not uniformly valid. It is therefore necessary to construct a local approximation to $\mathbf{X}(z)$ near such points using different techniques. We refer to these boundary points separating bands from gaps as *band edges*. We want to stress that band edges are to be distinguished from transition points making up the set Y_N defined in §4.1. Our method will be to define in a disc of fixed size near each band edge a matrix that exactly satisfies the jump conditions of $\mathbf{X}(z)$ and that matches well onto the outer asymptotics given by $\dot{\mathbf{X}}(z)$ at the boundary of the disc.

The distinguishing characteristic of a band edge $z = z_0$ is that in the adjacent gap Γ the function $\rho(z)$ is identically zero since the equilibrium measure $\mu_{\min}^c$ realizes the lower constraint for $z \in \Gamma$ if $\Gamma \subset \Sigma_0^\nabla$, or the upper constraint for $z \in \Gamma$ if $\Gamma \subset \Sigma_0^\triangle$, and meanwhile in the adjacent band $\rho(z)$ is a nonzero analytic function that vanishes at the band edge. The nature of the vanishing of $\rho(z)$ at the band edge must be understood before a local approximation can be constructed. Consider $\delta E_c/\delta\mu - \ell_c$, where the variational derivative is evaluated on the equilibrium measure. In the band, this quantity is identically zero according to the equilibrium condition (2.14). On the other hand, if $\rho^0(\cdot)$ and $V(\cdot)$ are analytic functions, then the function $\Psi(z)$ defined for $z \in (a, b)$ by

$$\Psi(z) := V(z) + \int_{\Sigma_0^\nabla} \log|z-x|\rho^0(x)\,dx - \int_{\Sigma_0^\triangle} \log|z-x|\rho^0(x)\,dx$$

extends analytically into the upper half-plane (it is analytic in a neighborhood of z_0 as long as z_0 is in the interior of either Σ_0^∇ or $\Sigma_0^\triangle$). Since

$$\Psi(z) + 2(d_N - c)\int_a^b \log|z-x|\rho(x)\,dx = \frac{\delta E_c}{\delta\mu}(z), \quad \text{for } z \in (a, b),$$

where the variational derivative is evaluated on the equilibrium measure, we have

$$0 \equiv \Psi(z) + 2(d_N - c) \int_a^b \log(z-x)\rho(x)\,dx - \ell_c - 2\pi i(d_N - c)\int_{z_0}^b \rho(x)\,dx + 2\pi i(d_N - c)\int_{z_0}^z \rho(x)\,dx, \quad (5.3)$$

for z near z_0 with $\Im(z) > 0$. Only the last integral involves contour integration off the real axis, and the integrand denotes the analytic function $\rho(\cdot)$ of the band. At the same time, the quantity $\delta E_c/\delta\mu - \ell_c$ extends into the upper half-plane from the gap Γ as

$$\left.\frac{\delta E_c}{\delta\mu} - \ell_c\right|_{z\in\Gamma} = \Psi(z) + 2(d_N - c)\int_a^b \log(z-x)\rho(x)\,dx - \ell_c - 2\pi i(d_N - c)\int_{z_0}^b \rho(x)\,dx$$

since $\rho(\cdot) \equiv 0$ for $z \in \Gamma$. We therefore deduce that

$$\left.\frac{\delta E_c}{\delta\mu} - \ell_c\right|_{z\in\Gamma} = -2\pi i(d_N - c)\left[\int_{z_0}^z \rho(x)\,dx\right]_+, \quad (5.4)$$

where on the right-hand side the integrand is the continuation of the analytic function $\rho(\cdot)$ defined in the adjacent band and the subscript denotes the boundary value taken on the gap Γ from the upper half-plane. Using virtually the same arguments but continuing all quantities into the lower half-plane, we find that

$$\left.\frac{\delta E_c}{\delta\mu} - \ell_c\right|_{z\in\Gamma} = 2\pi i(d_N - c)\left[\int_{z_0}^z \rho(x)\,dx\right]_-. \quad (5.5)$$

Combining (5.4) and (5.5) reveals the identity

$$\left[\int_{z_0}^z \rho(x)\,dx\right]_+ + \left[\int_{z_0}^z \rho(x)\,dx\right]_- = 0, \quad (5.6)$$

which holds for all z in the gap when the integrand $\rho(\cdot)$ is analytically extended about z_0 from the band. Differentiating this identity with respect to z, we discover that $\rho(z)^2$ extends from the band to a complex annulus surrounding z_0 as a single-valued analytic function that vanishes as $z \to z_0$ within the band (at least). Moreover, it follows from (5.3) that $\rho(z)^2$ is analytic at z_0 as well and so is necessarily of the form $\rho(z)^2 = (z - z_0)^p e^{f(z)}$, where $p = 1, 2, 3, \ldots$ and $f(z)$ is analytic at z_0.

Clearly, only odd values of the positive integer p are consistent with (5.6). However, even more is true. If the band edge point satisfies $z_0 \in \Sigma_0^\nabla$, then the combination $(c - d_N)\rho(x)$ can be seen by (4.5) to be strictly positive for x in the band adjacent to z_0, and furthermore the adjacent gap is a *void*; thus from (2.13) we see that the common left-hand side of (5.4) and (5.5) is strictly positive for $z \in \Gamma$. Similarly, if the band edge point satisfies $z_0 \in \Sigma_0^\Delta$, then the combination $(c - d_N)\rho(x)$ is strictly negative for x in the band adjacent to z_0, and the adjacent gap is a *saturated region* so that (2.16) makes the common left-hand side of (5.4) and (5.5) strictly negative for $z \in \Gamma$. In both cases, we can easily see that the equations (5.4) and (5.5) will be consistent with the assumption that $\rho(z)^2 = (z-z_0)^p e^{f(z)}$ for analytic $f(z)$ and $p = 1, 3, 5, 7, \ldots$ only if we discard the values $p = 3, 7, 11, \ldots$.

Therefore, using only the assumption that $\rho^0(\cdot)$ and $V(\cdot)$ are analytic functions, we have shown that at each band edge z_0 in the interior of Σ_0^∇ or Σ_0^Δ the positive analytic function $\rho(\cdot)$ vanishes like $(z - z_0)^{p/2}$, where p is of the form $p = 1 + 4m$ for $m = 0, 1, 2, 3, \ldots$. This is the general character of the vanishing of $\rho(\cdot)$ at band edges when $V(\cdot)$ and $\rho^0(\cdot)$ are analytic functions, and it is quite similar to the characterization of the local behavior of the equilibrium measure (without upper constraint) near band edges as explained in [DeiKM98].

As mentioned in §2.1.2 (see in particular (2.6) and (2.7)), for simplicity we will consider only the generic situation when $p = 1$ at all band edges. There are four cases. Let $h < 1$ be an arbitrary fixed positive parameter.

Left band edge with $z_0 = \alpha \in \Sigma_0^\nabla$ (lower constraint)

Let Γ denote the void to the left of α; then $e^{iN\theta(\alpha)} = e^{iN\theta_\Gamma}$. Let I denote the band to the right of α. Consider $D_\Gamma^{\nabla,L}$ to be an open disc centered at $z = \alpha$ of radius $h\epsilon$. Note that for ϵ sufficiently small, this

ASYMPTOTIC ANALYSIS   91

radius will be less than half the distance to the nearest distinct band edge and $D_\Gamma^{\nabla,L}$ will be disjoint from the endpoints $\{a,b\}$. We divide $D_\Gamma^{\nabla,L} \setminus (D_\Gamma^{\nabla,L} \cap \Sigma_{\text{SD}})$ into open quadrants:

$$D_{\Gamma,\text{I}}^{\nabla,L} = D_\Gamma^{\nabla,L} \cap \left\{ z \,\Big|\, z \neq \alpha, \, 0 < \arg(z-\alpha) < \frac{\pi}{2} \right\},$$

$$D_{\Gamma,\text{II}}^{\nabla,L} = D_\Gamma^{\nabla,L} \cap \left\{ z \,\Big|\, z \neq \alpha, \, \frac{\pi}{2} < \arg(z-\alpha) < \pi \right\},$$

$$D_{\Gamma,\text{III}}^{\nabla,L} = D_\Gamma^{\nabla,L} \cap \left\{ z \,\Big|\, z \neq \alpha, \, -\pi < \arg(z-\alpha) < -\frac{\pi}{2} \right\},$$

$$D_{\Gamma,\text{IV}}^{\nabla,L} = D_\Gamma^{\nabla,L} \cap \left\{ z \,\Big|\, z \neq \alpha, \, -\frac{\pi}{2} < \arg(z-\alpha) < 0 \right\}.$$

Now we introduce a local change of variables in $D_\Gamma^{\nabla,L}$. We set

$$\mathbf{Z}_\Gamma^{\nabla,L}(z) := \begin{cases} \mathbf{X}(z) e^{(\gamma - \eta(z) + 2\kappa g(z))\sigma_3/2} e^{-iN\theta_\Gamma \sigma_3/2}, & \text{for } z \in D_{\Gamma,\text{I}}^{\nabla,L}, \\ \mathbf{X}(z) T_\nabla(z)^{\sigma_3/2} e^{(\gamma - \eta(z) + 2\kappa g(z))\sigma_3/2} e^{-iN\theta_\Gamma \sigma_3/2}, & \text{for } z \in D_{\Gamma,\text{II}}^{\nabla,L}, \\ \mathbf{X}(z) T_\nabla(z)^{\sigma_3/2} e^{(\gamma - \eta(z) + 2\kappa g(z))\sigma_3/2} e^{iN\theta_\Gamma \sigma_3/2}, & \text{for } z \in D_{\Gamma,\text{III}}^{\nabla,L}, \\ \mathbf{X}(z) e^{(\gamma - \eta(z) + 2\kappa g(z))\sigma_3/2} e^{iN\theta_\Gamma \sigma_3/2}, & \text{for } z \in D_{\Gamma,\text{IV}}^{\nabla,L}. \end{cases} \quad (5.7)$$

According to (2.6), the equation $\zeta = \tau_\Gamma^{\nabla,L}(z)$ defined by (2.17) gives an invertible conformal mapping, taking for ϵ sufficiently small, the fixed disc $D_\Gamma^{\nabla,L}$ to a neighborhood of $\zeta = 0$ in the ζ-plane that scales like $N^{2/3}$. The transformation $\tau_\Gamma^{\nabla,L}(z)$ maps $\mathbb{R} \cap D_\Gamma^{\nabla,L}$ to $\mathbb{R}$, taking $z = \alpha$ to $\zeta = 0$, and is orientation-preserving since $d\tau_\Gamma^{\nabla,L}/dz(\alpha)$ is real and positive. The segments $\arg(z-\alpha) = \pm\pi/2$ in $D_\Gamma^{\nabla,L}$ are mapped to arcs in the ζ-plane that are tangent to the imaginary axis at $\zeta = 0$ and converge to the rays $\arg(\zeta) = \pm\pi/2$ as $N \to \infty$ uniformly for ζ in compact sets. The exact jump conditions satisfied by the boundary values of $\mathbf{Z}_\Gamma^{\nabla,L}(z)$ on $\Sigma_{\text{SD}} \cap D_\Gamma^{\nabla,L}$ may be written in terms of the new coordinate ζ as follows:

$$\mathbf{Z}_{\Gamma+}^{\nabla,L}(z) = \mathbf{Z}_{\Gamma-}^{\nabla,L}(z) \begin{pmatrix} 1 & ie^{-(-\zeta)^{3/2}} \\ 0 & 1 \end{pmatrix}, \quad \text{for } z \in \Gamma \cap D_\Gamma^{\nabla,L},$$

$$\mathbf{Z}_{\Gamma+}^{\nabla,L}(z) = \mathbf{Z}_{\Gamma-}^{\nabla,L}(z) \begin{pmatrix} 0 & -i \\ -i & 0 \end{pmatrix}, \quad \text{for } z \in I \cap D_\Gamma^{\nabla,L},$$

$$\mathbf{Z}_{\Gamma+}^{\nabla,L}(z) = \mathbf{Z}_{\Gamma-}^{\nabla,L}(z) \begin{pmatrix} 1 & 0 \\ -ie^{i\zeta^{3/2}} & 1 \end{pmatrix}, \quad \text{for } z \in \Sigma_{0+}^\nabla \cap D_\Gamma^{\nabla,L},$$

$$\mathbf{Z}_{\Gamma+}^{\nabla,L}(z) = \mathbf{Z}_{\Gamma-}^{\nabla,L}(z) \begin{pmatrix} 1 & 0 \\ -ie^{-i\zeta^{3/2}} & 1 \end{pmatrix}, \quad \text{for } z \in \Sigma_{0-}^\nabla \cap D_\Gamma^{\nabla,L}.$$

Here the subscripts $+$ and $-$, respectively, refer to boundary values taken on $\Sigma_{\text{SD}} \cap D_\Gamma^{\nabla,L}$ from the left and right relative to the orientation of Σ_{SD}.

At the same time, we can define a *comparison matrix* $\dot{\mathbf{Z}}_\Gamma^{\nabla,L}(z)$ from $\dot{\mathbf{X}}(z)$ by the relation

$$\dot{\mathbf{Z}}_\Gamma^{\nabla,L}(z) := \dot{\mathbf{X}}(z) e^{(\gamma - \eta(z) + 2\kappa g(z))\sigma_3/2} e^{-iN\mathrm{sgn}(\Im(z))\theta_\Gamma \sigma_3/2}, \quad \text{for } z \in D_\Gamma^{\nabla,L} \setminus (D_\Gamma^{\nabla,L} \cap \Sigma_{\text{SD}}). \quad (5.8)$$

Note the difference (a factor of $T_\nabla(z)^{\sigma_3/2}$ in quadrants II and III) between the transformation (5.8) and the transformation (5.7). This matrix extends to an analytic function in $D_\Gamma^{\nabla,L}$ with the exception of $z \in I \cap D_\Gamma^{\nabla,L}$, where it satisfies

$$\dot{\mathbf{Z}}_{\Gamma+}^{\nabla,L}(z) = \dot{\mathbf{Z}}_{\Gamma-}^{\nabla,L}(z) \begin{pmatrix} 0 & -i \\ -i & 0 \end{pmatrix}, \quad \text{for } z \in I \cap D_\Gamma^{\nabla,L}.$$

Again, the subscripts indicate boundary values consistent with the orientation of Σ_{SD}, with $+$ indicating approach from the left and $-$ indicating approach from the right. Because the matrix elements of $\dot{\mathbf{Z}}(z)$ blow up no worse than $(z-\alpha)^{-1/4}$, it is easy to see that $\dot{\mathbf{Z}}_\Gamma^{\nabla,L}(z)$ can be represented in the form

$$\dot{\mathbf{Z}}_\Gamma^{\nabla,L}(z) = \mathbf{H}_\Gamma^{\nabla,L}(z) \cdot \frac{1}{\sqrt{2}} (-\tau_\Gamma^{\nabla,L}(z))^{\sigma_3/4} \begin{pmatrix} 1 & 1 \\ -1 & 1 \end{pmatrix}, \quad (5.9)$$

where $\mathbf{H}_\Gamma^{\nabla,L}(z)$ is analytic in $D_\Gamma^{\nabla,L}$. The relations (5.9) and (5.8) together with (2.17) serve as a definition of $\mathbf{H}_\Gamma^{\nabla,L}(z)$ in terms of the solution $\dot{\mathbf{X}}(z)$ of Riemann-Hilbert Problem 5.1.

Since the image of the boundary of $D_\Gamma^{\nabla,L}$ in the ζ-plane expands as $N \to \infty$ with ϵ held fixed, and since on the boundary $\mathbf{Z}_\Gamma^{\nabla,L}(z)$ and $\dot{\mathbf{Z}}_\Gamma^{\nabla,L}(z)$ should be comparable, we propose to concretely determine an approximation of $\mathbf{Z}_\Gamma^{\nabla,L}(z)$ for $z \in D_\Gamma^{\nabla,L}$ by solving the following Riemann-Hilbert problem.

Riemann-Hilbert Problem 5.4. *Let C_+ be a contour connecting the origin to infinity lying entirely within a sector symmetric about the positive imaginary axis of opening angle strictly less than $\pi/3$. Let C_- denote the complex conjugate of C_+. Find a 2×2 matrix $\hat{\mathbf{Z}}^{\nabla,L}(\zeta)$ with the following properties:*

1. **Analyticity**: $\hat{\mathbf{Z}}^{\nabla,L}(\zeta)$ *is an analytic function of ζ for $\zeta \in \mathbb{C} \setminus (\mathbb{R} \cup C_+ \cup C_-)$.*

2. **Normalization**: *As $\zeta \to \infty$,*

$$\hat{\mathbf{Z}}^{\nabla,L}(\zeta) \cdot \frac{1}{\sqrt{2}} \begin{pmatrix} 1 & -1 \\ 1 & 1 \end{pmatrix} (-\zeta)^{-\sigma_3/4} = \mathbb{I} + O\left(\frac{1}{\zeta}\right) \qquad (5.10)$$

uniformly with respect to direction.

3. **Jump Conditions**: $\hat{\mathbf{Z}}^{\nabla,L}(\zeta)$ *takes continuous boundary values from each sector of its analyticity. The boundary values satisfy*

$$\hat{\mathbf{Z}}_+^{\nabla,L}(\zeta) = \hat{\mathbf{Z}}_-^{\nabla,L}(\zeta) \begin{pmatrix} 1 & ie^{-(-\zeta)^{3/2}} \\ 0 & 1 \end{pmatrix}, \qquad \text{for } \zeta \in \mathbb{R} \text{ and } \zeta < 0,$$

$$\hat{\mathbf{Z}}_+^{\nabla,L}(\zeta) = \hat{\mathbf{Z}}_-^{\nabla,L}(\zeta) \begin{pmatrix} 0 & -i \\ -i & 0 \end{pmatrix}, \qquad \text{for } \zeta \in \mathbb{R} \text{ and } \zeta > 0,$$

$$\hat{\mathbf{Z}}_+^{\nabla,L}(\zeta) = \hat{\mathbf{Z}}_-^{\nabla,L}(\zeta) \begin{pmatrix} 1 & 0 \\ -ie^{i\zeta^{3/2}} & 1 \end{pmatrix}, \qquad \text{for } \zeta \in C_+,$$

$$\hat{\mathbf{Z}}_+^{\nabla,L}(\zeta) = \hat{\mathbf{Z}}_-^{\nabla,L}(\zeta) \begin{pmatrix} 1 & 0 \\ -ie^{-i\zeta^{3/2}} & 1 \end{pmatrix}, \qquad \text{for } \zeta \in C_-.$$

To determine the boundary values, the contours on the real ζ-axis are oriented away from the origin and the contours C_+ and C_- are oriented toward the origin. As usual, $+$ indicates approach from the left and $-$ indicates approach from the right.

Note that the asymptotic behavior of $\hat{\mathbf{Z}}^{\nabla,L}(\zeta)$ is chosen to match the explicit terms in $\dot{\mathbf{Z}}_\Gamma^{\nabla,L}(z)$ with the exception of the holomorphic prefactor $\mathbf{H}_\Gamma^{\nabla,L}(z)$, the effect of which will be included after solving for $\hat{\mathbf{Z}}^{\nabla,L}(\zeta)$. The solution of Riemann-Hilbert Problem 5.4 was first described in [DeiZ95], and we provide it in the notation of our problem for completeness.

Proposition 5.5 (Deift and Zhou). *The unique solution of Riemann-Hilbert Problem 5.4 is given by the following explicit formulae. Let*

$$w := \left(\frac{3}{4}\right)^{2/3} \zeta.$$

For ζ between the positive real axis and the contour C_+,

$$\hat{\mathbf{Z}}^{\nabla,L}(\zeta) := \begin{pmatrix} e^{\frac{2\pi i}{3}} \sqrt{2\pi} \left(\frac{3}{4}\right)^{-\frac{1}{6}} e^{\frac{2iw^{3/2}}{3}} \operatorname{Ai}'\left(e^{\frac{\pi i}{3}} w\right) & e^{\frac{5\pi i}{6}} \sqrt{2\pi} \left(\frac{3}{4}\right)^{-\frac{1}{6}} e^{-\frac{2iw^{3/2}}{3}} \operatorname{Ai}'\left(e^{-\frac{\pi i}{3}} w\right) \\ e^{-\frac{2\pi i}{3}} \sqrt{2\pi} \left(\frac{3}{4}\right)^{\frac{1}{6}} e^{\frac{2iw^{3/2}}{3}} \operatorname{Ai}\left(e^{\frac{\pi i}{3}} w\right) & e^{\frac{\pi i}{6}} \sqrt{2\pi} \left(\frac{3}{4}\right)^{\frac{1}{6}} e^{-\frac{2iw^{3/2}}{3}} \operatorname{Ai}\left(e^{-\frac{\pi i}{3}} w\right) \end{pmatrix}.$$

ASYMPTOTIC ANALYSIS

For ζ between the positive real axis and the contour C_-,

$$\hat{\mathbf{Z}}^{\nabla,L}(\zeta) := \begin{pmatrix} e^{-\frac{2\pi i}{3}}\sqrt{2\pi}\left(\frac{3}{4}\right)^{-\frac{1}{6}} e^{-\frac{2iw^{3/2}}{3}}\operatorname{Ai}'\left(e^{-\frac{\pi i}{3}}w\right) & e^{-\frac{5\pi i}{6}}\sqrt{2\pi}\left(\frac{3}{4}\right)^{-\frac{1}{6}} e^{\frac{2iw^{3/2}}{3}}\operatorname{Ai}'\left(e^{\frac{\pi i}{3}}w\right) \\ e^{\frac{2\pi i}{3}}\sqrt{2\pi}\left(\frac{3}{4}\right)^{\frac{1}{6}} e^{-\frac{2iw^{3/2}}{3}}\operatorname{Ai}\left(e^{-\frac{\pi i}{3}}w\right) & e^{-\frac{\pi i}{6}}\sqrt{2\pi}\left(\frac{3}{4}\right)^{\frac{1}{6}} e^{\frac{2iw^{3/2}}{3}}\operatorname{Ai}\left(e^{\frac{\pi i}{3}}w\right) \end{pmatrix}.$$

For ζ between the contour C_+ and the negative real axis,

$$\hat{\mathbf{Z}}^{\nabla,L}(\zeta) := \begin{pmatrix} -\sqrt{2\pi}\left(\frac{3}{4}\right)^{-\frac{1}{6}} e^{\frac{2(-w)^{3/2}}{3}}\operatorname{Ai}'(-w) & e^{\frac{5\pi i}{6}}\sqrt{2\pi}\left(\frac{3}{4}\right)^{-\frac{1}{6}} e^{-\frac{2(-w)^{3/2}}{3}}\operatorname{Ai}'\left(e^{-\frac{\pi i}{3}}w\right) \\ -\sqrt{2\pi}\left(\frac{3}{4}\right)^{\frac{1}{6}} e^{\frac{2(-w)^{3/2}}{3}}\operatorname{Ai}(-w) & e^{\frac{\pi i}{6}}\sqrt{2\pi}\left(\frac{3}{4}\right)^{\frac{1}{6}} e^{-\frac{2(-w)^{3/2}}{3}}\operatorname{Ai}\left(e^{-\frac{\pi i}{3}}w\right) \end{pmatrix}. \tag{5.11}$$

Finally, for ζ between the contour C_- and the negative real axis,

$$\hat{\mathbf{Z}}^{\nabla,L}(\zeta) := \begin{pmatrix} -\sqrt{2\pi}\left(\frac{3}{4}\right)^{-\frac{1}{6}} e^{\frac{2(-w)^{3/2}}{3}}\operatorname{Ai}'(-w) & e^{-\frac{5\pi i}{6}}\sqrt{2\pi}\left(\frac{3}{4}\right)^{-\frac{1}{6}} e^{-\frac{2(-w)^{3/2}}{3}}\operatorname{Ai}'\left(e^{\frac{\pi i}{3}}w\right) \\ -\sqrt{2\pi}\left(\frac{3}{4}\right)^{\frac{1}{6}} e^{\frac{2(-w)^{3/2}}{3}}\operatorname{Ai}(-w) & e^{-\frac{\pi i}{6}}\sqrt{2\pi}\left(\frac{3}{4}\right)^{\frac{1}{6}} e^{-\frac{2(-w)^{3/2}}{3}}\operatorname{Ai}\left(e^{\frac{\pi i}{3}}w\right) \end{pmatrix}.$$

Proof. The jump conditions are easily verified with the help of the identity

$$\operatorname{Ai}(z) + e^{\frac{2\pi i}{3}}\operatorname{Ai}(e^{\frac{2\pi i}{3}}z) + e^{-\frac{2\pi i}{3}}\operatorname{Ai}(e^{-\frac{2\pi i}{3}}z) = 0.$$

The asymptotics are verified with the use of the steepest-descent asymptotic formulae

$$\begin{aligned}\operatorname{Ai}(z) &= \frac{1}{2\sqrt{\pi}} z^{-1/4} e^{-2z^{3/2}/3}(1 + O(z^{-3/2})), \\ \operatorname{Ai}'(z) &= -\frac{1}{2\sqrt{\pi}} z^{1/4} e^{-2z^{3/2}/3}(1 + O(z^{-3/2})),\end{aligned} \tag{5.12}$$

both of which hold as $z \to \infty$ with $-\pi < \arg(z) < \pi$. In fact, these calculations show that the $O(\zeta^{-1})$ error term in the normalization condition (5.10) is of a more precise form, namely,

$$\hat{\mathbf{Z}}^{\nabla,L}(\zeta) \cdot \frac{1}{\sqrt{2}}\begin{pmatrix}1 & -1 \\ 1 & 1\end{pmatrix}(-\zeta)^{-\sigma_3/4} = \begin{pmatrix} 1 + O(\zeta^{-3/2}) & O(\zeta^{-1}) \\ O(\zeta^{-2}) & 1 + O(\zeta^{-3/2})\end{pmatrix}. \tag{5.13}$$

In this sense the decay rate to the identity matrix of $1/\zeta$ is sharp in only one of the matrix elements, with the remaining matrix elements exhibiting more rapid decay. Uniqueness of the solution follows from Liouville's Theorem. □

The contours $C_\pm$ in Riemann-Hilbert Problem 5.4 are chosen so that in $\tau_\Gamma^{\nabla,L}(D_\Gamma^{\nabla,L})$ they agree with the images under $\tau_\Gamma^{\nabla,L}$ of the segments $\Sigma_{0\pm}^\nabla \cap D_\Gamma^{\nabla,L}$. Thus the sectorial condition on $C_\pm$ can be satisfied by taking the contour parameter ϵ controlling the radius of $D_\Gamma^{\nabla,L}$ to be sufficiently small. We now define a *local parametrix* for $\mathbf{X}(z)$ by the formula

$$\hat{\mathbf{X}}_\Gamma^{\nabla,L}(z) := \begin{cases} \mathbf{H}_\Gamma^{\nabla,L}(z)\hat{\mathbf{Z}}^{\nabla,L}(\tau_\Gamma^{\nabla,L}(z))e^{(\eta(z)-\gamma-2\kappa g(z))\sigma_3/2}e^{iN\theta_\Gamma\sigma_3/2}, & \text{for } z \in D_{\Gamma,\mathrm{I}}^{\nabla,L}, \\ \mathbf{H}_\Gamma^{\nabla,L}(z)\hat{\mathbf{Z}}^{\nabla,L}(\tau_\Gamma^{\nabla,L}(z))T_\nabla(z)^{-\sigma_3/2}e^{(\eta(z)-\gamma-2\kappa g(z))\sigma_3/2}e^{iN\theta_\Gamma\sigma_3/2}, & \text{for } z \in D_{\Gamma,\mathrm{II}}^{\nabla,L}, \\ \mathbf{H}_\Gamma^{\nabla,L}(z)\hat{\mathbf{Z}}^{\nabla,L}(\tau_\Gamma^{\nabla,L}(z))T_\nabla(z)^{-\sigma_3/2}e^{(\eta(z)-\gamma-2\kappa g(z))\sigma_3/2}e^{-iN\theta_\Gamma\sigma_3/2}, & \text{for } z \in D_{\Gamma,\mathrm{III}}^{\nabla,L}, \\ \mathbf{H}_\Gamma^{\nabla,L}(z)\hat{\mathbf{Z}}^{\nabla,L}(\tau_\Gamma^{\nabla,L}(z))e^{(\eta(z)-\gamma-2\kappa g(z))\sigma_3/2}e^{-iN\theta_\Gamma\sigma_3/2}, & \text{for } z \in D_{\Gamma,\mathrm{IV}}^{\nabla,L}. \end{cases} \tag{5.14}$$

Note that in this formula, the transformation $\tau_\Gamma^{\nabla,L}(\cdot)$ and the matrix $\mathbf{H}_\Gamma^{\nabla,L}(z)$ will be different in neighborhoods $D_\Gamma^{\nabla,L}$ corresponding to different left band edges in Σ_0^∇, being defined locally by (2.17), (5.8), and (5.9).

Right band edge with $z_0 = \beta \in \Sigma_0^\triangledown$ (lower constraint)

With Γ denoting the void to the right of the band edge β and I denoting the adjacent band on the left of β, we let $D_\Gamma^{\triangledown,R}$ be a disc centered at $z = \beta$ with radius $h\epsilon$. The four open quadrants of $D_\Gamma^{\triangledown,R} \setminus (D_\Gamma^{\triangledown,R} \cap \Sigma_{\text{SD}})$ are defined as

$$D_{\Gamma,\text{I}}^{\triangledown,R} = D_\Gamma^{\triangledown,R} \cap \left\{z \,\Big|\, z \neq \beta,\, 0 < \arg(z-\beta) < \frac{\pi}{2}\right\},$$

$$D_{\Gamma,\text{II}}^{\triangledown,R} = D_\Gamma^{\triangledown,R} \cap \left\{z \,\Big|\, z \neq \beta,\, \frac{\pi}{2} < \arg(z-\beta) < \pi\right\},$$

$$D_{\Gamma,\text{III}}^{\triangledown,R} = D_\Gamma^{\triangledown,R} \cap \left\{z \,\Big|\, z \neq \beta,\, -\pi < \arg(z-\beta) < -\frac{\pi}{2}\right\},$$

$$D_{\Gamma,\text{IV}}^{\triangledown,R} = D_\Gamma^{\triangledown,R} \cap \left\{z \,\Big|\, z \neq \beta,\, -\frac{\pi}{2} < \arg(z-\beta) < 0\right\}.$$

We introduce the local change of dependent variable

$$\mathbf{Z}_\Gamma^{\triangledown,R}(z) := \begin{cases} \mathbf{X}(z)T_\triangledown(z)^{\sigma_3/2}e^{(\gamma-\eta(z)+2\kappa g(z))\sigma_3/2}e^{-iN\theta_\Gamma \sigma_3/2}, & \text{for } z \in D_{\Gamma,\text{I}}^{\triangledown,R}, \\ \mathbf{X}(z)e^{(\gamma-\eta(z)+2\kappa g(z))\sigma_3/2}e^{-iN\theta_\Gamma \sigma_3/2}, & \text{for } z \in D_{\Gamma,\text{II}}^{\triangledown,R}, \\ \mathbf{X}(z)e^{(\gamma-\eta(z)+2\kappa g(z))\sigma_3/2}e^{iN\theta_\Gamma \sigma_3/2}, & \text{for } z \in D_{\Gamma,\text{III}}^{\triangledown,R}, \\ \mathbf{X}(z)T_\triangledown(z)^{\sigma_3/2}e^{(\gamma-\eta(z)+2\kappa g(z))\sigma_3/2}e^{iN\theta_\Gamma \sigma_3/2}, & \text{for } z \in D_{\Gamma,\text{IV}}^{\triangledown,R}, \end{cases}$$

(recall that $e^{iN\theta_\Gamma} = e^{iN\theta(\beta)}$) and the local conformal change of independent variable $\zeta = \tau_\Gamma^{\triangledown,R}(z)$ defined by (2.18). The mapping is orientation-reversing, taking $z < \beta$ to $\zeta > 0$ and $z > \beta$ to $\zeta < 0$. By taking ϵ sufficiently small, the radius $h\epsilon$ of $D_\Gamma^{\triangledown,R}$ will be small enough that the images under $\tau_\Gamma^{\triangledown,R}$ of the segments $\arg(z-\beta) = \pm\pi/2$ in $D_\Gamma^{\triangledown,R}$ lie within a symmetric sector of the imaginary ζ-axis of opening angle strictly less than $\pi/3$. The exact jump conditions satisfied by $\mathbf{Z}_\Gamma^{\triangledown,R}(z)$ in $D_\Gamma^{\triangledown,R}$ may be written in terms of ζ as

$$\begin{aligned} \mathbf{Z}_{\Gamma+}^{\triangledown,R}(z) &= \mathbf{Z}_{\Gamma-}^{\triangledown,R}(z) \begin{pmatrix} 1 & ie^{-(-\zeta)^{3/2}} \\ 0 & 1 \end{pmatrix}, & \text{for } z \in \Gamma \cap D_\Gamma^{\triangledown,R}, \\ \mathbf{Z}_{\Gamma+}^{\triangledown,R}(z) &= \mathbf{Z}_{\Gamma-}^{\triangledown,R}(z) \begin{pmatrix} 0 & -i \\ -i & 0 \end{pmatrix}, & \text{for } z \in I \cap D_\Gamma^{\triangledown,R}, \\ \mathbf{Z}_{\Gamma+}^{\triangledown,R}(z) &= \mathbf{Z}_{\Gamma-}^{\triangledown,R}(z) \begin{pmatrix} 1 & 0 \\ -ie^{i\zeta^{3/2}} & 1 \end{pmatrix}, & \text{for } z \in \Sigma_{0-}^\triangledown \cap D_\Gamma^{\triangledown,R}, \\ \mathbf{Z}_{\Gamma+}^{\triangledown,R}(z) &= \mathbf{Z}_{\Gamma-}^{\triangledown,R}(z) \begin{pmatrix} 1 & 0 \\ -ie^{-i\zeta^{3/2}} & 1 \end{pmatrix}, & \text{for } z \in \Sigma_{0+}^\triangledown \cap D_\Gamma^{\triangledown,R}. \end{aligned} \quad (5.15)$$

The subscripts $+$ and $-$ indicate, respectively, boundary values taken from the left and right of Σ_{SD} with respect to its orientation. The comparison matrix

$$\dot{\mathbf{Z}}_\Gamma^{\triangledown,R}(z) := \dot{\mathbf{X}}(z)e^{(\gamma-\eta(z)+2\kappa g(z))\sigma_3/2}e^{-iN\operatorname{sgn}(\Im(z))\theta_\Gamma \sigma_3/2}, \quad \text{for } z \in D_\Gamma^{\triangledown,R} \setminus (D_\Gamma^{\triangledown,R} \cap \Sigma_{\text{SD}}),$$

satisfies the same jump condition for $z \in I \cap D_\Gamma^{\triangledown,R}$ as $\mathbf{Z}_\Gamma^{\triangledown,R}(z)$ but is otherwise analytic in $D_\Gamma^{\triangledown,R}$ and can be written in the form

$$\dot{\mathbf{Z}}_\Gamma^{\triangledown,R}(z) := \mathbf{H}_\Gamma^{\triangledown,R}(z) \cdot \frac{1}{\sqrt{2}}(-\tau_\Gamma^{\triangledown,R}(z))^{\sigma_3/4}\begin{pmatrix} i & -i \\ -i & -i \end{pmatrix},$$

where $\mathbf{H}_\Gamma^{\triangledown,R}(z)$ is a holomorphic factor for $z \in D_\Gamma^{\triangledown,R}$. To come up with a matrix satisfying the jump conditions of $\mathbf{Z}_\Gamma^{\triangledown,R}(z)$ that is a good match to $\dot{\mathbf{Z}}_\Gamma^{\triangledown,R}(z)$ on the boundary of $D_\Gamma^{\triangledown,R}$, we consider the solution $\hat{\mathbf{Z}}^{\triangledown,L}(\zeta)$ of Riemann-Hilbert Problem 5.4 with the contours $C_\pm$ chosen such that $C_\pm \cap \tau_\Gamma^{\triangledown,R}(D_\Gamma^{\triangledown,R}) = \tau_\Gamma^{\triangledown,R}(\Sigma_{0\mp}^\triangledown)$ and we set

$$\hat{\mathbf{Z}}^{\triangledown,R}(\zeta) := \hat{\mathbf{Z}}^{\triangledown,L}(\zeta) \cdot i\sigma_3. \quad (5.16)$$

ASYMPTOTIC ANALYSIS

Proposition 5.6. *The matrix* $\hat{\mathbf{Z}}^{\nabla,R}(\zeta)$ *defined by (5.16) is an analytic function of* ζ *for* $\zeta \in \mathbb{C} \setminus (\mathbb{R} \cup C_+ \cup C_-)$ *that satisfies the normalization condition*

$$\hat{\mathbf{Z}}^{\nabla,R}(\zeta) \cdot \frac{1}{\sqrt{2}} \begin{pmatrix} -i & i \\ i & i \end{pmatrix} (-\zeta)^{-\sigma_3/4} = \mathbb{I} + O\left(\frac{1}{\zeta}\right)$$

as $\zeta \to \infty$ *uniformly with respect to direction. Moreover,* $\hat{\mathbf{Z}}^{\nabla,R}(\zeta)$ *takes continuous boundary values from each sector of its analyticity that with* $\zeta = \tau_\Gamma^{\nabla,R}(z)$ *satisfy the exact same set of relations (5.15) as* $\mathbf{Z}_\Gamma^{\nabla,R}(z)$.

We may construct a local parametrix for $\mathbf{X}(z)$ in $D_\Gamma^{\nabla,R}$ as follows:

$$\hat{\mathbf{X}}_\Gamma^{\nabla,R}(z) := \begin{cases} \mathbf{H}_\Gamma^{\nabla,R}(z)\hat{\mathbf{Z}}^{\nabla,R}(\tau_\Gamma^{\nabla,R}(z))T_\nabla(z)^{-\sigma_3/2}e^{(\eta(z)-\gamma-2\kappa g(z))\sigma_3/2}e^{iN\theta_\Gamma\sigma_3/2}, & \text{for } z \in D_{\Gamma,\mathrm{I}}^{\nabla,R}, \\ \mathbf{H}_\Gamma^{\nabla,R}(z)\hat{\mathbf{Z}}^{\nabla,R}(\tau_\Gamma^{\nabla,R}(z))e^{(\eta(z)-\gamma-2\kappa g(z))\sigma_3/2}e^{iN\theta_\Gamma\sigma_3/2}, & \text{for } z \in D_{\Gamma,\mathrm{II}}^{\nabla,R}, \\ \mathbf{H}_\Gamma^{\nabla,R}(z)\hat{\mathbf{Z}}^{\nabla,R}(\tau_\Gamma^{\nabla,R}(z))e^{(\eta(z)-\gamma-2\kappa g(z))\sigma_3/2}e^{-iN\theta_\Gamma\sigma_3/2}, & \text{for } z \in D_{\Gamma,\mathrm{III}}^{\nabla,R}, \\ \mathbf{H}_\Gamma^{\nabla,R}(z)\hat{\mathbf{Z}}^{\nabla,R}(\tau_\Gamma^{\nabla,R}(z))T_\nabla(z)^{-\sigma_3/2}e^{(\eta(z)-\gamma-2\kappa g(z))\sigma_3/2}e^{-iN\theta_\Gamma\sigma_3/2}, & \text{for } z \in D_{\Gamma,\mathrm{IV}}^{\nabla,R}. \end{cases}$$
(5.17)

Again, the transformation $\tau_\Gamma^{\nabla,R}(\cdot)$ and the matrix $\mathbf{H}_\Gamma^{\nabla,R}(z)$ will be different in neighborhoods $D_\Gamma^{\nabla,R}$ corresponding to different right band edges in Σ_0^∇.

Left band edge with $z_0 = \alpha \in \Sigma_0^\Delta$ **(upper constraint)**

Letting Γ denote the saturated region to the left of α, I denote the band to the right, and $D_\Gamma^{\Delta,L}$ denote a disc centered at $z = \alpha$ with radius $h\epsilon$, we partition the disc into quadrants:

$$D_{\Gamma,\mathrm{I}}^{\Delta,L} = D_\Gamma^{\Delta,L} \cap \left\{ z \,\Big|\, z \neq \alpha,\, 0 < \arg(z-\alpha) < \frac{\pi}{2} \right\},$$

$$D_{\Gamma,\mathrm{II}}^{\Delta,L} = D_\Gamma^{\Delta,L} \cap \left\{ z \,\Big|\, z \neq \alpha,\, \frac{\pi}{2} < \arg(z-\alpha) < \pi \right\},$$

$$D_{\Gamma,\mathrm{III}}^{\Delta,L} = D_\Gamma^{\Delta,L} \cap \left\{ z \,\Big|\, z \neq \alpha,\, -\pi < \arg(z-\alpha) < -\frac{\pi}{2} \right\},$$

$$D_{\Gamma,\mathrm{IV}}^{\Delta,L} = D_\Gamma^{\Delta,L} \cap \left\{ z \,\Big|\, z \neq \alpha,\, -\frac{\pi}{2} < \arg(z-\alpha) < 0 \right\}.$$

Next we set

$$\mathbf{Z}_\Gamma^{\Delta,L}(z) := \begin{cases} \mathbf{X}(z)e^{(\gamma-\eta(z)+2\kappa g(z))\sigma_3/2}e^{-iN\theta_\Gamma\sigma_3/2}, & \text{for } z \in D_{\Gamma,\mathrm{I}}^{\Delta,L}, \\ \mathbf{X}(z)T_\Delta(z)^{-\sigma_3/2}e^{(\gamma-\eta(z)+2\kappa g(z))\sigma_3/2}e^{-iN\theta_\Gamma\sigma_3/2}, & \text{for } z \in D_{\Gamma,\mathrm{II}}^{\Delta,L}, \\ \mathbf{X}(z)T_\Delta(z)^{-\sigma_3/2}e^{(\gamma-\eta(z)+2\kappa g(z))\sigma_3/2}e^{iN\theta_\Gamma\sigma_3/2}, & \text{for } z \in D_{\Gamma,\mathrm{III}}^{\Delta,L}, \\ \mathbf{X}(z)e^{(\gamma-\eta(z)+2\kappa g(z))\sigma_3/2}e^{iN\theta_\Gamma\sigma_3/2}, & \text{for } z \in D_{\Gamma,\mathrm{IV}}^{\Delta,L}, \end{cases}$$

where we recall that $e^{iN\theta_\Gamma} = e^{iN\theta(\alpha)}$, and consider the conformal mapping $\zeta = \tau_\Gamma^{\Delta,L}(z)$ defined by (2.19). We choose the parameter ϵ controlling the radius of $D_\Gamma^{\Delta,L}$ to be sufficiently small that the images $\tau_\Gamma^{\Delta,L}(\Sigma_{0\pm}^\Delta \cap D_\Gamma^{\Delta,L})$ lie within a symmetric sector of the imaginary ζ-axis of opening angle strictly less than $\pi/3$. The exact jump conditions satisfied by the matrix $\mathbf{Z}_\Gamma^{\Delta,L}(z)$ may be written in terms of ζ in a simple way:

$$\begin{aligned} \mathbf{Z}_{\Gamma+}^{\Delta,L}(z) &= \mathbf{Z}_{\Gamma-}^{\Delta,L}(z) \begin{pmatrix} 1 & 0 \\ ie^{-(-\zeta)^{3/2}} & 1 \end{pmatrix}, & \text{for } z \in \Gamma \cap D_\Gamma^{\Delta,L}, \\ \mathbf{Z}_{\Gamma+}^{\Delta,L}(z) &= \mathbf{Z}_{\Gamma-}^{\Delta,L}(z) \begin{pmatrix} 0 & -i \\ -i & 0 \end{pmatrix}, & \text{for } z \in I \cap D_\Gamma^{\Delta,L}, \\ \mathbf{Z}_{\Gamma+}^{\Delta,L}(z) &= \mathbf{Z}_{\Gamma-}^{\Delta,L}(z) \begin{pmatrix} 1 & -ie^{i\zeta^{3/2}} \\ 0 & 1 \end{pmatrix}, & \text{for } z \in \Sigma_{0+}^\Delta \cap D_\Gamma^{\Delta,L}, \\ \mathbf{Z}_{\Gamma+}^{\Delta,L}(z) &= \mathbf{Z}_{\Gamma-}^{\Delta,L}(z) \begin{pmatrix} 1 & -ie^{-i\zeta^{3/2}} \\ 0 & 1 \end{pmatrix}, & \text{for } z \in \Sigma_{0-}^\Delta \cap D_\Gamma^{\Delta,L}. \end{aligned}$$
(5.18)

The subscripts $+$ and $-$, respectively, refer to boundary values taken on the oriented contour Σ_{SD} from the left and right. The comparison matrix defined by the formula

$$\dot{\mathbf{Z}}_\Gamma^{\Delta,L}(z) := \dot{\mathbf{X}}(z)e^{(\gamma-\eta(z)+2\kappa g(z))\sigma_3/2}e^{-iN\text{sgn}(\Im(z))\theta_\Gamma\sigma_3/2}$$

satisfies the same jump condition for $z \in I \cap D_\Gamma^{\Delta,L}$ as $\mathbf{Z}_\Gamma^{\Delta,L}(z)$ but is otherwise analytic in $D_\Gamma^{\Delta,L}$ and thus may be written in the form

$$\dot{\mathbf{Z}}_\Gamma^{\Delta,L}(z) = \mathbf{H}_\Gamma^{\Delta,L}(z) \cdot \frac{1}{\sqrt{2}}(-\tau_\Gamma^{\Delta,L}(z))^{\sigma_3/4}\begin{pmatrix} i & i \\ i & -i \end{pmatrix}.$$

The quotient matrix $\mathbf{H}_\Gamma^{\Delta,L}(z)$ is holomorphic in $D_\Gamma^{\Delta,L}$. Finding a matrix with the same jump conditions as $\mathbf{Z}_\Gamma^{\Delta,L}(z)$ and matching onto $\dot{\mathbf{Z}}_\Gamma^{\Delta,L}(z)$ at the boundary of $D_\Gamma^{\Delta,L}$ leads us to recall the matrix $\hat{\mathbf{Z}}^{\nabla,L}(\zeta)$ solving Riemann-Hilbert Problem 5.4 with the contours $C_\pm$ taken to be such that for each N, $C_\pm \cap \tau_\Gamma^{\Delta,L}(D_\Gamma^{\Delta,L}) = \tau_\Gamma^{\Delta,L}(\Sigma_{0\pm}^\Delta)$ and to set

$$\hat{\mathbf{Z}}^{\Delta,L}(\zeta) := \hat{\mathbf{Z}}^{\nabla,L}(\zeta) \cdot i\sigma_1. \tag{5.19}$$

Proposition 5.7. *The matrix $\hat{\mathbf{Z}}^{\Delta,L}(\zeta)$ defined by (5.19) is an analytic function of ζ for $\zeta \in \mathbb{C} \setminus (\mathbb{R} \cup C_+ \cup C_-)$ that satisfies the normalization condition*

$$\hat{\mathbf{Z}}^{\Delta,L}(\zeta) \cdot \frac{1}{\sqrt{2}}\begin{pmatrix} -i & -i \\ -i & i \end{pmatrix}(-\zeta)^{-\sigma_3/4} = \mathbb{I} + O\left(\frac{1}{\zeta}\right)$$

as $\zeta \to \infty$ uniformly with respect to direction. Moreover, $\hat{\mathbf{Z}}^{\Delta,L}(\zeta)$ takes continuous boundary values from each sector of its analyticity that with $\zeta = \tau_\Gamma^{\Delta,L}(z)$ satisfy the exact same set of relations (5.18) as $\mathbf{Z}_\Gamma^{\Delta,L}(z)$.

We construct a local parametrix for $\mathbf{X}(z)$ with the formula

$$\hat{\mathbf{X}}_\Gamma^{\Delta,L}(z) := \begin{cases} \mathbf{H}_\Gamma^{\Delta,L}(z)\hat{Z}^{\Delta,L}(\tau_\Gamma^{\Delta,L}(z))e^{(\eta(z)-\gamma-2\kappa g(z))\sigma_3/2}e^{iN\theta_\Gamma\sigma_3/2}, & \text{for } z \in D_{\Gamma,\text{I}}^{\Delta,L}, \\ \mathbf{H}_\Gamma^{\Delta,L}(z)\hat{Z}^{\Delta,L}(\tau_\Gamma^{\Delta,L}(z))T_\Delta(z)^{\sigma_3/2}e^{(\eta(z)-\gamma-2\kappa g(z))\sigma_3/2}e^{iN\theta_\Gamma\sigma_3/2}, & \text{for } z \in D_{\Gamma,\text{II}}^{\Delta,L}, \\ \mathbf{H}_\Gamma^{\Delta,L}(z)\hat{Z}^{\Delta,L}(\tau_\Gamma^{\Delta,L}(z))T_\Delta(z)^{\sigma_3/2}e^{(\eta(z)-\gamma-2\kappa g(z))\sigma_3/2}e^{-iN\theta_\Gamma\sigma_3/2}, & \text{for } z \in D_{\Gamma,\text{III}}^{\Delta,L}, \\ \mathbf{H}_\Gamma^{\Delta,L}(z)\hat{Z}^{\Delta,L}(\tau_\Gamma^{\Delta,L}(z))e^{(\eta(z)-\gamma-2\kappa g(z))\sigma_3/2}e^{-iN\theta_\Gamma\sigma_3/2}, & \text{for } z \in D_{\Gamma,\text{IV}}^{\Delta,L}. \end{cases} \tag{5.20}$$

As before, the transformation $\tau_\Gamma^{\Delta,L}(\cdot)$ and the matrix $\mathbf{H}_\Gamma^{\Delta,L}(z)$ will be different in different neighborhoods $D_\Gamma^{\Delta,L}$ corresponding to different left band edges in Σ_0^Δ.

Right band edge with $z_0 = \beta \in \Sigma_0^\Delta$ (upper constraint)

With Γ denoting the saturated region to the right of β and I denoting the band to the left, we work in a disc $D_\Gamma^{\Delta,R}$ centered at $z = \beta$ with radius $h\epsilon$ and partition the disc into quadrants:

$$D_{\Gamma,\text{I}}^{\Delta,R} = D_\Gamma^{\Delta,R} \cap \left\{ z \,\Big|\, z \neq \beta, 0 < \arg(z-\beta) < \frac{\pi}{2} \right\},$$

$$D_{\Gamma,\text{II}}^{\Delta,R} = D_\Gamma^{\Delta,R} \cap \left\{ z \,\Big|\, z \neq \beta, \frac{\pi}{2} < \arg(z-\beta) < \pi \right\},$$

$$D_{\Gamma,\text{III}}^{\Delta,R} = D_\Gamma^{\Delta,R} \cap \left\{ z \,\Big|\, z \neq \beta, -\pi < \arg(z-\beta) < -\frac{\pi}{2} \right\},$$

$$D_{\Gamma,\text{IV}}^{\Delta,R} = D_\Gamma^{\Delta,R} \cap \left\{ z \,\Big|\, z \neq \beta, -\frac{\pi}{2} < \arg(z-\beta) < 0 \right\}.$$

We then set

$$\mathbf{Z}_\Gamma^{\Delta,R}(z) := \begin{cases} \mathbf{X}(z)T_\Delta(z)^{-\sigma_3/2}e^{(\gamma-\eta(z)+2\kappa g(z))\sigma_3/2}e^{-iN\theta_\Gamma\sigma_3/2}, & \text{for } z \in D_{\Gamma,\text{I}}^{\Delta,R}, \\ \mathbf{X}(z)e^{(\gamma-\eta(z)+2\kappa g(z))\sigma_3/2}e^{-iN\theta_\Gamma\sigma_3/2}, & \text{for } z \in D_{\Gamma,\text{II}}^{\Delta,R}, \\ \mathbf{X}(z)e^{(\gamma-\eta(z)+2\kappa g(z))\sigma_3/2}e^{iN\theta_\Gamma\sigma_3/2}, & \text{for } z \in D_{\Gamma,\text{III}}^{\Delta,R}, \\ \mathbf{X}(z)T_\Delta(z)^{-\sigma_3/2}e^{(\gamma-\eta(z)+2\kappa g(z))\sigma_3/2}e^{iN\theta_\Gamma\sigma_3/2}, & \text{for } z \in D_{\Gamma,\text{IV}}^{\Delta,R}, \end{cases}$$

ASYMPTOTIC ANALYSIS

where we recall that $e^{iN\theta_\Gamma} = e^{iN\theta(\beta)}$, and consider the conformal mapping $\zeta = \tau_\Gamma^{\Delta,R}(z)$ defined by (2.20). This is an orientation-reversing transformation of the neighborhood $D_\Gamma^{\Delta,R}$ of $z = \beta$ in the z-plane to a neighborhood of the origin in the ζ-plane. By making ϵ small enough, the radius of $D_\Gamma^{\Delta,R}$ will be so small that the images $\tau_\Gamma^{\Delta,R}(\Sigma_{0\pm}^\Delta \cap D_\Gamma^{\Delta,R})$ lie within a symmetric sector of the imaginary ζ-axis of opening angle strictly less than $\pi/3$. The matrix $\mathbf{Z}_\Gamma^{\Delta,R}(z)$ then satisfies exactly the following jump conditions:

$$\begin{aligned}
\mathbf{Z}_{\Gamma+}^{\Delta,R}(z) &= \mathbf{Z}_{\Gamma-}^{\Delta,R}(z) \begin{pmatrix} 1 & 0 \\ ie^{-(-\zeta)^{3/2}} & 1 \end{pmatrix}, && \text{for } z \in \Gamma \cap D_\Gamma^{\Delta,R}, \\
\mathbf{Z}_{\Gamma+}^{\Delta,R}(z) &= \mathbf{Z}_{\Gamma-}^{\Delta,R}(z) \begin{pmatrix} 0 & -i \\ -i & 0 \end{pmatrix}, && \text{for } z \in I \cap D_\Gamma^{\Delta,R}, \\
\mathbf{Z}_{\Gamma+}^{\Delta,R}(z) &= \mathbf{Z}_{\Gamma-}^{\Delta,R}(z) \begin{pmatrix} 1 & -ie^{i\zeta^{3/2}} \\ 0 & 1 \end{pmatrix}, && \text{for } z \in \Sigma_{0-}^\Delta \cap D_\Gamma^{\Delta,R}, \\
\mathbf{Z}_{\Gamma+}^{\Delta,R}(z) &= \mathbf{Z}_{\Gamma-}^{\Delta,R}(z) \begin{pmatrix} 1 & -ie^{-i\zeta^{3/2}} \\ 0 & 1 \end{pmatrix}, && \text{for } z \in \Sigma_{0+}^\Delta \cap D_\Gamma^{\Delta,R}.
\end{aligned} \quad (5.21)$$

The subscripts $+$ and $-$, respectively, indicate boundary values taken on Σ_{SD} from the left and right. The comparison matrix

$$\dot{\mathbf{Z}}_\Gamma^{\Delta,R}(z) := \dot{\mathbf{X}}(z) e^{(\gamma - \eta(z) + 2\kappa g(z))\sigma_3/2} e^{-iN\mathrm{sgn}(\Im(z))\theta_\Gamma \sigma_3/2}$$

satisfies the same jump condition for $z \in I \cap D_\Gamma^{\Delta,R}$ as $\mathbf{Z}_\Gamma^{\Delta,R}(z)$ and is otherwise analytic in $D_\Gamma^{\Delta,R}$; it may be written in the form

$$\dot{\mathbf{Z}}_\Gamma^{\Delta,R}(z) = \mathbf{H}_\Gamma^{\Delta,R}(z) \cdot \frac{1}{\sqrt{2}} (-\tau_\Gamma^{\Delta,R}(z))^{\sigma_3/4} \begin{pmatrix} -1 & -1 \\ 1 & -1 \end{pmatrix}.$$

The quotient $\mathbf{H}_\Gamma^{\Delta,R}(z)$ is holomorphic in $D_\Gamma^{\Delta,R}$. A matrix that satisfies the same jump conditions as $\mathbf{Z}_\Gamma^{\Delta,R}(z)$ and matches well onto $\dot{\mathbf{Z}}_\Gamma^{\Delta,R}(z)$ may be obtained by considering the matrix $\hat{\mathbf{Z}}^{\nabla,L}(\zeta)$ satisfying Riemann-Hilbert Problem 5.4, with the contours $C_\pm$ chosen so that $C_\pm \cap \tau_\Gamma^{\Delta,R}(D_\gamma^{\Delta,R}) = \tau_\Gamma^{\Delta,R}(\Sigma_{0\mp}^\Delta)$, and setting

$$\hat{\mathbf{Z}}^{\Delta,R}(\zeta) := \hat{\mathbf{Z}}^{\nabla,L}(\zeta) \cdot \sigma_1 \sigma_3. \quad (5.22)$$

Proposition 5.8. *The matrix $\hat{\mathbf{Z}}^{\Delta,R}(\zeta)$ defined by (5.22) is an analytic function of ζ for $\zeta \in \mathbb{C} \setminus (\mathbb{R} \cup C_+ \cup C_-)$ that satisfies the normalization condition*

$$\hat{\mathbf{Z}}^{\Delta,R}(\zeta) \cdot \frac{1}{\sqrt{2}} \begin{pmatrix} 1 & 1 \\ -1 & 1 \end{pmatrix} (-\zeta)^{-\sigma_3/4} = \mathbb{I} + O\left(\frac{1}{\zeta}\right)$$

as $\zeta \to \infty$ uniformly with respect to direction. Moreover, $\hat{\mathbf{Z}}^{\Delta,R}(\zeta)$ takes continuous boundary values from each sector of its analyticity that with $\zeta = \tau_\Gamma^{\Delta,R}(z)$ satisfy exactly the same set of relations (5.21) as $\mathbf{Z}_\Gamma^{\Delta,R}(z)$.

We use $\hat{\mathbf{Z}}^{\Delta,R}(\zeta)$ to construct a local parametrix for $\mathbf{X}(z)$ in $D_\Gamma^{\Delta,R}$ by the scheme

$$\hat{\mathbf{X}}_\Gamma^{\Delta,R}(z) := \begin{cases} \mathbf{H}_\Gamma^{\Delta,R}(z)\hat{\mathbf{Z}}^{\Delta,R}(\tau_\Gamma^{\Delta,R}(z))T_\Delta(z)^{\sigma_3/2} e^{(\eta(z)-\gamma-2\kappa g(z))\sigma_3/2} e^{iN\theta_\Gamma \sigma_3/2}, & \text{for } z \in D_{\Gamma,\mathrm{I}}^{\Delta,R}, \\ \mathbf{H}_\Gamma^{\Delta,R}(z)\hat{\mathbf{Z}}^{\Delta,R}(\tau_\Gamma^{\Delta,R}(z)) e^{(\eta(z)-\gamma-2\kappa g(z))\sigma_3/2} e^{iN\theta_\Gamma \sigma_3/2}, & \text{for } z \in D_{\Gamma,\mathrm{II}}^{\Delta,R}, \\ \mathbf{H}_\Gamma^{\Delta,R}(z)\hat{\mathbf{Z}}^{\Delta,R}(\tau_\Gamma^{\Delta,R}(z)) e^{(\eta(z)-\gamma-2\kappa g(z))\sigma_3/2} e^{-iN\theta_\Gamma \sigma_3/2}, & \text{for } z \in D_{\Gamma,\mathrm{III}}^{\Delta,R}, \\ \mathbf{H}_\Gamma^{\Delta,R}(z)\hat{\mathbf{Z}}^{\Delta,R}(\tau_\Gamma^{\Delta,R}(z))T_\Delta(z)^{\sigma_3/2} e^{(\eta(z)-\gamma-2\kappa g(z))\sigma_3/2} e^{-iN\theta_\Gamma \sigma_3/2}, & \text{for } z \in D_{\Gamma,\mathrm{IV}}^{\Delta,R}. \end{cases} \quad (5.23)$$

Once again, the transformation $\tau_\Gamma^{\Delta,R}(\cdot)$ and the matrix $\mathbf{H}_\Gamma^{\Delta,R}(z)$ will be different in different neighborhoods $D_\Gamma^{\Delta,R}$ corresponding to different right band edges in Σ_0^Δ.

Common properties of the four local approximations

The important properties of the local approximations are summarized in the following proposition.

Proposition 5.9. *Although originally defined in the four open quadrants within each disc, each function $\hat{\mathbf{X}}_\Gamma^{\nabla,L}(z)\mathbf{X}(z)^{-1}$, $\hat{\mathbf{X}}_\Gamma^{\nabla,R}(z)\mathbf{X}(z)^{-1}$, $\hat{\mathbf{X}}_\Gamma^{\Delta,L}(z)\mathbf{X}(z)^{-1}$, and $\hat{\mathbf{X}}_\Gamma^{\Delta,R}(z)\mathbf{X}(z)^{-1}$ has a continuous and hence analytic extension to the full interior of the corresponding disc. For each sufficiently small $\epsilon > 0$, there is a constant $M_\epsilon > 0$ such that on the boundary of each disc centered at a band edge $z = z_0$ we have*

$$\sup_{|z-z_0|=h\epsilon} \|\hat{\mathbf{X}}_\Gamma^{*,*}(z)\dot{\mathbf{X}}(z)^{-1} - \mathbb{I}\| \leq \frac{M_\epsilon}{N}, \qquad (5.24)$$

for sufficiently large N. Here $\hat{\mathbf{X}}_\Gamma^{*,*}(z)$ refers to any of the four different types of local parametrices.

Proof. The analyticity of $\hat{\mathbf{X}}_\Gamma^{*,*}(z)\mathbf{X}(z)^{-1}$ throughout $D_\Gamma^{*,*}$ follows directly from the construction in each case, in that there is no approximation of the jump matrix.

To prove (5.24), first note that since each band edge point z_0 is bounded away from all transition points $y_{k,N} \in Y_N$ and from the endpoints $\{a,b\}$, Proposition 4.3 guarantees that for $|z - z_0| \leq h\epsilon$,

$$\hat{\mathbf{X}}_\Gamma^{*,*}(z) = \mathbf{H}_\Gamma^{*,*}(z)\hat{\mathbf{Z}}^{*,*}(\tau_\Gamma^{*,*}(z))e^{(\eta(z)-\gamma-2\kappa g(z))\sigma_3/2}e^{iN\mathrm{sgn}(\Im(z))\theta_\Gamma\sigma_3/2}\left(\mathbb{I}+\mathbf{G}(z)\right),$$

where for some constant $J_\epsilon > 0$,

$$\sup_{|z-z_0|<h\epsilon} \|\mathbf{G}(z)\| \leq \frac{J_\epsilon}{N}.$$

Since according to Proposition 5.2, $\dot{\mathbf{X}}(z)$ is uniformly bounded for $|z - z_0| = h\epsilon$ and has determinant 1, it follows that a related constant $\tilde{J}_\epsilon > 0$ exists such that a similar estimate holds:

$$\sup_{|z-z_0|=h\epsilon} \|\dot{\mathbf{X}}(z)\mathbf{G}(z)\dot{\mathbf{X}}(z)^{-1}\| \leq \frac{\tilde{J}_\epsilon}{N},$$

for all sufficiently large N. Next, we recall the formula for the holomorphic prefactors $\mathbf{H}_\Gamma^{*,*}(z)$:

$$\mathbf{H}_\Gamma^{*,*}(z) = \dot{\mathbf{X}}(z)e^{(\gamma-\eta(z)+2\kappa g(z))\sigma_3/2}e^{-iN\mathrm{sgn}(\Im(z))\theta_\Gamma\sigma_3/2}\mathbf{C}^{*,*}[-\tau_\Gamma^{*,*}(z)]^{-\sigma_3/4}, \qquad (5.25)$$

where the constant matrices $\mathbf{C}^{*,*}$ are given by

$$\mathbf{C}^{\nabla,L} := \frac{1}{\sqrt{2}}\begin{pmatrix} 1 & -1 \\ 1 & 1 \end{pmatrix}, \quad \mathbf{C}^{\nabla,R} := \frac{1}{\sqrt{2}}\begin{pmatrix} -i & i \\ i & i \end{pmatrix},$$

$$\mathbf{C}^{\Delta,L} := \frac{1}{\sqrt{2}}\begin{pmatrix} -i & -i \\ -i & i \end{pmatrix}, \quad \mathbf{C}^{\Delta,R} := \frac{1}{\sqrt{2}}\begin{pmatrix} -1 & 1 \\ -1 & -1 \end{pmatrix}.$$

Thus we have

$$\hat{\mathbf{X}}_\Gamma^{*,*}(z)\dot{\mathbf{X}}(z)^{-1} = \mathbf{H}_\Gamma^{*,*}(z)\hat{\mathbf{Z}}^{*,*}(\tau_\Gamma^{*,*}(z))\mathbf{C}^{*,*}[-\tau_\Gamma^{*,*}(z)]^{-\sigma_3/4}\mathbf{H}_\Gamma^{*,*}(z)^{-1}\left(\mathbb{I}+\dot{\mathbf{X}}(z)\mathbf{G}(z)\dot{\mathbf{X}}(z)^{-1}\right).$$

Using (5.25) again, we can write this as

$$\hat{\mathbf{X}}_\Gamma^{*,*}(z)\dot{\mathbf{X}}(z)^{-1} = \mathbf{W}^{*,*}(z)$$
$$\cdot [-\tau_\Gamma^{*,*}(z)]^{-\sigma_3/4} \cdot \hat{\mathbf{Z}}^{*,*}(\tau_\Gamma^{*,*}(z))\mathbf{C}^{*,*}[-\tau_\Gamma^{*,*}(z)]^{-\sigma_3/4} \cdot [-\tau_\Gamma^{*,*}(z)]^{\sigma_3/4}$$
$$\cdot \mathbf{W}^{*,*}(z)^{-1}\left(\mathbb{I}+\dot{\mathbf{X}}(z)\mathbf{G}(z)\dot{\mathbf{X}}(z)^{-1}\right),$$

where

$$\mathbf{W}^{*,*}(z) := \dot{\mathbf{X}}(z)e^{(\gamma-\eta(z)+2\kappa g(z))\sigma_3/2}e^{-iN\mathrm{sgn}(\Im(z))\theta_\Gamma\sigma_3/2}\mathbf{C}^{*,*}$$

is a matrix that is, according to Proposition 5.2, uniformly bounded when $|z - z_0| = h\epsilon$. But, from (5.13), we get that for $|z - z_0| = h\epsilon$, which corresponds to $\tau_\Gamma^{*,*}(z)$ of size $N^{2/3}$,

$$[-\tau_\Gamma^{*,*}(z)]^{-\sigma_3/4}\hat{\mathbf{Z}}^{*,*}(\tau_\Gamma^{*,*}(z))\mathbf{C}^{*,*}[-\tau_\Gamma^{*,*}(z)]^{-\sigma_3/4}[-\tau_\Gamma^{*,*}(z)]^{\sigma_3/4}$$
$$= [-\tau_\Gamma^{*,*}(z)]^{-\sigma_3/4}\left[\mathbb{I}+\begin{pmatrix} O(\tau_\Gamma^{*,*}(z)^{-3/2}) & O(\tau_\Gamma^{*,*}(z)^{-1}) \\ O(\tau_\Gamma^{*,*}(z)^{-2}) & O(\tau_\Gamma^{*,*}(z)^{-3/2}) \end{pmatrix}\right][-\tau_\Gamma^{*,*}(z)]^{\sigma_3/4}$$
$$= \mathbb{I} + O(\tau_\Gamma^{*,*}(z)^{-3/2}),$$

which is, as desired, of order $1/N$ when $|z - z_0| = h\epsilon$. This establishes (5.24). □

ASYMPTOTIC ANALYSIS   99

5.1.3 Definition of the global parametrix $\hat{\mathbf{X}}(z)$

The global parametrix $\hat{\mathbf{X}}(z)$ is an explicit approximation of $\mathbf{X}(z)$, the uniform validity of which we will establish in §5.2. It is defined for $z \in \mathbb{C} \setminus (\Sigma_{\mathrm{SD}} \cup \{\text{disc boundaries}\})$ as follows. About each left band edge $\alpha \in (a, b)$ where the lower constraint becomes active in a void Γ, we have placed a disc $D_\Gamma^{\nabla, L}$. For $z \in D_\Gamma^{\nabla, L} \cap (\mathbb{C} \setminus \Sigma_{\mathrm{SD}})$, we set

$$\hat{\mathbf{X}}(z) := \hat{\mathbf{X}}_\Gamma^{\nabla, L}(z). \tag{5.26}$$

About each right band edge $\beta \in (a, b)$ where the lower constraint becomes active in a void Γ, we have placed a disc $D_\Gamma^{\nabla, R}$. For $z \in D_\Gamma^{\nabla, R} \cap (\mathbb{C} \setminus \Sigma_{\mathrm{SD}})$, we set

$$\hat{\mathbf{X}}(z) := \hat{\mathbf{X}}_\Gamma^{\nabla, R}(z). \tag{5.27}$$

About each left band edge $\alpha \in (a, b)$ where the upper constraint becomes active in a saturated region Γ, we have placed a disc $D_\Gamma^{\Delta, L}$. For $z \in D_\Gamma^{\Delta, L} \cap (\mathbb{C} \setminus \Sigma_{\mathrm{SD}})$, we set

$$\hat{\mathbf{X}}(z) := \hat{\mathbf{X}}_\Gamma^{\Delta, L}(z). \tag{5.28}$$

About each right band edge $\beta \in (a, b)$ where the upper constraint becomes active in a saturated region Γ, we have placed a disc $D_\Gamma^{\Delta, R}$. For $z \in D_\Gamma^{\Delta, R} \cap (\mathbb{C} \setminus \Sigma_{\mathrm{SD}})$, we set

$$\hat{\mathbf{X}}(z) := \hat{\mathbf{X}}_\Gamma^{\Delta, R}(z). \tag{5.29}$$

Finally, for all $z \in \mathbb{C} \setminus \Sigma_{\mathrm{SD}}$ lying outside the closure of all discs, we set

$$\hat{\mathbf{X}}(z) := \dot{\mathbf{X}}(z). \tag{5.30}$$

5.2 ERROR ESTIMATION

To compare the (unknown) solution $\mathbf{X}(z)$ of Riemann-Hilbert Problem 4.6 to the explicit global parametrix $\hat{\mathbf{X}}(z)$, we consider the error matrix $\mathbf{E}(z)$ defined by

$$\mathbf{E}(z) := \mathbf{X}(z)\hat{\mathbf{X}}(z)^{-1}. \tag{5.31}$$

A direct calculation shows that this matrix has a continuous (and thus analytic) extension to each band I and also to the interior of each disc $D_\Gamma^{*,*}$. In other words, $\mathbf{E}(z)$ is analytic for $z \in \mathbb{C} \setminus \Sigma_E$, where Σ_E is the contour pictured in Figure 5.2. We want to deduce, for sufficiently small positive ϵ, an estimate for

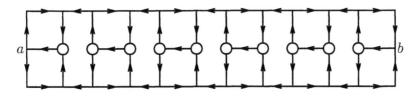

Figure 5.2 *The contour Σ_E lies in the region $a \leq \Re(z) \leq b$ and $|\Im(z)| \leq \epsilon$. The circles of radius $h\epsilon$ ($h < 1$) are all oriented in the clockwise direction.*

$\mathbf{E}(z) - \mathbb{I}$ that is valid in a neighborhood of an arbitrary point of $[a, b]$. In order to do this, it is useful to first introduce an intermediate matrix $\mathbf{F}(z)$ which will differ from $\mathbf{E}(z)$ only near all gaps Γ and near the endpoints a and b.

For each void interval Γ that lies between two consecutive bands (Γ is necessarily an interior gap), let x and y be the points where Γ meets the boundaries of the discs $D_\Gamma^{\nabla, R}$ and $D_\Gamma^{\nabla, L}$ and let L_Γ^∇ denote the open chord (*i.e.*, the part of a disc bounded by a circular arc of the boundary and the straight-line segment joining the endpoints of the arc) determined by the points x, $(x+y)/2 - ih\epsilon$, and y. If the lower constraint is satisfied at $z = a$ and Γ is the corresponding void interval that meets the boundary of the disc $D_\Gamma^{\nabla,L}$ at a point x, then we let L_Γ^∇ denote the open triangle with vertices a, x, and $a - ih\epsilon$. If the lower constraint is satisfied at $z = b$ and Γ is the corresponding void interval that meets the boundary of the disc $D_\Gamma^{\nabla,R}$ at a point x, then we let L_Γ^∇ denote the open triangle with vertices b, x, and $b - ih\epsilon$. The various regions L_Γ^∇ lie in the range $a < \Re(z) < b$ and are illustrated with darker shading in Figure 5.3. We make the change of variables

$$\mathbf{F}(z) = \mathbf{E}(z)\dot{\mathbf{X}}(z) \begin{pmatrix} 1 & -iT_\nabla(z)e^{\gamma - \eta(z) + 2\kappa g(z)}e^{iN\theta_\Gamma}e^{-N\xi_\Gamma(z)} \\ 0 & 1 \end{pmatrix} \dot{\mathbf{X}}(z)^{-1}, \qquad \text{for } z \in L_\Gamma^\nabla. \tag{5.32}$$

For each saturated region Γ that lies between two consecutive bands (making Γ an interior gap), let x and y be the points where Γ meets the boundaries of the discs $D_\Gamma^{\Delta,R}$ and $D_\Gamma^{\Delta,L}$ and let L_Γ^Δ denote the open chord determined by the points x, $(x+y)/2 - ih\epsilon$, and y. If the upper constraint is satisfied at $z = a$ and Γ is the corresponding saturated region that meets the boundary of the disc $D_\Gamma^{\Delta,L}$ at a point x, then we let L_Γ^Δ denote the open triangle with vertices a, x, and $a - ih\epsilon$. If the upper constraint is satisfied at $z = b$ and Γ is the corresponding saturated region that meets the boundary of the disc $D_\Gamma^{\Delta,R}$ at a point x, then we let L_Γ^Δ denote the open triangle with vertices b, x, and $b - ih\epsilon$. The various regions L_Γ^Δ lie in the range $a < \Re(z) < b$ and are illustrated with lighter shading in Figure 5.3. We make the change of variables

$$\mathbf{F}(z) = \mathbf{E}(z)\dot{\mathbf{X}}(z) \begin{pmatrix} 1 & 0 \\ -iT_\Delta(z)e^{\eta(z) - \gamma - 2\kappa g(z)}e^{-iN\theta_\Gamma}e^{-N\xi_\Gamma(z)} & 1 \end{pmatrix} \dot{\mathbf{X}}(z)^{-1}, \qquad \text{for } z \in L_\Gamma^\Delta. \tag{5.33}$$

Next, we define two open half-discs: $D_a = \{z | \Re(z) < a, |z - a| < h\epsilon\}$ and $D_b = \{z | \Re(z) > b, |z - b| < h\epsilon\}$. In each of these half-discs centered at an endpoint where the lower constraint is active (indicated with darker shading in Figure 5.3), we set

$$\mathbf{F}(z) = \mathbf{E}(z)\dot{\mathbf{X}}(z) \begin{pmatrix} 1 & -ie^{\gamma - \eta(z)} \dfrac{\prod_{n \in \Delta}(z - x_{N,n})}{\prod_{n \in \nabla}(z - x_{N,n})} e^{N(\ell_c - V(z) - i\theta^0(z)/2)} e^{(k - \#\Delta)g(z)} \\ 0 & 1 \end{pmatrix} \dot{\mathbf{X}}(z)^{-1}, \tag{5.34}$$

and in each half-disc centered at an endpoint where the upper constraint is active (indicated with lighter shading in Figure 5.3), we set

$$\mathbf{F}(z) = \mathbf{E}(z)\dot{\mathbf{X}}(z) \begin{pmatrix} 1 & 0 \\ -ie^{\eta(z) - \gamma} \dfrac{\prod_{n \in \nabla}(z - x_{N,n})}{\prod_{n \in \Delta}(z - x_{N,n})} e^{N(V(z) - \ell_c - i\theta^0(z)/2)} e^{(\#\Delta - k)g(z)} & 1 \end{pmatrix} \dot{\mathbf{X}}(z)^{-1}. \tag{5.35}$$

It is important to observe that the matrix relating $\mathbf{F}(z)$ and $\mathbf{E}(z)$ in (5.34) and (5.35) is always an analytic function of z in the half-disc under consideration. Indeed, the poles are all located in $[a, b]$, $e^{g(z)}$ is analytic for $z \in \mathbb{C} \setminus [a, b]$, and $k - \#\Delta \in \mathbb{Z}$. For all remaining $z \in \mathbb{C} \setminus \Sigma_E$, we set $\mathbf{F}(z) = \mathbf{E}(z)$.

Lemma 5.10. *The matrix $\mathbf{F}(z)$ admits a continuous and hence analytic extension to the upper boundaries of all regions L_Γ^∇ and L_Γ^Δ, as well as to the segments $\Re(z) = a$ and $\Re(z) = b$ with $|\Im(z)| < h\epsilon$.*

Proof. This proof is rather straightforward once one makes the following observations. First, in the quarter-discs $D_a \cap \{z | \Im(z) > 0\}$ and $D_b \cap \{z | \Im(z) > 0\}$ centered at endpoints where the lower constraint holds, we have the identity

$$-ie^{\gamma - \eta(z)} \dfrac{\prod_{n \in \Delta}(z - x_{N,n})}{\prod_{n \in \nabla}(z - x_{N,n})} e^{N(\ell_c - V(z) - i\theta^0(z)/2)} e^{(k - \#\Delta)g(z)} = -iY(z)e^{\gamma - \eta(z) + 2\kappa g(z)} e^{-iN\theta_\Gamma} e^{-iN\theta^0(z)} e^{-N\xi_\Gamma(z)}.$$

$$\tag{5.36}$$

ASYMPTOTIC ANALYSIS

Here Γ refers to the void that is adjacent to the endpoint. If the upper constraint is active, we have in the same region the identity

$$-ie^{\eta(z)-\gamma}\frac{\prod_{n\in\nabla}(z-x_{N,n})}{\prod_{n\in\Delta}(z-x_{N,n})}e^{N(V(z)-\ell_c-i\theta^0(z)/2)}e^{(\#\Delta-k)g(z)} = -iY(z)^{-1}e^{\eta(z)-\gamma-2\kappa g(z)}e^{iN\theta_\Gamma}e^{-iN\theta^0(z)}e^{-N\xi_\Gamma(z)}.$$
(5.37)

On the other hand, in the quarter-discs $D_a \cap \{z|\Im(z) < 0\}$ and $D_b \cap \{z|\Im(z) < 0\}$ centered at endpoints where the lower constraint holds, we have

$$-ie^{\gamma-\eta(z)}\frac{\prod_{n\in\Delta}(z-x_{N,n})}{\prod_{n\in\nabla}(z-x_{N,n})}e^{N(\ell_c-V(z)-i\theta^0(z)/2)}e^{(k-\#\Delta)g(z)} = -iY(z)e^{\gamma-\eta(z)+2\kappa g(z)}e^{iN\theta_\Gamma}e^{-N\xi_\Gamma(z)}. \quad (5.38)$$

If the upper constraint is active, then in the same region,

$$-ie^{\eta(z)-\gamma}\frac{\prod_{n\in\nabla}(z-x_{N,n})}{\prod_{n\in\Delta}(z-x_{N,n})}e^{N(V(z)-\ell_c-i\theta^0(z)/2)}e^{(\#\Delta-k)g(z)} = -iY(z)^{-1}e^{\eta(z)-\gamma-2\kappa g(z)}e^{-iN\theta_\Gamma}e^{-N\xi_\Gamma(z)}. \quad (5.39)$$

The claimed continuity of $\mathbf{F}(z)$ follows from these identities upon using the definition $\mathbf{E}(z) = \mathbf{X}(z)\dot{\mathbf{X}}(z)^{-1}$ (since $\hat{\mathbf{X}}(z) = \dot{\mathbf{X}}(z)$ in for all relevant z in the current context), the jump conditions satisfied by $\mathbf{X}(z)$ and $\dot{\mathbf{X}}(z)$, and the relations (4.20) connecting $T_\Delta(z)$, $T_\nabla(z)$, and $Y(z)$ for $\Im(z) < 0$. □

The contour Σ_F where $\mathbf{F}(z)$ fails to be analytic is shown in Figure 5.3. In order to estimate $\mathbf{E}(z)^{-1}\mathbf{F}(z) - \mathbb{I}$,

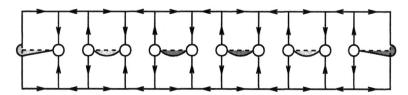

Figure 5.3 The contour Σ_F. Dashed lines indicate contour segments of Σ_E to which $\mathbf{F}(z)$ has a continuous and hence analytic extension. As with Σ_E, the disc boundaries are oriented in the clockwise direction. The circular boundaries of the half-discs D_a and D_b are also oriented in the clockwise direction. The lower boundaries of all regions L_Γ^∇ and L_Γ^Δ are oriented from right to left.

and subsequently to estimate $\mathbf{F}(z) - \mathbb{I}$, we will now need to recall the behavior of the functions $T_\nabla(z)$ and $T_\Delta(z)$ in the asymptotic limit $N \to \infty$.

It follows from Proposition 4.3 that in each region L_Γ^∇ (respectively, L_Γ^Δ), $T_\nabla(z)$ (respectively, $T_\Delta(z)$) is uniformly bounded. Furthermore, in any half-disc D_a or D_b centered at an endpoint where the lower constraint is active, the function $Y(z)$ is uniformly bounded, and in any half-disc centered at an endpoint where the upper constraint is active, the function $Y(z)^{-1}$ is uniformly bounded. Using the identities (5.36)–(5.39) and the variational inequalities (2.13) and (2.16) that control $\Re(\xi_\Gamma(z))$ in these regions (and noting that in particular $\Re(\xi_\Gamma(a)) > 0$ and $\Re(\xi_\Gamma(b)) > 0$ by assumption — see §2.1.2), we have the following result.

Lemma 5.11. *Let the contour parameter $\epsilon > 0$ be sufficiently small. Then there are constants $C_{1,\epsilon} > 0$ and $C_{2,\epsilon} > 0$ such that for all sufficiently large N,*

$$\sup_{z\in\mathbb{C}\setminus(\Sigma_E\cup\Sigma_F)} \|\mathbf{E}(z)^{-1}\mathbf{F}(z) - \mathbb{I}\| \leq C_{1,\epsilon}e^{-C_{2,\epsilon}N}.$$

Here $\|\cdot\|$ denotes an arbitrary matrix norm.

Being obtained from $\mathbf{X}(z)$ satisfying Riemann-Hilbert Problem 4.6 by explicit transformations involving the global parametrix $\dot{\mathbf{X}}(z)$ as well as the explicit relations (5.32)–(5.35), the (unknown) matrix $\mathbf{F}(z)$ is the solution of a Riemann-Hilbert problem as well.

Riemann-Hilbert Problem 5.12. *Find a 2×2 matrix $\mathbf{F}(z)$ with the following properties:*

1. **Analyticity**: $\mathbf{F}(z)$ *is an analytic function of z for $z \in \mathbb{C} \setminus \Sigma_F$.*

2. **Normalization**: *As $z \to \infty$,*
$$\mathbf{F}(z) = \mathbb{I} + O\left(\frac{1}{z}\right).$$

3. **Jump Conditions**: *$\mathbf{F}(z)$ takes uniformly continuous boundary values on Σ_F from each connected component of $\mathbb{C} \setminus \Sigma_F$. For each non-self-intersection point $z \in \Sigma_F$, we denote by $\mathbf{F}_+(z)$ ($\mathbf{F}_-(z)$) the limit of $\mathbf{F}(w)$ as $w \to z$ from the left (right). The boundary values satisfy the jump condition $\mathbf{F}_+(z) = \mathbf{F}_-(z)\mathbf{v}_\mathbf{F}(z)$, where for z on the lower boundary of a region L_Γ^∇ below a void $\Gamma \subset \Sigma_0^\nabla$,*
$$\mathbf{v}_\mathbf{F}(z) = \dot{\mathbf{X}}(z) \begin{pmatrix} 1 & iT_\nabla(z)e^{\gamma - \eta(z) + 2\kappa g(z)}e^{iN\theta_\Gamma}e^{-N\xi_\Gamma(z)} \\ 0 & 1 \end{pmatrix} \dot{\mathbf{X}}(z)^{-1}.$$

For z on the lower boundary of a region L_Γ^Δ below a saturated region $\Gamma \subset \Sigma_0^\Delta$,
$$\mathbf{v}_\mathbf{F}(z) = \dot{\mathbf{X}}(z) \begin{pmatrix} 1 & 0 \\ iT_\Delta(z)e^{\eta(z) - \gamma - 2\kappa g(z)}e^{-iN\theta_\Gamma}e^{-N\xi_\Gamma(z)} & 1 \end{pmatrix} \dot{\mathbf{X}}(z)^{-1}.$$

For z in any vertical segment $\Sigma_{0\pm}^\nabla \cap \Sigma_F$ meeting the boundary of a disc centered at an endpoint z_0 of a band I,
$$\mathbf{v}_\mathbf{F}(z) = \dot{\mathbf{X}}(z) \begin{pmatrix} T_\nabla(z)^{\pm 1/2} & 0 \\ -iT_\nabla(z)^{-1/2}e^{\eta(z) - \gamma - 2\kappa g(z)}e^{\pm iN\theta(z_0)}\exp\left(\pm 2\pi iNc \int_{z_0}^z \psi_I(s)\,ds\right) & T_\nabla(z)^{\mp 1/2} \end{pmatrix} \dot{\mathbf{X}}(z)^{-1}. \tag{5.40}$$

For z in any vertical segment $\Sigma_{0\pm}^\Delta \cap \Sigma_F$ meeting the boundary of a disc centered at an endpoint z_0 of a band I,
$$\mathbf{v}_\mathbf{F}(z) = \dot{\mathbf{X}}(z) \begin{pmatrix} T_\Delta(z)^{\mp 1/2} & -iT_\Delta(z)^{-1/2}e^{\gamma - \eta(z) + 2\kappa g(z)}e^{\mp iN\theta(z_0)}\exp\left(\pm 2\pi iNc \int_{z_0}^z \overline{\psi}_I(z)\,ds\right) \\ 0 & T_\Delta(z)^{\pm 1/2} \end{pmatrix} \dot{\mathbf{X}}(z)^{-1}. \tag{5.41}$$

For z in any segment $\Sigma_{\Gamma\pm}^\nabla \cap \Sigma_F$ parallel to a void $\Gamma \subset \Sigma_0^\nabla$ or with $\Re(z) = a$ or $\Re(z) = b$,
$$\mathbf{v}_\mathbf{F}(z) = \dot{\mathbf{X}}(z) \begin{pmatrix} 1 & iY(z)e^{\gamma - \eta(z) + 2\kappa g(z)}e^{\mp iN\theta_\Gamma}e^{\mp iN\theta^0(z)}e^{-N\xi_\Gamma(z)} \\ 0 & 1 \end{pmatrix} \dot{\mathbf{X}}(z)^{-1},$$

and for z in the semicircular boundary of a half-disc D_a or D_b centered at an endpoint where the lower constraint is active,
$$\mathbf{v}_\mathbf{F}(z) = \dot{\mathbf{X}}(z) \begin{pmatrix} 1 & ie^{\gamma - \eta(z)}\dfrac{\prod_{n \in \Delta}(z - x_{N,n})}{\prod_{n \in \nabla}(z - x_{N,n})}e^{N(\ell_c - V(z) - i\theta^0(z)/2)}e^{(k - \#\Delta)g(z)} \\ 0 & 1 \end{pmatrix} \dot{\mathbf{X}}(z)^{-1}.$$

For z in any segment $\Sigma_{\Gamma\pm}^\Delta$ parallel to a saturated region $\Gamma \subset \Sigma_0^\Delta$ or with $\Re(z) = a$ or $\Re(z) = b$,
$$\mathbf{v}_\mathbf{F}(z) = \dot{\mathbf{X}}(z) \begin{pmatrix} 1 & 0 \\ iY(z)^{-1}e^{\eta(z) - \gamma - 2\kappa g(z)}e^{\pm iN\theta_\Gamma}e^{\mp iN\theta^0(z)}e^{-N\xi_\Gamma(z)} & 1 \end{pmatrix} \dot{\mathbf{X}}(z)^{-1},$$

and for z in the semicircular boundary of a half-disc D_a or D_b centered at an endpoint where the upper constraint is active,

$$\mathbf{v_F}(z) = \dot{\mathbf{X}}(z)\begin{pmatrix} 1 & 0 \\ ie^{\eta(z)-\gamma}\dfrac{\prod_{n\in\nabla}(z-x_{N,n})}{\prod_{n\in\Delta}(z-x_{N,n})}e^{N(V(z)-\ell_c-i\theta^0(z)/2)}e^{(\#\Delta-k)g(z)} & 1 \end{pmatrix}\dot{\mathbf{X}}(z)^{-1}.$$

With a sequence $\{y_N\}_{N=0}^{\infty}$ determined as in the formulation of Riemann-Hilbert Problem 4.6, we have that for z in any segment $\Sigma_{I\pm}$ parallel to any band I,

$$\mathbf{v_F}(z) = \dot{\mathbf{X}}(z)\begin{pmatrix} T_\Delta(z)^{-1/2} & v_{12}^\pm(z) \\ v_{21}^\pm(z) & T_\nabla(z)^{-1/2} \end{pmatrix}^{\pm 1}\dot{\mathbf{X}}(z)^{-1}, \qquad (5.42)$$

where

$$v_{12}^\pm(z) := \mp i T_\Delta(z)^{-1/2} e^{\gamma-\eta(z)+2\kappa g(z)} e^{\mp iN\theta(y_N)} \exp\left(\pm 2\pi i N c \int_{y_N}^z \overline{\psi}_I(s)\,ds\right),$$

$$v_{21}^\pm(z) := \mp i T_\nabla(z)^{-1/2} e^{\eta(z)-\gamma-2\kappa g(z)} e^{\pm iN\theta(y_N)} \exp\left(\pm 2\pi i N c \int_{y_N}^z \psi_I(s)\,ds\right).$$

Finally, for z in the clockwise-oriented boundary of any disc $D_\Gamma^{\nabla,L}$,

$$\mathbf{v_F}(z) = \hat{\mathbf{X}}_\Gamma^{\nabla,L}(z)\dot{\mathbf{X}}(z)^{-1},$$

for z in the clockwise-oriented boundary of any disc $D_\Gamma^{\nabla,R}$,

$$\mathbf{v_F}(z) = \hat{\mathbf{X}}_\Gamma^{\nabla,R}(z)\dot{\mathbf{X}}(z)^{-1},$$

for z in the clockwise-oriented boundary of any disc $D_\Gamma^{\Delta,L}$,

$$\mathbf{v_F}(z) = \hat{\mathbf{X}}_\Gamma^{\Delta,L}(z)\dot{\mathbf{X}}(z)^{-1},$$

and for z in the clockwise-oriented boundary of any disc $D_\Gamma^{\Delta,R}$,

$$\mathbf{v_F}(z) = \hat{\mathbf{X}}_\Gamma^{\Delta,R}(z)\dot{\mathbf{X}}(z)^{-1}.$$

We have the following characterization of the jump matrix for $\mathbf{F}(z)$.

Lemma 5.13. *Let the parameter $\epsilon > 0$ of the contour Σ_F be sufficiently small. Then there is a constant $C_\epsilon > 0$ such that*

$$\sup_{z\in\Sigma_F}\|\mathbf{v_F}(z) - \mathbb{I}\| \leq \frac{C_\epsilon}{N}$$

holds for sufficiently large N.

Proof. The estimates on the boundaries of the discs $D_{*,*}^{*,*}$ all follow from Proposition 5.9. For the remaining parts of Σ_F, we note that by Proposition 5.2 $\dot{\mathbf{X}}(z)$ and $\dot{\mathbf{X}}(z)^{-1}$ are both uniformly bounded for $z\in\Sigma_F$; thus it suffices to prove an estimate of the same order for $\dot{\mathbf{X}}(z)^{-1}\mathbf{v_F}(z)\dot{\mathbf{X}}(z)$. Now using Proposition 4.3, one sees that the diagonal elements in (5.40)–(5.42) all differ from 1 by a quantity of order $1/N$. All off-diagonal matrix elements are exponentially small as $N\to\infty$ for two different reasons. First, we recall the variational inequalities (2.13) and (2.16) that hold on the real axis in the voids and saturated regions, respectively; these control the off-diagonal elements of $\dot{\mathbf{X}}(z)^{-1}\mathbf{v_F}(z)\dot{\mathbf{X}}(z)$ involving a factor $e^{-N\xi_\Gamma(z)}$ for ϵ sufficiently small that the inequality $\Re(\xi_\Gamma(z)) > 0$ holds for z on relevant portions of Σ_F, as it does for z in the gap $\Gamma \subset [a,b]$. Second, we recall the inequalities $\psi_I(x) > 0$ and $\overline{\psi}_I(x) > 0$ that hold for x in each band I together with the presumed square-root vanishing of $\psi_I(x)$ at band edges where the lower constraint becomes active and of $\overline{\psi}_I(x)$ at band edges where the upper constraint becomes active; these facts control the off-diagonal elements of $\dot{\mathbf{X}}(z)^{-1}\mathbf{v_F}(z)\dot{\mathbf{X}}(z)$ in the segments $\Sigma_{0\pm}^\nabla \cap \Sigma_F$, $\Sigma_{0\pm}^\Delta \cap \Sigma_F$, and $\Sigma_{I\pm}$. □

Lemma 5.14. *Let the contour parameter $\epsilon > 0$ be sufficiently small. Then Riemann-Hilbert Problem 5.12 has a unique solution for sufficiently large N, and the solution has the Cauchy integral representation*

$$\mathbf{F}(z) = \mathbb{I} + \int_{\Sigma_F} (z-s)^{-1} \mathbf{m}(s) \, ds, \tag{5.43}$$

where $\mathbf{m}(\cdot)$ is an arcwise-continuous matrix function in $L^2(\Sigma_F)$. There is a constant $L_\epsilon > 0$ such that

$$\|\mathbf{m}\|_2 \leq \frac{L_\epsilon}{N} \tag{5.44}$$

holds for all sufficiently large N. Also, $\det(\mathbf{F}(z)) = 1$, for all $z \in \mathbb{C} \setminus \Sigma_\mathbf{F}$.

Proof. This is essentially a consequence of the theory of matrix Riemann-Hilbert problems with L^2 boundary values and uniformly near-identity jump matrices (see, for example, [Zho89]). The key idea is that it is possible to convert the Riemann-Hilbert problem into a system of singular integral equations of the form $(1-B)\mathbf{u} = \mathbb{I}$, where B is a singular integral operator acting on matrix functions $\mathbf{u}(z)$ defined on Σ_F; then the desired density $\mathbf{m}(z)$ is proportional to both $\mathbf{u}(z)$ and $\mathbf{v}_F(z) - \mathbb{I}$. The operator B can be written as a composition of multiplication by $\mathbf{v}_F(z) - \mathbb{I}$ and a singular integral operator with Cauchy kernel. It is a profound result of modern harmonic analysis [CoiMM82] that the Cauchy kernel singular integral operators are bounded in L^2 on contours that may be decomposed as finite unions of graphs of Lipschitz functions (an appropriate Lipschitz condition must also be satisfied at each self-intersection point). The norm of B in $L^2(\Sigma_F)$ is proportional to the product of $\|\mathbf{v}_F(z) - \mathbb{I}\|_\infty$, which we know can be made arbitrarily small according to Lemma 5.13, and the L^2 norm of a Cauchy integral over Σ_F, which is finite if ϵ is taken to be sufficiently small (this makes all self-intersections of Σ_F nontangential). Thus for sufficiently large N, we will have $\|B\|_2 < 1$, and the integral equation for $\mathbf{u}(z)$ can be solved in $L^2(\Sigma_F)$ by a Neumann series: $\mathbf{u}(z) = \mathbb{I} + B\mathbb{I} + B^2\mathbb{I} + \cdots$.

We therefore have the invertibility of the operator $1-B$ for sufficiently large N, with $\|(1-B)^{-1}\|_2$ bounded independently of N, and thus the existence of $\mathbf{u} \in L^2(\Sigma_F)$. Moreover, since the total length of Σ_F is independent of N, we have $\|\mathbf{u}\|_2 = \|(1-B)^{-1}\mathbb{I}\|_2$ being bounded uniformly with respect to N as well. This proves (5.44) since $\mathbf{m}(z)$ is proportional to the product of $\mathbf{u}(z)$ and $\mathbf{v}_F(z) - \mathbb{I}$.

The fact that the boundary values taken by the solution $\mathbf{F}(z)$ supplied by the L^2 theory are in fact uniformly continuous along the boundary of each connected component of $\mathbb{C} \setminus \Sigma_F$ warrants some additional explanation. Indeed, the L^2 theory guarantees a solution of the Riemann-Hilbert problem taking boundary values only in the L^2 sense. However, since the jump matrix $\mathbf{v}_F(z)$ is analytic on each arc of Σ_F, it follows that both $\mathbf{F}_+(z)$ and $\mathbf{F}_-(z)$ may be continued analytically through to the opposite side of each arc, and then from Morera's Theorem we deduce not only that is $\mathbf{F}(z)$ continuous up to the boundary but also that both boundary values are analytic functions of z. That uniform continuity extends even to self-intersection points of Σ_F can be shown using the compatibility of the limiting values of $\mathbf{v}_F(z)$ along all arcs meeting at such a point; namely, the cyclic product of the limiting values is the identity matrix for all self-intersection points. Thus the unique L^2 solution is in fact a classical solution of the Riemann-Hilbert problem. □

Thus we arrive at the main result of this section.

Proposition 5.15. *Let the contour parameter ϵ be sufficiently small. Then for each closed set $K \subset \mathbb{C} \setminus \Sigma_F$, not necessarily bounded, there is a constant $Q_{K,\epsilon} > 0$ such that*

$$\sup_{z \in K} \|\mathbf{E}(z) - \mathbb{I}\| \leq \frac{Q_{K,\epsilon}}{N}$$

holds for all sufficiently large N. Recall that $\mathbf{E}(z) = \mathbf{X}(z)\hat{\mathbf{X}}(z)^{-1}$.

Proof. From (5.43) and (5.44) we obtain the desired estimate for the matrix $\mathbf{F}(z)$. To complete the proof, we recall Lemma 5.11. □

Chapter Six

Discrete Orthogonal Polynomials: Proofs of Theorems Stated in §2.3

In this chapter, we start with the exact formula for $\mathbf{X}(z)$ valid in the entire complex z-plane:

$$\mathbf{X}(z) = \mathbf{E}(z)\hat{\mathbf{X}}(z). \tag{6.1}$$

Thus $\mathbf{X}(z)$ is written in terms of the explicit global parametrix and the matrix $\mathbf{E}(z)$, which while not explicit is characterized by Proposition 5.15. We then work backward to the matrix $\mathbf{P}(z; N, k)$ and thereby obtain exact formulae for the monic polynomials $\pi_{N,k}(z)$ valid in the whole complex plane, as well as the normalization constants $\gamma_{N,k}$ and recurrence coefficients $a_{N,k}$ and $b_{N,k}$, in terms of the matrix elements of $\hat{\mathbf{X}}(z)$ and $\mathbf{E}(z)$, and their asymptotics for large z. Then, under various conditions on z we extract simple asymptotic formulae by direct asymptotic expansion of the exact formulae. In particular, we will obtain Plancherel-Rotach-type asymptotics of the monic polynomials $\pi_{N,k}(z)$ for real z in the interval $[a, b]$ of accumulation of the discrete nodes of support of the weights.

6.1 ASYMPTOTIC ANALYSIS OF $\mathbf{P}(z; N, k)$ FOR $z \in \mathbb{C} \setminus [a, b]$

6.1.1 Asymptotic behavior of $\pi_{N,k}(z)$ for $z \in \mathbb{C} \setminus [a, b]$: the proof of Theorem 2.7

Let $K \subset \mathbb{C} \setminus [a, b]$ be a fixed closed set, not necessarily bounded. The parameter ϵ in the contour Σ_F may then be fixed at such a sufficiently small positive value that $K \cap \Sigma_F = \emptyset$ and K is contained in the unbounded component of $\mathbb{C} \setminus \Sigma_F$. For $z \in K$, from (4.45), (4.46), (5.30), and (6.1), we thus have

$$\mathbf{P}(z; N, k) = e^{-(N\ell_c+\gamma)\sigma_3/2}\mathbf{E}(z)\dot{\mathbf{X}}(z)e^{(N\ell_c+\gamma)\sigma_3/2}e^{(k-\#\Delta)g(z)\sigma_3}\prod_{n\in\Delta}(z-x_{N,n})^{\sigma_3}.$$

Since $z \in K$ is bounded away from $[a, b]$, we use the midpoint rule to obtain

$$e^{(k-\#\Delta)g(z)}\prod_{n\in\Delta}(z-x_{N,n}) = e^{(k-\#\Delta)g(z)}\exp\left(N\int_{\Sigma_0^\Delta}\log(z-x)\rho^0(x)\,dx\right)\cdot\left(1+O\left(\frac{1}{N}\right)\right),$$

where the error term is uniform for $z \in K$. Combining this result with (4.5) and (4.7) and recalling that $k = cN + \kappa$, we have

$$e^{(k-\#\Delta)g(z)}\prod_{n\in\Delta}(z-x_{N,n}) = e^{\kappa g(z)}e^{NL_c(z)}\cdot\left(1+O\left(\frac{1}{N}\right)\right),$$

where $L_c(z)$ is defined in (2.10). Note that the product $e^{\kappa g(z)}e^{NL_c(z)}$ is analytic for $z \in \mathbb{C} \setminus [a, b]$. In particular, this analysis leads to the following formula:

$$P_{11}(z; N, k) = \left[E_{11}(z)\dot{X}_{11}(z)e^{\kappa g(z)} + E_{12}(z)\dot{X}_{21}(z)e^{\kappa g(z)}\right]e^{NL_c(z)}\cdot\left(1+O\left(\frac{1}{N}\right)\right).$$

We use Proposition 5.15 to estimate $\mathbf{E}(z) - \mathbb{I}$, and Proposition 5.2 to characterize $\dot{X}_{11}(z)$. The proof is complete upon noting that $W(z) = \dot{X}_{11}(z)e^{\kappa g(z)}$, using the formulae for $\dot{X}_{11}(z)$ obtained in Appendix A, and recalling from Proposition 1.3 that $P_{11}(z; N, k) = \pi_{N,k}(z)$.

6.1.2 Asymptotic behavior of the leading coefficients $\gamma_{N,k}$ and of the recurrence coefficients $a_{N,k}$ and $b_{N,k}$: the proof of Theorem 2.8

Taking the set K in the proof of Theorem 2.7 above to be unbounded allows us to consider $z \to \infty$. For arbitrary fixed N, we have the expansion

$$e^{(k-\#\Delta)g(z)} \prod_{n \in \Delta}(z - x_{N,n}) = e^{\kappa(g(z)-\log(z))} z^k \left(1 + \frac{H_{k,N}}{z} + O\left(\frac{1}{z^2}\right)\right)$$

as $z \to \infty$, where

$$H_{k,N} := N \int_{\Sigma_0^\Delta} x\rho^0(x)\, dx - \sum_{n \in \Delta} x_{N,n} - Nc \int_a^b x\, d\mu^c_{\min}(x)\,.$$

The matrices $\mathbf{E}(z)$ and $\dot{\mathbf{X}}(z) e^{\kappa(g(z)-\log(z))\sigma_3}$ have asymptotic expansions for large z of the form

$$\mathbf{E}(z) = \mathbb{I} + \frac{1}{z}\mathbf{E}^{(1)} + \frac{1}{z^2}\mathbf{E}^{(2)} + O\left(\frac{1}{z^3}\right),$$

$$\dot{\mathbf{X}}(z) e^{\kappa(g(z)-\log(z))\sigma_3} = \mathbb{I} + \frac{1}{z}\mathbf{B}^{(1)} + \frac{1}{z^2}\mathbf{B}^{(2)} + O\left(\frac{1}{z^3}\right)$$

as $z \to \infty$. In terms of these coefficients we thus have for each fixed N the expansions

$$z^{-k} P_{11}(z) = 1 + \frac{H_{k,N} + B_{11}^{(1)} + E_{11}^{(1)}}{z} + O\left(\frac{1}{z^2}\right),$$

$$z^{-k} P_{21}(z) = \frac{e^{N\ell_c + \gamma}}{z}(B_{21}^{(1)} + E_{21}^{(1)}) + O\left(\frac{1}{z^2}\right),$$

$$z^k P_{12}(z) = \frac{e^{-N\ell_c - \gamma}}{z}(B_{12}^{(1)} + E_{12}^{(1)})$$
$$+ \frac{e^{-N\ell_c - \gamma}}{z^2}(B_{12}^{(2)} + E_{12}^{(2)} + E_{11}^{(1)} B_{12}^{(1)} + E_{12}^{(1)} B_{22}^{(1)} - H_{k,N} B_{12}^{(1)} - H_{k,N} E_{12}^{(1)})$$
$$+ O\left(\frac{1}{z^3}\right).$$

Comparing with (1.27), we therefore have the following exact formulae in which $H_{k,N}$ does not appear:

$$\gamma_{N,k} = \frac{e^{(N\ell_c + \gamma)/2}}{\sqrt{B_{12}^{(1)} + E_{12}^{(1)}}}$$

$$\gamma_{N,k-1} = e^{(N\ell_c + \gamma)/2} \sqrt{B_{21}^{(1)} + E_{21}^{(1)}}\,,$$

$$b_{N,k-1} = \sqrt{(B_{12}^{(1)} + E_{12}^{(1)})(B_{21}^{(1)} + E_{21}^{(1)})}\,,$$

$$a_{N,k} = B_{11}^{(1)} + E_{11}^{(1)} + \frac{B_{12}^{(2)} + E_{12}^{(2)} + E_{11}^{(1)} B_{12}^{(1)} + E_{12}^{(1)} B_{22}^{(1)}}{B_{12}^{(1)} + E_{12}^{(1)}}\,.$$

Now for sufficiently large z, we have $\mathbf{E}(z) = \mathbf{F}(z)$, and therefore Lemma 5.14 and in particular the Cauchy integral representation (5.43) of $\mathbf{F}(z)$ implies that the coefficients $E_{jk}^{(1)}$ and $E_{jk}^{(2)}$ are all of order $1/N$ as $N \to \infty$. Furthermore, $\dot{\mathbf{X}}(z) e^{\kappa(g(z)-\log(z))}$ is a matrix that for some fixed $R > 0$ is analytic and uniformly bounded (independently of N) for $|z| > R$, which implies that the coefficients $B_{jk}^{(1)}$ and $B_{jk}^{(2)}$ remain bounded as $N \to \infty$. In fact, for sufficiently large N, $B_{12}^{(1)}$ and $B_{21}^{(1)}$ are bounded away from zero, and thus

$$\gamma_{N,k} = \frac{e^{(N\ell_c + \gamma)/2}}{\sqrt{B_{12}^{(1)}}} \left(1 + O\left(\frac{1}{N}\right)\right),$$

$$\gamma_{N,k-1} = e^{(N\ell_c + \gamma)/2} \sqrt{B_{21}^{(1)}} \left(1 + O\left(\frac{1}{N}\right)\right),$$

$$b_{N,k-1} = \sqrt{B_{12}^{(1)} B_{21}^{(1)}} \left(1 + O\left(\frac{1}{N}\right)\right),$$

$$a_{N,k} = B_{11}^{(1)} + \frac{B_{12}^{(2)}}{B_{12}^{(1)}} + O\left(\frac{1}{N}\right).$$

Using the formulae obtained in Proposition A.4 established in Appendix A then completes the proof. It should be remarked that the quantities $B_{12}^{(1)}$ and $B_{21}^{(1)}$ are necessarily positive since ℓ_c and γ are real.

6.2 ASYMPTOTIC BEHAVIOR OF $\pi_{N,k}(z)$ FOR z NEAR A VOID OF $[a,b]$: THE PROOF OF THEOREM 2.9

The variational inequality (2.13) holds strictly throughout the closed interval $J \subset [a,b]$, and while it is possible for either a or b to be an endpoint of J, neither endpoint of J may be a band edge. We choose the contour parameter ϵ to be sufficiently small that Proposition 5.15 controls $\mathbf{E}(z) - \mathbb{I}$ and then take δ to be small enough that $K_J^\delta \cap \Sigma_F = \emptyset$. Then, for all $z \in K_J^\delta$, regardless of whether $\Im(z)$ is positive or negative or of whether $\Re(z) \in (a,b)$ or not, we have the exact formula

$$\pi_{N,k}(z) = \left[E_{11}(z)\dot{X}_{11}(z) + E_{12}(z)\dot{X}_{21}(z)\right] e^{(k - \#\Delta)g(z)} \prod_{n \in \Delta} (z - x_{N,n}). \tag{6.2}$$

This follows from (4.45)–(4.47), (5.30), (6.1), and Proposition 1.3. It is not hard to verify that the right-hand side extends analytically to the whole compact set K_J^δ. Since each node $x_{N,n}$ with $n \in \Delta$ is bounded away from K_J^δ, we may approximate the product to within a relative error of order $1/N$ uniform in K_J^δ to find

$$\pi_{N,k}(z) = \left[E_{11}(z)\dot{X}_{11}(z) + E_{12}(z)\dot{X}_{21}(z)\right] e^{\kappa g(z)} e^{NL_c(z)} \left(1 + O\left(\frac{1}{N}\right)\right).$$

Here we have used (4.5) and (4.7) and $k = Nc + \kappa$, and $L_c(z)$ is defined by (2.10). Finally, using Proposition 5.15 to estimate $\mathbf{E}(z) - \mathbb{I}$ and Proposition 5.2 to uniformly bound $\dot{X}_{11}(z)$ and $\dot{X}_{21}(z)$ for $z \in K_J^\delta$, we arrive at

$$\pi_{N,k}(z) = e^{NL_c(z)} \left[\dot{X}_{11}(z) e^{\kappa g(z)} + O\left(\frac{1}{N}\right)\right]. \tag{6.3}$$

We recall the definition (2.11) of the analytic function $\overline{L}_c^\Gamma(z)$ and note that $W(z) = \dot{X}_{11}(z) e^{\kappa g(z)}$.

The estimate (2.49) follows from (6.3) because $e^{N(L_c(z) - \overline{L}_c^\Gamma(z))}$ is uniformly bounded in K_J^δ. Indeed, we have

$$L_c(z) - \overline{L}_c^\Gamma(z) = -\frac{i\theta_\Gamma}{2} \cdot \operatorname{sgn}(\Im(z)). \tag{6.4}$$

Thus the right-hand side of (6.4) is simply a different imaginary constant in each half-plane. This also establishes the uniform boundedness of $A_\Gamma^\nabla(z)$ when we use Proposition 5.2 to bound $\dot{X}_{11}(z)$. The analyticity of $A_\Gamma^\nabla(z)$ in K_J^δ is a consequence of the jump condition satisfied by $\dot{\mathbf{X}}(z)$ in the void Γ; using $+$ $(-)$ to denote boundary values taken on the real axis from above (below), we have for real $z \in K_J^\delta$,

$$\begin{aligned}
A_{\Gamma+}^\nabla(z) &= e^{-iN\theta_\Gamma/2} \dot{X}_{11+}(z) e^{\kappa g_+(z)} \\
&= e^{iN\theta_\Gamma/2} \dot{X}_{11-}(z) e^{i\phi_\Gamma} e^{\kappa g_+(z)} \\
&= e^{iN\theta_\Gamma/2} \dot{X}_{11-}(z) e^{\kappa g_-(z)} \\
&= A_{\Gamma-}^\nabla(z)
\end{aligned}$$

6.3 ASYMPTOTIC BEHAVIOR OF $\pi_{N,k}(z)$ FOR z NEAR A SATURATED REGION OF $[a,b]$

6.3.1 Asymptotics valid away from hard edges: the proof of Theorem 2.10

Because the closed interval $J \subset \Gamma$ is bounded away from all of the points $a, \alpha_0, \beta_0, \ldots, \alpha_G, \beta_G, b$, we may fix the parameter $\epsilon > 0$ sufficiently small that Proposition 5.15 controls $\mathbf{E}(z) - \mathbb{I}$ and then select $\delta > 0$ small enough that $K_J^\delta \cap \Sigma_F = \emptyset$, where the compact set K_J^δ is defined by (2.48). For $z \in K_J^\delta$, we thus have the following exact formula:

$$\pi_{N,k}(z) = \left[E_{11}(z)\dot{X}_{11}(z) + E_{12}(z)\dot{X}_{21}(z)\right] e^{(k-\#\Delta)g(z)} \prod_{n\in\Delta}(z - x_{N,n})$$
$$+ \left[E_{11}(z)\dot{X}_{12}(z) + E_{12}(z)\dot{X}_{22}(z)\right] i\,\mathrm{sgn}(\Im(z))e^{\eta(z)-\gamma}e^{N(V(z)-\ell_c-i\mathrm{sgn}(\Im(z))\theta^0(z)/2)}e^{(\#\Delta-k)g(z)}\prod_{n\in\nabla}(z-x_{N,n}).$$

This formula follows from (4.45), (4.48), (5.30), (6.1), and Proposition 1.3, and the right-hand side extends analytically to the whole set K_J^δ. Using the definition (4.11) of the function $T_\Delta(z)$ and its characterization for nonreal z in Proposition 4.2, and recalling the definition (2.10), we can rewrite this formula as

$$\pi_{N,k}(z) = \left[\exp\left(-N\int_{\Sigma_0^\nabla}\log(z-x)\rho^0(x)\,dx\right)\prod_{n\in\nabla}(z-x_{N,n})\right]e^{NL_c(z)-iN\mathrm{sgn}(\Im(z))\theta^0(z)/2}$$
$$\cdot\left(\left[E_{11}(z)\dot{X}_{11}(z)e^{\kappa g(z)} + E_{12}(z)\dot{X}_{21}(z)e^{\kappa g(z)}\right]T_\Delta(z)^{-1}2\cos\left(\frac{N\theta^0(z)}{2}\right)\right.$$
$$\left.+\left[E_{11}(z)\dot{X}_{12}(z)e^{-\kappa g(z)} + E_{12}(z)\dot{X}_{22}(z)e^{-\kappa g(z)}\right]i\,\mathrm{sgn}(\Im(z))e^{\eta(z)-\gamma}e^{-N\xi_\Gamma(z)}e^{iN\mathrm{sgn}(\Im(z))(\theta_\Gamma-\theta^0(z)/2)}\right).$$
(6.5)

Since K_J^δ is bounded away from any nodes $x_{N,n}$, for $n \in \nabla$, the product on the first line of (6.5) may be approximated in terms of an exponential of an integral up to a relative error of order $1/N$ uniformly in K_J^δ. From Proposition 4.3 it follows that $T_\Delta(z)^{-1} - 1$ is also of order $1/N$ uniformly in K_J^δ. Finally, using Proposition 5.15 to estimate $\mathbf{E}(z) - \mathbb{I}$ and Proposition 5.2 to bound $\dot{\mathbf{X}}(z)$ uniformly in K_J^δ, we see that

$$\pi_{N,k}(z) = e^{NL_c(z)-iN\mathrm{sgn}(\Im(z))\theta^0(z)/2}$$
$$\cdot\left(\left(\dot{X}_{11}(z)e^{\kappa g(z)} + O\left(\frac{1}{N}\right)\right)2\cos\left(\frac{N\theta^0(z)}{2}\right) + O\left(\exp\left(-N\inf_{z\in K_J^\delta}\Re(\xi_\Gamma(z))\right)\right)\right). \quad (6.6)$$

The exponential estimate holds because $\mathrm{sgn}(\Re(i\theta^0(z))) = \mathrm{sgn}(\Im(z))$ for all $z \in K_J^\delta$. Also, since $\Re(\xi_\Gamma(z))$ is strictly positive for all $z \in K_J^\delta$ (which is equivalent to the inequality (2.16) being strict in J and δ being sufficiently small), this term is exponentially small as $N \to \infty$. We note that $W(z) = \dot{X}_{11}(z)e^{\kappa g(z)}$.

The estimates (2.52) follow from (6.6) because $e^{N(L_c(z)-\overline{L}_c^\Gamma(z)-i\mathrm{sgn}(\Im(z))\theta^0(z)/2)}$ is uniformly bounded in K_J^δ. Indeed, we have

$$L_c(z) - \overline{L}_c^\Gamma(z) - \frac{i}{2}\mathrm{sgn}(\Im(z))\theta^0(z) = -\frac{i\theta_\Gamma}{2}\cdot\mathrm{sgn}(\Im(z)).$$

The rest of the proof follows that of Theorem 2.9.

6.3.2 Asymptotics uniformly valid near hard edges: the proof of Theorem 2.11

We will analyze the case where the saturated region is $\Gamma = (a, \alpha_0)$ and $J = [a, t]$, with $t \in \Gamma$, in detail. The analysis in a saturated region near $z = b$ is similar.

The upper constraint is active throughout J, and the variational inequality (2.16) holds strictly for all $z \in J$. We take the fixed parameter ϵ to be sufficiently small that Proposition 5.15 controls $\mathbf{E}(z) - \mathbb{I}$ and then choose $\delta > 0$ small enough that $K_J^\delta \cap \Sigma_F = \emptyset$, where K_J^δ is defined by (2.48). The set K_J^δ is the closure of the union of two open sets: $K_{J,\text{out}}^\delta$ consisting of the points in the interior of K_J^δ with $\Re(z) < a$, and $K_{J,\text{in}}^\delta$ consisting of the points in the interior of K_J^δ with $\Re(z) > a$.

For $z \in K_{J,\text{in}}^\delta$, the exact formula (6.5) for $\pi_{N,k}(z)$ is valid. Since $K_{J,\text{in}}^\delta$ is not bounded away from $z = a$, we may no longer neglect $T_\Delta(z)^{-1} - 1$. However, we may substitute from Proposition 4.3 an asymptotic formula for $T_\Delta(z)^{-1}$ that is uniformly valid in $K_{J,\text{in}}^\delta$. The remaining approximations we make for $z \in K_{J,\text{in}}^\delta$ are exactly the same as in the proof of Theorem 2.10.

On the other hand, for $z \in K_{J,\text{out}}^\delta$, the exact formula (6.2) holds. Using (4.18), we may write this in the form

$$\pi_{N,k}(z) = \left[E_{11}(z)\dot{X}_{11}(z) + E_{12}(z)\dot{X}_{21}(z) \right] e^{(k-\#\Delta)g(z)} Y(z)$$

$$\cdot \left[\exp\left(-N \int_{\Sigma_0^\nabla} \log(z-x)\rho^0(x)\,dx \right) \prod_{n \in \nabla} (z - x_{N,n}) \right] \exp\left(N \int_{\Sigma_0^\Delta} \log(z-x)\rho^0(x)\,dx \right).$$

The terms in the large square brackets may be estimated using the midpoint rule to approximate the integral in the exponent; these terms are thus of the form $1 + O(1/N)$ uniformly for $z \in K_{J,\text{out}}^\delta$. From Proposition 4.3 we may substitute an asymptotic formula for $Y(z)$ that is uniformly valid in $K_{J,\text{out}}^\delta$. Using Proposition 5.15 to estimate $\mathbf{E}(z) - \mathbb{I}$ uniformly for $z \in K_{J,\text{out}}^\delta$, and Proposition 5.2 to uniformly bound $\dot{X}_{11}(z)$ and $\dot{X}_{12}(z)$ in the same region, we obtain an asymptotic expression for $\pi_{N,k}(z)$ that is valid in $K_{J,\text{out}}^\delta$. To write this expression, we note that $W(z) = \dot{X}_{11}(z) e^{\kappa g(z)}$.

The two asymptotic formulae so-obtained are uniformly valid right up to the line $\Re(z) = a$ that divides K_J^δ into two parts. Moreover, it is straightforward to check that the formulae agree for $\Re(z) = a$. In this way, we obtain a uniform approximation for $\pi_{N,k}(z)$ for z near a that is an analytic function, and the proof is complete.

6.3.3 Asymptotics of the zeros of $\pi_{N,k}(z)$ in saturated regions: the proof of Theorem 2.12

The zeros of the cosine function in (2.51) and (2.53) are exactly the nodes of orthogonalization making up the set X_N. We may thus expect that there should be a zero of $\pi_{N,k}(z)$ very close to each node $x_{N,n}$ in a saturated region. To make this idea precise, we now study how the zeros of the leading term in (2.51) are perturbed by the term $\delta_N(z)$. Neither $\varepsilon_N(z)$ nor $\delta_N(z)$ in (2.51) is purely real for real z (although the imaginary part of $\varepsilon_N(z)$ is necessarily exponentially small for real z to balance with that of $\delta_N(z)$ since $\pi_{N,k}(z)$ is a real polynomial). However, from (6.5) we get the exact formula

$$\Re(\delta_N(z)) = \left(B_\Gamma^\Delta(z) \sin\left(\frac{N\theta^0(z)}{2} \right) + \sigma_N(z) \right) e^{\eta(z) - \gamma - N\xi_\Gamma(z)},$$

where $\sigma_N(z)$ is uniformly of order $1/N$ for $z \in \mathbb{R} \cap K_J^\delta$, with K_J^δ defined by (2.48) for δ small enough, and

$$B_\Gamma^\Delta(z) := e^{N(L_c(z) - \overline{L}_c^\Gamma(z))} e^{-iN\operatorname{sgn}(\Im(z))\theta^0(z)/2} \dot{X}_{12}(z) e^{-\kappa g(z)} e^{iN\operatorname{sgn}(\Im(z))\theta_\Gamma}. \tag{6.7}$$

Note that (6.7) apparently defines $B_\Gamma^\Delta(z)$ for $\Im(z) \neq 0$, but it is easy to check that this definition extends analytically to a real function for real z.

Now if the saturated region is the interval $\Gamma = (a, \alpha_0)$, then Proposition 5.3 guarantees that $A_\Gamma^\Delta(z)$ and $B_\Gamma^\Delta(z)$ are bounded away from zero and have opposite signs. Since $\theta^0(z)$ is a strictly decreasing function of z for $z \in (a, b)$, it follows that there is a zero of $\pi_{N,k}(z)$ exponentially close to but strictly greater than each node $x_{N,n}$ in the interval J (and no other zeros in J). Similarly, if the saturated region is the interval

$\Gamma = (\beta_G, b)$, then Proposition 5.3 guarantees that $A_\Gamma^\triangle(z)$ and $B_\Gamma^\triangle(z)$ are bounded away from zero and have the same sign. From this it follows that there is a zero of $\pi_{N,k}(z)$ exponentially close to but strictly less than each node $x_{N,n}$ in the interval J, and no other zeros in J. Note that with the use of the asymptotic formulae given in Theorem 2.11, it follows that these conclusions hold true even if the interval J under consideration contains either $z = a$ or $z = b$ as an endpoint.

If the saturated region is $\Gamma = \Gamma_j = (\beta_{j-1}, \alpha_j)$ for some $j = 1, \ldots, G$, then Proposition 5.3 implies that the product $A_\Gamma^\triangle(z) B_\Gamma^\triangle(z)$ vanishes at exactly one point $z = z_j$ in $\Gamma = \Gamma_j$. If $z_j < \min J$, then $A_\Gamma^\triangle(z)$ and $B_\Gamma^\triangle(z)$ are bounded away from zero and have opposite signs for $z \in J$, and thus there is a zero of $\pi_{N,k}(z)$ exponentially close to but strictly greater than each node $x_{N,n}$ in J, and no other zeros in J. If $z_j > \max J$, then $A_\Gamma^\triangle(z)$ and $B_\Gamma^\triangle(z)$ have the same sign, and thus there is a zero of $\pi_{N,k}(z)$ exponentially close to but strictly less than each node $x_{N,n}$ in J, and no other zeros in J.

Continuing with the case $\Gamma = \Gamma_j = (\beta_{j-1}, \alpha_j)$, suppose that z_j lies in the interior of $J \subset \Gamma_j$. If it is $B_\Gamma^\triangle(z)$ that vanishes at $z = z_j$, then it is clear that $\pi_{N,k}(z)$ has a zero exponentially close to each node $x_{N,n}$ in J, and no other zeros in J. Moreover, in this case there is a neighborhood of z_j of length proportional to $1/N$ outside of which $\Re(\delta_N(z))$ has the same sign as its leading term; thus with the possible exception of a bounded number of nodes surrounding $z = z_j$, the zeros exponentially localized near the nodes lying to the left (right) of $z = z_j$ lie to the left (right) of the nearest node. In fact, Proposition 1.1 guarantees that this situation persists inward from the left and right to a single interval between two consecutive nodes $[x_{N,m}, x_{N,m+1}]$ that contains no zeros of $\pi_{N,k}(z)$ at all, and such that there is a zero exponentially close to but to the left of $x_{N,m}$ and another zero exponentially close to but to the right of $x_{N,m+1}$. Thus in this situation, the interval J contains precisely one less than the maximum possible number of zeros of $\pi_{N,k}(z)$ since there are exactly two consecutive nodes that do not have a zero between them.

On the other hand, if it is $A_\Gamma^\triangle(z)$ that vanishes at $z = z_j$ in the interior of J, then in addition to the zeros of the cosine function, there is a single zero of $A_\Gamma^\triangle(z) + \Re(\varepsilon_N(z))$, say $z = z_{j,N}$, that is subjected to perturbation. The zeros of the cosine lying to the left (right) of $z = z_{j,N}$ are easily seen (using Proposition 5.3 to analyze the relative signs of $A_\Gamma^\triangle(z)$ and $B_\Gamma^\triangle(z)$) to move under perturbation an exponentially small amount to the left (right). The spurious zero $z_{j,N}$ is also perturbed an exponentially small amount, and it is easy to see that the closer $z_{j,N}$ lies to a node in X_N, the more it is repelled by the perturbation. Even in the degenerate case where $z_{j,N}$ coincides exactly with a node, it is easy to see that the perturbation always serves to unfold the double zero into two real zeros of $\pi_{N,k}(z)$, both exponentially close to the same node with one on either side. Thus in this situation, the interval J always contains precisely the maximum possible number of zeros of $\pi_{N,k}(z)$ (one zero between each consecutive pair of nodes), all exponentially localized to nodes in X_N with the possible exception of exactly one that necessarily corresponds to the zero $z_{j,N}$ of $A_\Gamma^\triangle(z) + \Re(\varepsilon_N(z))$. This completes the proof.

6.4 ASYMPTOTIC BEHAVIOR OF $\pi_{N,k}(z)$ FOR z NEAR A BAND

6.4.1 Asymptotic behavior of $\pi_{N,k}(z)$: the proof of Theorem 2.13

The closed interval J is necessarily bounded away from the two nearest band edge points $z = \alpha_j$ and $z = \beta_j$. Therefore, given $\epsilon > 0$ sufficiently small that Proposition 5.15 controls $\mathbf{E}(z) - \mathbb{I}$, we may choose $\delta > 0$ small enough that the set K_J^δ defined by (2.48) satisfies $K_J^\delta \cap \Sigma_F = \emptyset$.

Suppose first that the band I containing J is not a transition band but rather is completely contained in Σ_0^∇. Then, from Proposition 1.3, (4.45), (4.49), (4.50), (5.30), and (6.1), we have the following exact

DISCRETE ORTHOGONAL POLYNOMIALS: PROOFS OF THEOREMS STATED IN §2.3   111

formula for $\pi_{N,k}(z)$ in K_J^δ:

$$\pi_{N,k}(z) = T_\nabla(z)^{-1/2} \left[\exp\left(-N \int_{\Sigma_0^\Delta} \log(z-x)\rho^0(x)\,dx\right) \prod_{n\in\Delta}(z-x_{N,n}) \right] e^{N\overline{L}_c^I(z)}(-1)^{M_I^\nabla}$$

$$\cdot \left[\left(E_{11}(z)\dot{X}_{11}(z) + E_{12}(z)\dot{X}_{21}(z)\right) e^{\kappa g(z)} e^{-iN\mathrm{sgn}(\Im(z))\theta_I^\nabla(z)/2} \right. \quad (6.8)$$

$$\left. + i\,\mathrm{sgn}(\Im(z))e^{\eta(z)-\gamma}\left(E_{11}(z)\dot{X}_{12}(z) + E_{12}(z)\dot{X}_{22}(z)\right) e^{-\kappa g(z)} e^{iN\mathrm{sgn}(\Im(z))\theta_I^\nabla(z)/2} \right],$$

where $\overline{L}_c^I(z)$ is defined by (2.12) and

$$M_I^\nabla := N \int_{y<x\in\Sigma_0^\Delta} \rho^0(x)\,dx,$$

where y is the nearest transition point to the right of $J \subset I$. It follows from (4.1) that $M_I^\nabla \in \mathbb{Z}$. The right-hand side of (6.8) extends analytically to the whole compact set K_J^δ. The terms in square brackets on the first line of (6.8) are seen to be $1 + O(1/N)$ uniformly for $z \in K_J^\delta$ by a midpoint-rule approximation argument (since K_J^δ is in this case bounded away from any component of Σ_0^Δ). Similarly, $T_\nabla(z)^{-1/2} = 1 + O(1/N)$ uniformly for $z \in K_J^\delta$ by Proposition 4.3. Propositions 5.2 and 5.15 then imply that the terms in parentheses on the second line of (6.8) are simply $\dot{X}_{11}(z) + O(1/N)$ and that the terms in parentheses on the third line of (6.8) are just $\dot{X}_{12}(z) + O(1/N)$, with all errors uniform in K_J^δ. Thus one obtains an asymptotic formula for $\pi_{N,k}(z)$ valid uniformly in K_J^δ.

Next, suppose that the band I containing J is not a transition band but is rather completely contained in Σ_0^Δ. In this case, from Proposition 1.3, (4.45), (4.51), (4.52), (5.30), and (6.1), we have the following exact formula for $\pi_{N,k}(z)$ in K_J^δ:

$$\pi_{N,k}(z) = T_\Delta(z)^{-1/2} \left[\exp\left(-N \int_{\Sigma_0^\nabla} \log(z-x)\rho^0(x)\,dx\right) \prod_{n\in\nabla}(z-x_{N,n}) \right] e^{N\overline{L}_c^I(z)}(-1)^{M_I^\Delta}$$

$$\cdot \left[\left(E_{11}(z)\dot{X}_{11}(z) + E_{12}(z)\dot{X}_{21}(z)\right) e^{\kappa g(z)} e^{iN\mathrm{sgn}(\Im(z))[\theta^0(z)-\theta_I^\Delta(z)]/2} \right.$$

$$\left. + i\,\mathrm{sgn}(\Im(z))e^{\eta(z)-\gamma}\left(E_{11}(z)\dot{X}_{12}(z) + E_{12}(z)\dot{X}_{22}(z)\right) e^{\kappa g(z)} e^{-iN\mathrm{sgn}(\Im(z))[\theta^0(z)-\theta_I^\Delta(z)]/2} \right],$$
(6.9)

where

$$M_I^\Delta := N \int_{y<x\in\Sigma_0^\nabla} \rho^0(x)\,dx \in \mathbb{Z}$$

and y is the nearest transition point to the right of $J \subset I$. Once again, the right-hand side may be considered an analytic function in the set K_J^δ. Since K_J^δ is bounded away from Σ_0^∇ in this case, the expression (6.9) may be approximated in virtually the same way as (6.8) in order to obtain a uniformly valid asymptotic formula for $\pi_{N,k}(z)$.

Finally, suppose that the band I containing J is a transition band in which we must place a transition point $y \in Y_N$. Recall that J is bounded away from the endpoints α_j and β_j of $I = I_j$. Thus, without any loss of generality, we may choose the transition point $y \in I \cap Y_N$ such that either $y < \min J$ or $y > \max J$. This means that either $J \subset \Sigma_0^\nabla$ or $J \subset \Sigma_0^\Delta$, and we may analyze either (6.8) or (6.9), respectively, exactly as we did above.

We now wish to write the two exact formulae (6.8) and (6.9) in such a form that it is clear that the limit $N \to \infty$ yields an asymptotic formula that is independent of whether $J \subset \Sigma_0^\nabla$ or $J \subset \Sigma_0^\Delta$. In fact, a direct calculation using (4.38) and (4.39) along with the quantization condition (4.1) shows that

$$(-1)^{M_I^\nabla} e^{\mp iN\theta_I^\nabla(z)/2} = (-1)^{M_I^\Delta} e^{\pm iN[\theta^0(z)-\theta_I^\Delta(z)]/2} = \exp\left(\pm i\pi Nc\left[\mu_{\min}^c([x,b]) - \int_x^z \psi_I(s)\,ds\right]\right),$$

where x is any point or endpoint of the band I.

Therefore, in considering the limit $N \to \infty$, it remains to recall that Proposition 5.3 implies that $\dot{X}_{11+}(z)$ does not vanish at any point of I and that $W(z) = \dot{X}_{11}(z)e^{\kappa g(z)}$. This completes the proof.

6.4.2 Asymptotic behavior of the zeros: the proof of Theorem 2.14

Theorem 2.14 is a consequence of the estimate (2.57) established in Theorem 2.13, the strict inequalities $0 < d\mu_{\min}^c/dx < \rho^0(x)/c$ holding by definition for $J \subset I$ because I is a band, and from the strict inequality $A_I(x) > 0$ for $x \in J \subset \mathbb{R}$ stated in Theorem 2.13.

6.5 ASYMPTOTIC BEHAVIOR OF $\pi_{N,k}(z)$ FOR z NEAR A BAND EDGE

6.5.1 Band/void edges: the proof of Theorem 2.15

First consider a left band endpoint $z = \alpha$ between a band I (on the right) and a void Γ (on the left). We take the contour parameter ϵ sufficiently small that Proposition 5.15 controls $\mathbf{E}(z) - \mathbb{I}$ and then choose $r > 0$ sufficiently small that the disc $|z - \alpha| \leq r$ is contained in the disc $D_\Gamma^{\nabla,L}$. For such z, we thus have the following exact formula for $\pi_{N,k}(z)$:

$$\pi_{N,k}(z) = -\sqrt{2\pi}e^{(\eta(z)-\gamma)/2}e^{N\overline{L}_c^I(z)}$$

$$\cdot T_\nabla(z)^{-1/2}\left[\exp\left(-N\int_{\Sigma_0^\Delta}\log(z-x)\rho^0(x)\,dx\right)\prod_{n\in\Delta}(z-x_{N,n})\right]$$

$$\cdot \left[\left(\frac{3}{4}\right)^{1/6}\left(E_{11}(z)H_{\Gamma,12}^{\nabla,L}(z) + E_{12}(z)H_{\Gamma,22}^{\nabla,L}(z)\right)\operatorname{Ai}\left(-\left(\frac{3}{4}\right)^{2/3}\tau_\Gamma^{\nabla,L}(z)\right)\right.$$

$$\left.+ \left(\frac{3}{4}\right)^{-1/6}\left(E_{11}(z)H_{\Gamma,11}^{\nabla,L}(z) + E_{12}(z)H_{\Gamma,21}^{\nabla,L}(z)\right)\operatorname{Ai}'\left(-\left(\frac{3}{4}\right)^{2/3}\tau_\Gamma^{\nabla,L}(z)\right)\right]. \quad (6.10)$$

This follows from (4.45), (4.47), (4.49), (4.50), (5.14), (5.26), (6.1), and Proposition 1.3.

Recall from §5.1.2 that $\mathbf{H}_\Gamma^{\nabla,L}(z)$ is analytic throughout $D_\Gamma^{\nabla,L}$. From the definition of this function in terms of $\tau_\Gamma^{\nabla,L}(z)$ and $\dot{\mathbf{X}}(z)$, it follows that the first column (second column) of $\mathbf{H}_\Gamma^{\nabla,L}(z)$ is uniformly bounded in $D_\Gamma^{\nabla,L}$ by a quantity of order $N^{-1/6}$ (of order $N^{1/6}$). Also, Proposition 5.3 implies that the matrix elements of $\mathbf{H}_\Gamma^{\nabla,L}(z)$ are real for real z. We may now use an argument based on the midpoint rule for Riemann sums to approximate the terms in square brackets on the second line of (6.10), recall Proposition 4.3 to handle $T_\Delta(z)$, and use Proposition 5.15 to estimate $\mathbf{E}(z) - \mathbb{I}$.

Finally, we may observe from §5.1.2 the relations

$$H_{\Gamma,12}^{\nabla,L}(z) = -H_\Gamma^-(z)(-\tau_\Gamma^{\nabla,L}(z))^{1/4},$$

$$H_{\Gamma,11}^{\nabla,L}(z) = H_\Gamma^+(z)(-\tau_\Gamma^{\nabla,L}(z))^{-1/4},$$

where we have used the identities $W(z) = \dot{X}_{11}(z)e^{\kappa g(z)}$ and $Z(z) = \dot{X}_{12}(z)e^{-\kappa g(z)}$ and the functions $H_\Gamma^\pm(z)$ are defined by (2.47). This completes the proof of the asymptotic formula (2.58) and the corresponding error estimates.

Since $\tau_\Gamma^{\nabla,L}(z)$ is uniformly bounded independently of N for z in shrinking neighborhoods of the band edge $z = \alpha$ with radius of order $N^{-2/3}$, we immediately obtain the asymptotic formula (2.59) and the corresponding error estimate.

Next consider a right band endpoint $z = \beta$ between a band I (on the left) and a void Γ (on the right). Again taking ϵ small enough that Proposition 5.15 controls $\mathbf{E}(z) - \mathbb{I}$, we take the parameter r small enough

DISCRETE ORTHOGONAL POLYNOMIALS: PROOFS OF THEOREMS STATED IN §2.3   113

that the disc $|z-\beta| \leq r$ is contained in the disc $D_\Gamma^{\nabla,R}$. In this case, we have the exact formula:

$$\pi_{N,k}(z) = -i\sqrt{2\pi}e^{(\eta(z)-\gamma)/2}e^{N\overline{L}_c^I(z)}$$
$$\cdot T_\nabla(z)^{-1/2}\left[\exp\left(-N\int_{\Sigma_0^\Delta}\log(z-x)\rho^0(x)\,dx\right)\prod_{n\in\Delta}(z-x_{N,n})\right]$$
$$\cdot \left[\left(\frac{3}{4}\right)^{1/6}\left(E_{11}(z)H_{\Gamma,12}^{\nabla,R}(z)+E_{12}(z)H_{\Gamma,22}^{\nabla,R}(z)\right)\operatorname{Ai}\left(-\left(\frac{3}{4}\right)^{2/3}\tau_\Gamma^{\nabla,R}(z)\right)\right.$$
$$\left.+\left(\frac{3}{4}\right)^{-1/6}\left(E_{11}(z)H_{\Gamma,11}^{\nabla,R}(z)+E_{12}(z)H_{\Gamma,21}^{\nabla,R}(z)\right)\operatorname{Ai}'\left(-\left(\frac{3}{4}\right)^{2/3}\tau_\Gamma^{\nabla,R}(z)\right)\right].$$
(6.11)

This follows from (4.45), (4.47), (4.49), (4.50), (5.17), (5.27), (6.1), and Proposition 1.3. Once again, we see that the second line in (6.11) may be replaced by $1+O(1/N)$ uniformly for $|z-\beta| \leq r$. Since the first column of $\mathbf{H}_\Gamma^{\nabla,R}(z)$ is uniformly of order $N^{-1/6}$ and the second column of $\mathbf{H}_\Gamma^{\nabla,R}(z)$ is uniformly of order $N^{1/6}$, and since we have the exact representations (from §5.1.2)

$$H_{\Gamma,12}^{\nabla,R}(z) = iH_\Gamma^+(z)(-\tau_\Gamma^{\nabla,R}(z))^{1/4},$$
$$H_{\Gamma,11}^{\nabla,R}(z) = -iH_\Gamma^-(z)(-\tau_\Gamma^{\nabla,R}(z))^{-1/4},$$

we immediately obtain the asymptotic formula (2.60) and the corresponding error estimates with the use of Proposition 5.15. The asymptotic formula (2.61) and its error estimate then follow exactly as before since $\tau_\Gamma^{\nabla,R}(z)$ remains uniformly bounded as $N \to \infty$ if $|z-\beta| \leq rN^{-2/3}$. Note that in this case the matrix elements of $\mathbf{H}_\Gamma^{\nabla,R}$ are imaginary for real z.

6.5.2 Band/saturated region edges: the proof of Theorem 2.16

First consider the neighborhood of a left band edge $z = \alpha$ separating a band I (for $z > \alpha$) from a saturated region Γ (for $z < \alpha$). We choose the contour parameter ϵ sufficiently small that Proposition 5.15 controls the matrix $\mathbf{E}(z) - \mathbb{I}$. Then we choose $r > 0$ small enough that the disc $|z-\alpha| \leq r$ is contained within the disc $D_\Gamma^{\Delta,L}$. In this case, we have the following exact formula for $\pi_{N,k}(z)$:

$$\pi_{N,k}(z) = i\sqrt{2\pi}e^{(\eta(z)-\gamma)/2}e^{N\overline{L}_c^I(z)}$$
$$\cdot T_\Delta(z)^{-1/2}\left[\exp\left(-N\int_{\Sigma_0^\nabla}\log(z-x)\rho^0(x)\,dx\right)\prod_{n\in\nabla}(z-x_{N,n})\right]$$
$$\cdot \left[\left(\frac{3}{4}\right)^{1/6}\left(E_{11}(z)H_{\Gamma,12}^{\Delta,L}(z)+E_{12}(z)H_{\Gamma,22}^{\Delta,L}(z)\right)F_A^L(z)\right.$$
$$\left.+\left(\frac{3}{4}\right)^{-1/6}\left(E_{11}(z)H_{\Gamma,11}^{\Delta,L}(z)+E_{12}(z)H_{\Gamma,21}^{\Delta,L}(z)\right)F_B^L(z)\right],$$
(6.12)

where $F_A^L(z)$ and $F_B^L(z)$ are the combinations of trigonometric functions and Airy functions and their derivatives defined by (2.63). This formula follows from (4.45), (4.48), (4.51), (4.52), (5.20), (5.28), (6.1), and Proposition 1.3. The terms on the second line of the right-hand side in (6.12) are $1+O(1/N)$ as $N \to \infty$ uniformly for $|z-\alpha| \leq r$, as can be seen from a midpoint-rule approximation of the integral and by using Proposition 4.3. Proposition 5.15 is then used to control $\mathbf{E}(z) - \mathbb{I}$. Noting that the second column of $\mathbf{H}_\Gamma^{\Delta,L}(z)$ is uniformly of order $N^{1/6}$ while the first column of $\mathbf{H}_\Gamma^{\Delta,L}(z)$ is uniformly of order $N^{-1/6}$, and moreover recalling from §5.1.2 the explicit formulae

$$H_{\Gamma,12}^{\Delta,L}(z) = -iH_\Gamma^-(z)\left(-\tau_\Gamma^{\Delta,L}(z)\right)^{1/4},$$
$$H_{\Gamma,11}^{\Delta,L}(z) = -iH_\Gamma^+(z)\left(-\tau_\Gamma^{\Delta,L}(z)\right)^{-1/4},$$

which also rely on the identities $W(z) = \dot{X}_{11}(z)e^{\kappa g(z)}$ and $Z(z) = \dot{X}_{12}(z)e^{-\kappa g(z)}$, we obtain the asymptotic formula (2.62) along with the corresponding error estimates. The asymptotic formula (2.64) then follows along with its error estimate by noting that $\tau_\Gamma^{\Delta,L}(z)$ remains uniformly bounded as $N \to \infty$ if $|z - \alpha| \leq rN^{-2/3}$.

Next consider the neighborhood of a right band edge $z = \beta$ separating a band I (for $z < \beta$) from a saturated region Γ (for $z > \beta$). Again take the contour parameter ϵ sufficiently small that Proposition 5.15 provides a uniform estimate of $\mathbf{E}(z) - \mathbb{I}$ on appropriate closed sets and then choose $r > 0$ small enough that the disc $|z - \beta| \leq r$ is contained within the disc $D_\Gamma^{\Delta,R}$. Then we have for z, with $|z - \beta| \leq r$, the exact formula

$$\pi_{N,k}(z) = \sqrt{2\pi}e^{(\eta(z)-\gamma)/2}e^{N\overline{L}_c^I(z)}$$

$$\cdot T_\Delta(z)^{-1/2} \left[\exp\left(-N \int_{\Sigma_0^\nabla} \log(z-x)\rho^0(x)\,dx\right) \prod_{n \in \nabla}(z - x_{N,n}) \right]$$

$$\cdot \left[\left(\frac{3}{4}\right)^{1/6} \left(E_{11}(z)H_{\Gamma,12}^{\Delta,R}(z) + E_{12}(z)H_{\Gamma,22}^{\Delta,R}(z)\right) F_A^R(z) \right. \quad (6.13)$$

$$\left. + \left(\frac{3}{4}\right)^{-1/6} \left(E_{11}(z)H_{\Gamma,11}^{\Delta,R}(z) + E_{12}(z)H_{\Gamma,21}^{\Delta,R}(z)\right) F_B^R(z) \right],$$

where $F_A^R(z)$ and $F_B^R(z)$ are the expressions defined by (2.66). This formula follows from (4.45), (4.48), (4.51), (4.52), (5.23), (5.29), (6.1), and Proposition 1.3. Once again, the terms on the second line of the right-hand side of (6.13) can be approximated uniformly for $|z - \beta| \leq r$ as $1 + O(1/N)$ as $N \to \infty$. Proposition 5.15 again guarantees that, uniformly for $|z - \beta| \leq r$, we have $\mathbf{E}(z) - \mathbb{I} = O(1/N)$, and then noting that $\mathbf{H}_\Gamma^{\Delta,R}(z)N^{\sigma_3/6}$ remains uniformly bounded as $N \to \infty$ and more specifically that

$$H_{\Gamma,12}^{\Delta,R}(z) = H_\Gamma^-(z)\left(-\tau_\Gamma^{\Delta,R}(z)\right)^{1/4},$$

$$H_{\Gamma,11}^{\Delta,R}(z) = -H_\Gamma^+(z)\left(-\tau_\Gamma^{\Delta,R}(z)\right)^{-1/4},$$

we complete the proof of the asymptotic formula (2.65) and its corresponding error estimates. Since $\tau_\Gamma^{\Delta,R}(z)$ is uniformly bounded as $N \to \infty$ with $|z - \beta| \leq rN^{-2/3}$, we then obtain immediately the asymptotic formula (2.67) and its corresponding error estimate.

Chapter Seven

Universality: Proofs of Theorems Stated in §3.3

7.1 RELATION BETWEEN CORRELATION FUNCTIONS OF DUAL ENSEMBLES

Since the holes are also governed by a discrete orthogonal polynomial ensemble, the correlation functions for holes are again represented as determinants involving the reproducing kernel, this time corresponding to the dual weights. It turns out that there is a simple relation between the correlation functions for particles and those for holes.

7.1.1 Probabilistic approach

Let $\overline{R}_m^{(N,\bar{k})}$ be the m-point correlation function of the dual orthogonal polynomial ensemble for the holes. Hence $\overline{R}_m^{(N,\bar{k})}$ is defined as in (3.2), with the replacement of $p^{(N,k)}$ by $\overline{p}^{(N,\bar{k})}$. Let $\overline{K}_{N,\bar{k}}$ denote the reproducing kernel of the dual ensemble. Then (3.3) implies that

$$\overline{R}_m^{(N,\bar{k})}(x_1,\ldots,x_m) = \det\bigl(\overline{K}_{N,\bar{k}}(x_i,x_j)\bigr)_{1\leq i,j\leq m},$$

for nodes $x_1,\ldots,x_m$. Now, given nodes $x_1,\ldots,x_m$,

$\mathbb{P}(\text{there are particles at each of the nodes } x_1,\ldots,x_m)$
$$= \mathbb{P}(\text{there are no holes at any of the nodes } x_1,\ldots,x_m)$$
$$= 1 - \sum_{i=1}^m \mathbb{P}(\text{there is a hole at the node } x_i)$$
$$+ \sum_{1\leq i<j\leq m} \mathbb{P}(\text{there are holes at both of the nodes } x_i \text{ and } x_j)$$
$$- \sum_{1\leq i<j<k\leq m} \mathbb{P}(\text{there are holes at each of the nodes } x_i, x_j, x_k) + \cdots.$$

Thus from (3.2),

$$R_m^{(N,k)}(x_1,\ldots,x_m) = 1 - \sum_{i=1}^m \overline{R}_1^{(N,\bar{k})}(x_i) + \sum_{1\leq i<j\leq m} \overline{R}_2^{(N,\bar{k})}(x_i,x_j) - \sum_{1\leq i<j<k\leq m} \overline{R}_3^{(N,\bar{k})}(x_i,x_j,x_k) + \cdots \quad (7.1)$$

Therefore the determinantal formula (3.3) for the correlation functions implies the following.

Proposition 7.1. *Let $K_{N,k}$ be the reproducing kernel (3.4) for the discrete orthogonal polynomial ensemble and let $\overline{K}_{N,\bar{k}}$ be the reproducing kernel of the corresponding dual orthogonal polynomial ensemble. Then with $\bar{k} = N - k$,*

$$\det\bigl(K_{N,k}(x_i,x_j)\bigr)_{1\leq i,j\leq m} = \det\bigl(\delta_{ij} - \overline{K}_{N,\bar{k}}(x_i,x_j)\bigr)_{1\leq i,j\leq m}.$$

In particular, when $m = 1$, this result implies that for a node $x \in X_N$,

$$K_{N,k}(x,x) = 1 - \overline{K}_{N,\bar{k}}(x,x),$$

and then when $m = 2$, we further discover that for nodes $x \neq y$,

$$K_{N,k}(x,y)^2 = \overline{K}_{N,\bar{k}}(x,y)^2. \quad (7.2)$$

7.1.2 Direct approach

It is possible to establish these same results, and also to refine (7.2) by determining the relative signs of $K_{N,k}(x,y)$ and $\overline{K}_{N,\bar{k}}(x,y)$, by using Proposition 1.3 regarding the solution formula for Interpolation Problem 1.2 and the dual relation

$$\overline{\mathbf{P}}(z;N,\bar{k}) = \sigma_1 \mathbf{P}(z;N,k) \prod_{n=0}^{N-1} (z - x_{N,n})^{-\sigma_3} \sigma_1, \qquad \bar{k} = N - k. \tag{7.3}$$

Here $\mathbf{P}(z;N,k)$ is the solution of Interpolation Problem 1.2 with weights $\{w_{N,j}\}$ on the nodes X_N, and $\overline{\mathbf{P}}(z;N,\bar{k})$ is the solution of Interpolation Problem 1.2 with the dual weights $\{\overline{w}_{N,j}\}$ defined by (1.46) and with the exponent k in the normalization condition replaced by $\bar{k}$. Note that (7.3) implies in particular that if z and w are *not* nodes ($z, w \notin X_N$), then

$$\left(\overline{\mathbf{P}}(z;N,\bar{k})^{-1}\overline{\mathbf{P}}(w;N,\bar{k})\right)_{21} = \left(\mathbf{P}(z;N,k)^{-1}\mathbf{P}(w;N,k)\right)_{12} \prod_{n=0}^{N-1} (z - x_{N,n})(w - x_{N,n}). \tag{7.4}$$

Suppose first that $n \neq m$ are distinct indices. Then

$$\overline{K}_{N,\bar{k}}(x_{N,m}, x_{N,n}) = \frac{\sqrt{\overline{w}_{N,m}\overline{w}_{N,n}}}{x_{N,m} - x_{N,n}} \left(\overline{\mathbf{P}}(x_{N,m};N,\bar{k})^{-1}\overline{\mathbf{P}}(x_{N,n};N,\bar{k})\right)_{21}$$

$$= \frac{\sqrt{\overline{w}_{N,m}\overline{w}_{N,n}}}{x_{N,m} - x_{N,n}} \lim_{\substack{w \to x_{N,m} \\ z \to x_{N,n}}} \left(\overline{\mathbf{P}}(w;N,\bar{k})^{-1}\overline{\mathbf{P}}(z;N,\bar{k})\right)_{21}$$

$$= \frac{\sqrt{\overline{w}_{N,m}\overline{w}_{N,n}}}{x_{N,m} - x_{N,n}}$$

$$\cdot \lim_{\substack{w \to x_{N,m} \\ z \to x_{N,n}}} \left[\prod_{j=0}^{N-1} (w - x_{N,j})(z - x_{N,j}) \cdot \left(\mathbf{P}(w;N,k)^{-1}\mathbf{P}(z;N,k)\right)_{12} \right],$$

where in going from the second to the third line we use (7.4). The limiting operation is necessary because while $\left(\overline{\mathbf{P}}(w;N,\bar{k})^{-1}\overline{\mathbf{P}}(z;N,\bar{k})\right)_{21}$ is analytic in w and z near $w = x_{N,m}$ and $z = x_{N,n}$, $\left(\mathbf{P}(w;N,k)^{-1}\mathbf{P}(z;N,k)\right)_{12}$ has singularities at these points. Next, using the definition (1.46) of the dual weights, we obtain

$$\overline{K}_{N,\bar{k}}(x_{N,m}, x_{N,n}) = \frac{(-1)^{m+n}}{\sqrt{w_{N,m}w_{N,n}}} \cdot \frac{\lim_{\substack{w \to x_{N,m} \\ z \to x_{N,n}}} \left[(w - x_{N,m})(z - x_{N,n}) \left(\mathbf{P}(w;N,k)^{-1}\mathbf{P}(z;N,k)\right)_{12}\right]}{x_{N,m} - x_{N,n}}$$

$$= \frac{(-1)^{m+n}}{\sqrt{w_{N,m}w_{N,n}}} \cdot \frac{\left(\operatorname*{Res}_{w=x_{N,m}} \mathbf{P}(w;N,k)^{-1} \operatorname*{Res}_{z=x_{N,n}} \mathbf{P}(z;N,k)\right)_{12}}{x_{N,m} - x_{N,n}},$$

where we use the fact that $\det \mathbf{P}(z;N,k) = 1$, which implies that $\mathbf{P}(z;N,k)^{-1}$ has simple poles at the nodes just like $\mathbf{P}(z;N,k)$ does. Now again because $\det \mathbf{P}(z;N,k) = 1$, we obtain from (1.21) that

$$\operatorname*{Res}_{w=x_{N,m}} \mathbf{P}(w;N,k)^{-1} = \lim_{w \to x_{N,m}} \begin{pmatrix} 0 & -w_{N,m} \\ 0 & 0 \end{pmatrix} \mathbf{P}(w;N,k)^{-1}.$$

Using this equation together with (1.21), we arrive at

$$\overline{K}_{N,\bar{k}}(x_{N,m}, x_{N,n}) = \frac{(-1)^{m+n}}{\sqrt{w_{N,m}w_{N,n}}} \cdot \frac{-w_{N,m}w_{N,n} \left(\mathbf{P}(x_{N,m};N,k)^{-1}\mathbf{P}(x_{N,n};N,k)\right)_{21}}{x_{N,m} - x_{N,n}}$$

$$= (-1)^{m+n+1} \sqrt{w_{N,m}w_{N,n}} \frac{\left(\mathbf{P}(x_{N,m};N,k)^{-1}\mathbf{P}(x_{N,n};N,k)\right)_{21}}{x_{N,m} - x_{N,n}}$$

$$= (-1)^{m+n+1} K_{N,k}(x_{N,m}, x_{N,n}).$$

Thus we have proved the following, a more specific version of (7.2).

UNIVERSALITY: PROOFS OF THEOREMS STATED IN §3.3

Proposition 7.2. *For distinct nodes $x = x_{N,m}$ and $y = x_{N,n}$ in X_N,*
$$\overline{K}_{N,\bar{k}}(x,y) = (-1)^{m+n+1} K_{N,k}(x,y),$$
where $\bar{k} = N - k$.

Now we consider the reproducing kernel and its dual on the diagonal. We begin with
$$\overline{K}_{N,\bar{k}}(x_{N,m}, x_{N,m}) = \overline{w}_{N,m} \left(\frac{d}{dz} \overline{\mathbf{P}}(z; N, \bar{k})^{-1} \bigg|_{z=x_{N,m}} \overline{\mathbf{P}}(x_{N,m}; N; \bar{k}) \right)_{21}$$
$$= -\overline{w}_{N,m} \left(\overline{\mathbf{P}}(x_{N,m}; N, \bar{k})^{-1} \frac{d}{dz} \overline{\mathbf{P}}(z; N, \bar{k}) \bigg|_{z=x_{N,m}} \right)_{21}.$$

But, using (7.3), we see that
$$\overline{\mathbf{P}}(z; N, \bar{k})^{-1} \frac{d}{dz} \overline{\mathbf{P}}(z; N, \bar{k}) = \sigma_1 \prod_{j=0}^{N-1}(z - x_{N,j})^{\sigma_3} \cdot \left[\mathbf{P}(z; N, k)^{-1} \frac{d}{dz} \mathbf{P}(z; N, k) \right] \cdot \prod_{j=0}^{N-1}(z - x_{N,j})^{-\sigma_3} \sigma_1$$
$$+ \sigma_1 \prod_{j=1}^{N-1}(z - x_{N,j})^{\sigma_3} \cdot \frac{d}{dz}\left[\prod_{j=0}^{N-1}(z - x_{N,j})^{-\sigma_3} \right] \sigma_1,$$

and the second term is a diagonal matrix. Consequently,
$$\overline{K}_{N,\bar{k}}(x_{N,m}, x_{N,m}) = -\overline{w}_{N,m} \lim_{z \to x_{N,m}} \left[\prod_{j=0}^{N-1}(z - x_{N,m})^2 \cdot \left(\mathbf{P}(z; N, k)^{-1} \frac{d}{dz} \mathbf{P}(z; N, k) \right)_{12} \right].$$

From Proposition 1.3, we then have
$$\left(\mathbf{P}(z; N, k)^{-1} \frac{d}{dz} \mathbf{P}(z; N, k) \right)_{12}$$
$$= \sum_{n=0}^{N-1} \sum_{j=0}^{N-1} \frac{P_{11}(x_{N,n}; N, k)P_{21}(x_{N,j}; N, k) - P_{11}(x_{N,j}; N, k)P_{21}(x_{N,n}; N, k)}{(z - x_{N,n})(z - x_{N,j})^2} w_{N,n} w_{N,j}$$
$$= \sum_{n \neq j} \frac{P_{11}(x_{N,n}; N, k)P_{21}(x_{N,j}; N, k) - P_{11}(x_{N,j}; N, k)P_{21}(x_{N,n}; N, k)}{(z - x_{N,n})(z - x_{N,j})^2} w_{N,n} w_{N,j}.$$

Therefore
$$\overline{K}_{N,\bar{k}}(x_{N,m}, x_{N,m}) = -\overline{w}_{N,m} w_{N,m} \prod_{\substack{j=0 \\ j \neq m}}^{N-1}(x_{N,m} - x_{N,j})^2$$
$$\cdot \sum_{\substack{n=0 \\ n \neq m}}^{N-1} \frac{P_{11}(x_{N,n}; N, k)P_{21}(x_{N,m}; N, k) - P_{11}(x_{N,m}; N, k)P_{21}(x_{N,n}; N, k)}{x_{N,m} - x_{N,n}} w_{N,n},$$

and with the use of (1.46), this becomes
$$\overline{K}_{N,\bar{k}}(x_{N,m}, x_{N,m}) = -\sum_{\substack{n=0 \\ n \neq m}}^{N-1} \frac{P_{11}(x_{N,n}; N, k)P_{21}(x_{N,m}; N, k) - P_{11}(x_{N,m}; N, k)P_{21}(x_{N,n}; N, k)}{x_{N,m} - x_{N,n}} w_{N,n}.$$

Now for $z \in \mathbb{C} \setminus X_N$, we have $\det \mathbf{P}(z; N, k) = 1$, and taking the limit $z \to x_{N,m}$ with the use of the explicit formula for $\mathbf{P}(z; N, k)$ furnished by Proposition 1.3 yields the identity

$$w_{N,m} \left[P_{21}(x_{N,m}; N, k) \frac{d}{dz} P_{11}(z; N, k) \bigg|_{z=x_{N,m}} - P_{11}(x_{N,m}; N, k) \frac{d}{dz} P_{21}(z; N, k) \bigg|_{z=x_{N,m}} \right]$$
$$+ \sum_{\substack{n=0 \\ n \neq m}}^{N-1} \frac{P_{11}(x_{N,m}; N, k) P_{21}(x_{N,n}; N, k) - P_{11}(x_{N,n}; N, k) P_{21}(x_{N,m}; N, k)}{x_{N,m} - x_{N,n}} w_{N,n} = 1.$$

So we have (again using $\det \mathbf{P}(z; N, k) = 1$)

$$\overline{K}_{N,\bar{k}}(x_{N,m}, x_{N,m}) = 1 - w_{N,m} \left(\frac{d}{dz} \mathbf{P}(z; N, k)^{-1} \bigg|_{z=x_{N,m}} \mathbf{P}(x_{N,m}; N, k) \right)_{21} = 1 - K_{N,k}(x_{N,m}, x_{N,m}),$$

which completes the direct proof of the following.

Proposition 7.3. *For any node $x \in X_N$,*

$$\overline{K}_{N,\bar{k}}(x, x) = 1 - K_{N,k}(x, x),$$

where $\bar{k} = N - k$.

Combining Propositions 7.2 and 7.3, we therefore may write, for any given set of nodes $x_1, \ldots, x_m$,

$$\left(\overline{K}_{N,\bar{k}}(x_i, x_j) \right)_{1 \leq i,j \leq m} = \mathbf{D} \left(\delta_{ij} - K_{N,k}(x_i, x_j) \right)_{1 \leq i,j \leq m} \mathbf{D},$$

where $\mathbf{D} := \mathrm{diag}(1, -1, 1, -1, \ldots, (-1)^{m+1})$. Taking determinants then yields another independent proof of Proposition 7.1.

◁ **Remark:** The dual ensemble is useful for several reasons. Of course, the statistics of holes are often of independent interest. But even if one is interested only in particle statistics, the dual ensemble is very helpful in the analysis of statistics near saturated regions of the node space X_N where the upper constraint is active for the particle weights. It follows from Proposition 2.6 that each saturated region for the particle weights with k particles is a void for the (dual) hole weights with $\bar{k} = N - k$ holes. In this way, each calculation valid for the hole ensemble near a void automatically translates via Propositions 7.2 and 7.3 into a statement about particle statistics near saturated regions. ▷

7.2 EXACT FORMULAE FOR $K_{N,k}(x, y)$

The following result will be used often below to obtain formulae for $K_{N,k}(x, y)$ and $K_{N,k}(x, x)$ in various regions of $[a, b]$.

Lemma 7.4. *Let x be any node satisfying $x \in X_N \cap \Sigma_0^{\nabla}$. Then*

$$\sqrt{w(x)} e^{(N\ell_c + \gamma)/2} e^{(k - \#\Delta) g_+(x)} \prod_{n \in \Delta} |x - x_{N,n}|$$
$$= e^{(-\eta(x) + \gamma + 2\kappa g_+(x))/2} e^{-iN\theta(x)/2} \frac{e^{-\frac{1}{2} N \left[\frac{\delta E_c}{\delta \mu}(x) - \ell_c \right]}}{\sqrt{2\pi N \rho^0(x)}} T_{\nabla}(x)^{1/2}. \quad (7.5)$$

Here the variational derivative is evaluated on the equilibrium measure $\mu_{\min}^c$, and $g_+(x)$ denotes the boundary value taken by $g(z)$ as $z \to x$ with $\Im(z) > 0$.

UNIVERSALITY: PROOFS OF THEOREMS STATED IN §3.3

Proof. Let $x = x_{N,j} \in X_N \cap \Sigma_0^\nabla$. Hence $j \in \nabla$. Substituting for $w_N(\cdot)$ from (1.9) and (1.16) and using the fact that $x = x_{N,j} \in X_N$, we have

$$w(x)e^{N\ell_c+\gamma}e^{2(k-\#\Delta)g_+(x)}\prod_{n\in\Delta}(x-x_{N,n})^2$$
$$= (-1)^{N-1-j}e^{-NV(x)-\eta(x))+N\ell_c+\gamma+2(k-\#\Delta)g_+(x)}\frac{\prod_{n\in\Delta}(x_{N,j}-x_{N,n})}{\prod_{\substack{n\in\nabla \\ n\neq j}}(x_{N,j}-x_{N,n})}.$$

But, using (4.10), we have

$$\frac{\prod_{n\in\Delta}(x_{N,j}-x_{N,n})}{\prod_{\substack{n\in\nabla \\ n\neq j}}(x_{N,j}-x_{N,n})}$$

$$= \lim_{z\to x_{N,j}}(z-x_{N,j})\frac{\prod_{n\in\Delta}(z-x_{N,n})}{\prod_{n\in\nabla}(z-x_{N,n})}$$

$$= \lim_{z\to x_{N,j}}\frac{z-x_{N,j}}{2\cos\left(\frac{N\theta^0(z)}{2}\right)}T_\nabla(z)\exp\left(-N\left[\int_{\Sigma_0^\nabla}\log|z-s|\rho^0(s)\,ds - \int_{\Sigma_0^\Delta}\log|z-s|\rho^0(s)\,ds\right]\right).$$

The limit of the fraction can be found using l'Hôpital's rule, and the remaining factors are continuous for real z. Thus we arrive at

$$\frac{\prod_{n\in\Delta}(x_{N,j}-x_{N,n})}{\prod_{\substack{n\in\nabla \\ n\neq j}}(x_{N,j}-x_{N,n})} = \frac{T_\nabla(x_{N,j})\exp\left(-N\left[\int_{\Sigma_0^\nabla}\log|x_{N,j}-s|\rho^0(s)\,ds - \int_{\Sigma_0^\Delta}\log|x_{N,j}-s|\rho^0(s)\,ds\right]\right)}{2\pi N\rho^0(x_{N,j})\sin\left(\frac{N\theta^0(x_{N,j})}{2}\right)}.$$

From the definition of ρ^0 and (1.15),

$$\theta^0(x_{N,j}) = \pi\frac{2N-2j-1}{N}, \quad \text{so } \sin\left(\frac{N\theta^0(x_{N,j})}{2}\right) = (-1)^{N-j-1}.$$

Therefore, recalling the definition (4.7) and (4.5) of the complex phase function $g(z)$, the definition (2.9) of the variational derivative of the energy functional $E_c[\cdot]$, and the definition of $\theta(z)$ (4.9), we obtain an identity that is the square of (7.5). By directly comparing the arguments of both sides of (7.5), one verifies that the square root has been taken consistently. □

The following elementary lemma will be useful.

Lemma 7.5. *Let $f(x)$ and $M(x,y)$ be differentiable functions with $M(x,x) \equiv 0$. Then*

$$\frac{\partial}{\partial x}[f(x)f(y)M(x,y)]_{y=x} = f(x)^2\frac{\partial}{\partial x}M(x,y)\bigg|_{y=x}. \tag{7.6}$$

We will now use these results to express $K_{N,k}(x,y)$ in terms of the piecewise-analytic global parametrix $\hat{\mathbf{X}}(z)$ and the error matrix $\mathbf{E}(z)$, for x and y in different parts of the interval $[a,b]$ of accumulation of the nodes. The first result in this direction is the following.

Proposition 7.6. *Let x and y be distinct nodes in a band I, both lying in the same component of Σ_0^∇ and lying outside all discs $D_\Gamma^{\nabla,*}$. Then*

$$K_{N,k}(x,y) = \frac{1}{2\pi N \sqrt{\rho^0(x)\rho^0(y)}} \frac{\mathbf{v}^T e^{iN\theta(x)\sigma_3/2} \mathbf{B}(x)^{-1} \mathbf{B}(y) e^{-iN\theta(y)\sigma_3/2} \mathbf{w}}{x-y} \qquad (7.7)$$

and

$$K_{N,k}(x,x) = \frac{1}{2\pi N \rho^0(x)} \left[2\pi Nc \frac{d\mu_{\min}^c}{dx}(x) - \mathbf{v}^T e^{iN\theta(x)\sigma_3/2} \mathbf{B}(x)^{-1} \mathbf{B}'(x) e^{-iN\theta(x)\sigma_3/2} \mathbf{w} \right], \qquad (7.8)$$

where

$$\mathbf{v} := \begin{pmatrix} -i \\ 1 \end{pmatrix}, \qquad \mathbf{w} := \begin{pmatrix} 1 \\ i \end{pmatrix}$$

(note that $\mathbf{v}^T \mathbf{w} = 0$), and

$$\mathbf{B}(x) := \mathbf{E}_+(x) \dot{\mathbf{X}}_+(x) e^{(\kappa g_+(x) + \gamma/2 - \eta(x)/2)\sigma_3}. \qquad (7.9)$$

Here the subscript $+$ denotes the boundary value taken as $z \to x$ with $\Im(z) > 0$.

Proof. For distinct nodes x and y, we begin with

$$K_{N,k}(x,y) = \sqrt{w(x)w(y)} \frac{\left(\mathbf{P}(x;N,k)^{-1}\mathbf{P}(y;N,k)\right)_{21}}{x-y}$$

and define the quotient by l'Hôpital's rule:

$$K_{N,k}(x,x) = w(x) \frac{\partial}{\partial x} \left(\mathbf{P}(x;N,k)^{-1}\mathbf{P}(y;N,k)\right)_{21} \bigg|_{y=x}.$$

Now, for any real $x \in \Sigma_0^\nabla$, we have from (1.43) and (4.4) that

$$\mathbf{P}(x;N,k) = \mathbf{R}_+(x) \begin{pmatrix} 1 & ie^{-iN\theta^0(x)/2} e^{-NV_N(x)} \frac{\prod_{n \in \Delta}(x - x_{N,n})}{\prod_{n \in \nabla}(x - x_{N,n})} \\ 0 & 1 \end{pmatrix} \left[\prod_{n \in \Delta}(x - x_{N,n})\right]^{\sigma_3},$$

where $\mathbf{R}_+(x)$ denotes the boundary value taken from the upper half-plane (from the left-hand side of the contour Σ; see Figure 4.2). Thus

$$\left(\mathbf{P}(x;N,k)^{-1}\mathbf{P}(y;N,k)\right)_{21} = \left(\mathbf{R}_+(x)^{-1}\mathbf{R}_+(y)\right)_{21} \prod_{n \in \Delta}(x - x_{N,n})(y - x_{N,n})$$

$$= \left(\mathbf{S}_+(x)^{-1}\mathbf{S}_+(y)\right)_{21} e^{N\ell_c + \gamma} e^{(k-\#\Delta)(g_+(x)+g_+(y))} \prod_{n \in \Delta}(x - x_{N,n})(y - x_{N,n}),$$

where the second equality follows from (4.8). When we further suppose that x and y lie within the same component of Σ_0^∇, this formula may be rewritten as

$$\left(\mathbf{P}(x;N,k)^{-1}\mathbf{P}(y;N,k)\right)_{21} = \left(\mathbf{S}_+(x)^{-1}\mathbf{S}_+(y)\right)_{21} e^{N\ell_c + \gamma} e^{(k-\#\Delta)(g_+(x)+g_+(y))} \prod_{n \in \Delta}|x - x_{N,n}||y - x_{N,n}|. \qquad (7.10)$$

Letting x and y lie in a band $I \subset \Sigma_0^\nabla$, we have from (4.40), (5.30), and (5.31) that for $z = x$ or $z = y$,

$$\mathbf{S}_+(z) = \mathbf{E}(z) \dot{\mathbf{X}}_+(z) \begin{pmatrix} 1 & 0 \\ ie^{\eta(z) - \gamma - 2\kappa g_+(z)} e^{iN\theta(z)} & 1 \end{pmatrix} T_\nabla(z)^{-\sigma_3/2}, \qquad (7.11)$$

and thus

$$\left(\mathbf{S}_+(x)^{-1}\mathbf{S}_+(y)\right)_{21} = T_\nabla(x)^{-1/2} e^{-(\kappa g_+(x) + \gamma/2 - \eta(x)/2)} e^{iN\theta(x)/2}$$

$$\cdot T_\nabla(y)^{-1/2} e^{-(\kappa g_+(y) + \gamma/2 - \eta(y)/2)} e^{iN\theta(y)/2}$$

$$\cdot \mathbf{v}^T e^{iN\theta(x)\sigma_3/2} \mathbf{B}(x)^{-1} \mathbf{B}(y) e^{-iN\theta(y)\sigma_3/2} \mathbf{w}$$

since $\mathbf{E}(z)$ is analytic in the band so that $\mathbf{E}(z) = \mathbf{E}_+(z)$. Now we substitute into (7.10):

$$\left(\mathbf{P}(x;N,k)^{-1}\mathbf{P}(y;N,k)\right)_{21} = f(x)f(y)\mathbf{v}^T e^{iN\theta(x)\sigma_3/2}\mathbf{B}(x)^{-1}\mathbf{B}(y)e^{-iN\theta(y)\sigma_3/2}\mathbf{w}, \qquad (7.12)$$

where

$$f(z) := T_\nabla(z)^{-1/2} e^{-(\kappa g_+(z)-\eta(z)/2)} e^{N\ell_c/2} e^{(k-\#\Delta)g_+(z)} \prod_{n \in \Delta} |z - x_{N,n}|.$$

Now (7.12) holds for any x and y in the same band of Σ_0^∇, and when we specialize to nodes $x, y \in X_N$, we obtain formulae for the reproducing kernel. Therefore

$$K_{N,k}(x,y) = \sqrt{w(x)w(y)} f(x)f(y) \frac{\mathbf{v}^T e^{iN\theta(x)\sigma_3/2}\mathbf{B}(x)^{-1}\mathbf{B}(y)e^{-iN\theta(y)\sigma_3/2}\mathbf{w}}{x-y},$$

and using Lemma 7.5, we have

$$K_{N,k}(x,x) = w(x)f(x)^2 \frac{\partial}{\partial x}\left(\mathbf{v}^T e^{iN\theta(x)\sigma_3/2}\mathbf{B}(x)^{-1}\mathbf{B}(y)e^{-iN\theta(y)\sigma_3/2}\mathbf{w}\right)\bigg|_{y=x}. \qquad (7.13)$$

Since $x \in X_N$ and $y \in X_N$, we may use Lemma 7.4 along with the equilibrium condition (2.14) that holds for x and y in a band I to deduce

$$\sqrt{w(x)}f(x) = \frac{1}{\sqrt{2\pi N \rho^0(x)}} \quad \text{and} \quad \sqrt{w(y)}f(y) = \frac{1}{\sqrt{2\pi N \rho^0(y)}}, \quad \text{for } x \text{ and } y \text{ in } X_N \cap I \subset \Sigma_0^\nabla.$$

This proves (7.7). To complete the proof of (7.8), we carry out the differentiation in (7.13), noting that by definition (see (4.5) and (4.9)),

$$\theta'(x) = 2\pi c \frac{d\mu_{\min}^c}{dx}(x), \qquad x \in \Sigma_0^\nabla.$$

$\square$

Proposition 7.7. *Let x and y be distinct nodes in a void Γ lying outside all discs $D_\Gamma^{\nabla,*}$. Then*

$$K_{N,k}(x,y) = \frac{T_\nabla(x)^{1/2}T_\nabla(y)^{1/2} e^{-\frac{1}{2}N\left[\frac{\delta E_c}{\delta\mu}(x)-\ell_c\right]} e^{-\frac{1}{2}N\left[\frac{\delta E_c}{\delta\mu}(y)-\ell_c\right]}}{2\pi N \sqrt{\rho^0(x)\rho^0(y)}} \cdot \frac{\mathbf{a}^T e^{iN\theta_\Gamma \sigma_3/2}\mathbf{B}(x)^{-1}\mathbf{B}(y)e^{-iN\theta_\Gamma \sigma_3/2}\mathbf{b}}{x-y} \qquad (7.14)$$

and

$$K_{N,k}(x,x) = -\frac{T_\nabla(x) e^{-N\left[\frac{\delta E_c}{\delta\mu}(x)-\ell_c\right]}}{2\pi N \rho^0(x)} \cdot \mathbf{a}^T e^{iN\theta_\Gamma \sigma_3/2}\mathbf{B}(x)^{-1}\mathbf{B}'(x)e^{-iN\theta_\Gamma \sigma_3/2}\mathbf{b}, \qquad (7.15)$$

where

$$\mathbf{a} := \begin{pmatrix} 0 \\ 1 \end{pmatrix}, \qquad \mathbf{b} := \begin{pmatrix} 1 \\ 0 \end{pmatrix}$$

(note that $\mathbf{a}^T \mathbf{b} = 0$), $\mathbf{B}(x)$ is defined by (7.9), and the variational derivative is evaluated on the equilibrium measure $\mu_{\min}^c$.

Proof. Note that the two points x and y lying in the same void interval necessarily belong to the same component of Σ_0^∇. The proof follows that of Proposition 7.6 with only a few modifications. First, in place of (7.11) we have the simpler relation

$$\mathbf{S}_+(x) = \mathbf{E}_+(x)\dot{\mathbf{X}}_+(x).$$

Next, when we use Lemma 7.4, we must retain the exponentials involving the variational derivative since in place of (2.14) we have the variational inequality (2.13) because x and y are in a void Γ. Finally, we recall that the function $e^{iN\theta(x)}$ takes the constant value $e^{iN\theta_\Gamma}$ throughout Γ. $\square$

Recall the definition of the mappings $\tau_\Gamma^{\nabla,L}$ and $\tau_\Gamma^{\nabla,R}$ given in (2.17) and (2.18), respectively.

Proposition 7.8. *Let x and y be distinct nodes in a disc $D_\Gamma^{\nabla,L}$. Then*

$$K_{N,k}(x,y) = \frac{1}{N^{2/3}\sqrt{\rho^0(x)\rho^0(y)}} \cdot \frac{\mathbf{q}_\Gamma^{\nabla,L}(x)^T \mathbf{A}_\Gamma^{\nabla,L}(x)^{-1} \mathbf{A}_\Gamma^{\nabla,L}(y) \mathbf{r}_\Gamma^{\nabla,L}(y)}{x-y} \qquad (7.16)$$

and

$$K_{N,k}(x,x) = \frac{1}{N^{2/3}\rho^0(x)}\left[-\mathbf{q}_\Gamma^{\nabla,L}(x)^T \mathbf{A}_\Gamma^{\nabla,L}(x)^{-1}\frac{d\mathbf{A}_\Gamma^{\nabla,L}}{dx}(x)\mathbf{r}_\Gamma^{\nabla,L}(x) - \mathbf{q}_\Gamma^{\nabla,L}(x)^T\frac{d\mathbf{r}_\Gamma^{\nabla,L}}{dx}(x)\right], \qquad (7.17)$$

where

$$\mathbf{q}_\Gamma^{\nabla,L}(x) := \begin{pmatrix} -\mathrm{Ai}\left(-\left(\frac{3}{4}\right)^{2/3}\tau_\Gamma^{\nabla,L}(x)\right) \\ N^{-1/3}\mathrm{Ai}'\left(-\left(\frac{3}{4}\right)^{2/3}\tau_\Gamma^{\nabla,L}(x)\right) \end{pmatrix}, \quad \mathbf{r}_\Gamma^{\nabla,L}(x) := \begin{pmatrix} N^{-1/3}\mathrm{Ai}'\left(-\left(\frac{3}{4}\right)^{2/3}\tau_\Gamma^{\nabla,L}(x)\right) \\ \mathrm{Ai}\left(-\left(\frac{3}{4}\right)^{2/3}\tau_\Gamma^{\nabla,L}(x)\right) \end{pmatrix}$$

(note that $\mathbf{q}_\Gamma^{\nabla,L}(x)^T \mathbf{r}_\Gamma^{\nabla,L}(x) \equiv 0$), and

$$\mathbf{A}_\Gamma^{\nabla,L}(x) := \mathbf{E}(x)\mathbf{H}_\Gamma^{\nabla,L}(x)N^{\sigma_3/6}\left(\frac{3}{4}\right)^{-\sigma_3/6}. \qquad (7.18)$$

Similarly, if x and y are distinct nodes in a disc $D_\Gamma^{\nabla,R}$, then

$$K_{N,k}(x,y) = -\frac{1}{N^{2/3}\sqrt{\rho^0(x)\rho^0(y)}} \cdot \frac{\mathbf{q}_\Gamma^{\nabla,R}(x)\mathbf{A}_\Gamma^{\nabla,R}(x)^{-1}\mathbf{A}_\Gamma^{\nabla,R}(y)\mathbf{r}_\Gamma^{\nabla,R}(y)}{x-y} \qquad (7.19)$$

and

$$K_{N,k}(x,x) = -\frac{1}{N^{2/3}\rho^0(x)}\left[-\mathbf{q}_\Gamma^{\nabla,R}(x)^T \mathbf{A}_\Gamma^{\nabla,R}(x)^{-1}\frac{d\mathbf{A}_\Gamma^{\nabla,R}}{dx}(x)\mathbf{r}_\Gamma^{\nabla,R}(x) - \mathbf{q}_\Gamma^{\nabla,R}(x)^T\frac{d\mathbf{r}_\Gamma^{\nabla,R}}{dx}(x)\right], \qquad (7.20)$$

where

$$\mathbf{q}_\Gamma^{\nabla,R}(x) := \begin{pmatrix} -\mathrm{Ai}\left(-\left(\frac{3}{4}\right)^{2/3}\tau_\Gamma^{\nabla,R}(x)\right) \\ N^{-1/3}\mathrm{Ai}'\left(-\left(\frac{3}{4}\right)^{2/3}\tau_\Gamma^{\nabla,R}(x)\right) \end{pmatrix}, \quad \mathbf{r}_\Gamma^{\nabla,R}(x) := \begin{pmatrix} N^{-1/3}\mathrm{Ai}'\left(-\left(\frac{3}{4}\right)^{2/3}\tau_\Gamma^{\nabla,R}(x)\right) \\ \mathrm{Ai}\left(-\left(\frac{3}{4}\right)^{2/3}\tau_\Gamma^{\nabla,R}(x)\right) \end{pmatrix}$$

(note again that $\mathbf{q}_\Gamma^{\nabla,R}(x)^T \mathbf{r}_\Gamma^{\nabla,R}(x) \equiv 0$), and

$$\mathbf{A}_\Gamma^{\nabla,R}(x) := \mathbf{E}(x)\mathbf{H}_\Gamma^{\nabla,R}(x)N^{\sigma_3/6}\left(\frac{3}{4}\right)^{-\sigma_3/6}. \qquad (7.21)$$

Proof. Using the fact that x and y necessarily lie in the same component of Σ_0^∇, we still have the relation (7.10), where the subscript + indicates a boundary value taken from the upper half-plane. Now the matrix $\mathbf{S}(z)$ is analytic for all $z \in D_\Gamma^{\nabla,L} \cap \mathbb{C}_+$, so to obtain a formula for $\mathbf{S}(z)$ we may choose arbitrarily whether to consider z in quadrant I or quadrant II of $D_\Gamma^{\nabla,L}$ (in each case the answer is necessarily the same). For concreteness, we choose to evaluate $\mathbf{S}_+(x)$ by taking a limit from $D_{\Gamma,\mathrm{II}}^{\nabla,L}$ (above the void Γ). In this region, $\mathbf{S}(z) \equiv \mathbf{X}(z) \equiv \mathbf{E}(z)\hat{\mathbf{X}}_\Gamma^{\nabla,L}(z)$, so from (5.14) we see that for x and y in $D_{\Gamma,\mathrm{II}}^{\nabla,L}$,

$$\left(\mathbf{S}(x)^{-1}\mathbf{S}(y)\right)_{21} = T_\nabla(x)^{-1/2}e^{(\eta(x)-\gamma-2\kappa g(x))/2}e^{iN\theta_\Gamma/2} \cdot T_\nabla(y)^{-1/2}e^{(\eta(y)-\gamma-2\kappa g(y))/2}e^{iN\theta_\Gamma/2}$$

$$\cdot N^{-1/3}\left(\frac{3}{4}\right)^{1/3}\left(\mathbf{G}(x)^{-1}\mathbf{A}_\Gamma^{\nabla,L}(x)^{-1}\mathbf{A}_\Gamma^{\nabla,L}(y)\mathbf{G}(y)\right)_{21},$$

where
$$\mathbf{G}(z) := \left(\frac{3}{4}\right)^{\sigma_3/6} N^{-\sigma_3/6} \hat{\mathbf{Z}}^{\nabla,L}(\tau_\Gamma^{\nabla,L}(z)) N^{\sigma_3/6} \left(\frac{3}{4}\right)^{-\sigma_3/6}.$$

Using the explicit formula for $\hat{\mathbf{Z}}^{\nabla,L}(\zeta)$ furnished by (5.11) in Proposition 5.5, we then obtain
$$\left(\mathbf{G}(x)^{-1}\mathbf{A}_\Gamma^{\nabla,L}(x)^{-1}\mathbf{A}_\Gamma^{\nabla,L}(y)\mathbf{G}(y)\right)_{21} = 2\pi \left(\frac{3}{4}\right)^{-1/3} N^{2/3} e^{(-\tau_\Gamma^{\nabla,L}(x))^{3/2}/2} e^{(-\tau_\Gamma^{\nabla,L}(y))^{3/2}/2}$$
$$\cdot \mathbf{q}_\Gamma^{\nabla,L}(x)^T \mathbf{A}_\Gamma^{\nabla,L}(x)^{-1} \mathbf{A}_\Gamma^{\nabla,L}(y) \mathbf{r}_\Gamma^{\nabla,L}(y).$$

Substituting into (7.10) gives
$$\left(\mathbf{P}(x;N,k)^{-1}\mathbf{P}(y;N,k)\right)_{21} = N^{1/3} b(x) b(y) \mathbf{q}_\Gamma^{\nabla,L}(x)^T \mathbf{A}_\Gamma^{\nabla,L}(x)^{-1} \mathbf{A}_\Gamma^{\nabla,L}(y) \mathbf{r}_\Gamma^{\nabla,L}(y),$$
where
$$b(z) := \sqrt{2\pi} e^{(-\tau_\Gamma^{\nabla,L}(z))^{3/2}/2} T_\nabla(z)^{-1/2} e^{-(\kappa g_+(z) - \eta(z))/2} e^{iN\theta_\Gamma/2} e^{N\ell_c/2} e^{(k - \#\Delta)g_+(z)} \prod_{n \in \Delta} |z - x_{N,n}|.$$

Here, by $(-\tau_\Gamma^{\nabla,L}(z))^{3/2}$, we understand the boundary value taken on $\mathbb{R}$ from $\Im(z) > 0$, or equivalently $\Im(\tau_\Gamma^{\nabla,L}(z)) > 0$. The subscript $+$ on $g_+(z)$ denotes the same limit. Using Lemma 7.4, we see that for any node z in $D_\Gamma^{\nabla,L}$,
$$\sqrt{w(z)} b(z) = \frac{e^{(-\tau_\Gamma^{\nabla,L}(z))^{-3/2}/2} e^{-iN(\theta(z) - \theta_\Gamma)/2} e^{-\frac{1}{2}N\left[\frac{\delta E_c}{\delta\mu}(z) - \ell_c\right]}}{\sqrt{N\rho^0(z)}},$$
where the variational derivative is evaluated for $\mu = \mu_{\min}^c$. Now, if the node z lies in the void Γ, then $\theta(z) = \theta_\Gamma$ modulo $2\pi/N$, but from (5.4) and (2.17), we see that $(-\tau_\Gamma^{\nabla,L}(z))^{-3/2} = N[\delta E_c/\delta\mu(z) - \ell_c]$. On the other hand, if the node z lies in the adjacent band, then from (2.14) we have $\delta E_c/\delta\mu(z) - \ell_c = 0$, but again (2.17) gives the identity $(-\tau_\Gamma^{\nabla,L}(z))^{-3/2} = iN(\theta(z) - \theta_\Gamma)$ modulo $2\pi i$. Thus, for all nodes z in $D_\Gamma^{\nabla,L}$, we have $\sqrt{w(z)} b(z) = 1/\sqrt{N\rho^0(z)}$. This proves (7.16). Using Lemma 7.5, we also obtain (7.17).

The proofs of (7.19) and (7.20) are analogous. It is perhaps noteworthy that the origin of the leading minus sign in these formulae is the factor $i\sigma_3$ relating $\hat{\mathbf{Z}}^{\nabla,L}(\zeta)$ and $\hat{\mathbf{Z}}^{\nabla,R}(\zeta)$ (see (5.16)). □

◁ **Remark:** In each case we may verify after the fact that for a node $x \in X_N$,
$$K_{N,k}(x,x) = \lim_{\substack{z,w \to x \\ z,w \in \mathbb{C},\ z \neq w}} K_{N,k}(z,w); \tag{7.22}$$
that is, in each region $K_{N,k}(x,y)$ may be viewed as an analytic function of two complex variables sampled at the discrete nodes $X_N \times X_N$. This is not obvious from the definition. Indeed, the definition (3.4) of $K_{N,k}(x,y)$ can *a priori* be evaluated only when x and y are both nodes due to the factor $\sqrt{w(x)w(y)}$. If the weights were given in the form $w_{N,n} = w(x_{N,n})$ for some analytic function $w(x)$, there would be a direct interpretation of the limit process (7.22). However, the weights under consideration (given by (1.16)) do not have the exact form of an analytic function simply sampled at the nodes, because of the presence of a factor involving an essentially discrete product over nodes. Indeed, the derivation of the exact formulae above for $K_{N,k}(x,y)$ both on and off the diagonal made explicit use of the fact that x and y are discrete nodes via Lemma 7.4. ▷

The following result can also be extracted from the proofs of Propositions 7.7 and 7.8.

Proposition 7.9. *Let x and y be nodes in the same component of Σ_0^∇. If y lies in a disc $D_\Gamma^{\nabla,L}$ and x lies outside the disc but in the adjacent void Γ, then*
$$K_{N,k}(x,y) = -\frac{T_\nabla(x)^{1/2} e^{-\frac{1}{2}N\left[\frac{\delta E_c}{\delta\mu}(x) - \ell_c\right]}}{N^{5/6}\sqrt{2\pi\rho^0(x)\rho^0(y)}} \cdot \frac{\mathbf{a}^T e^{iN\theta_\Gamma\sigma_3/2} \mathbf{B}(x)^{-1} \mathbf{A}_\Gamma^{\nabla,L}(y) \mathbf{r}_\Gamma^{\nabla,L}(y)}{x - y}.$$
Similarly, if y lies in a disc $D_\Gamma^{\nabla,R}$ and x lies outside the disc but in the adjacent void Γ, then
$$K_{N,k}(x,y) = -i\frac{T_\nabla(x)^{1/2} e^{-\frac{1}{2}N\left[\frac{\delta E_c}{\delta\mu}(x) - \ell_c\right]}}{N^{5/6}\sqrt{2\pi\rho^0(x)\rho^0(y)}} \cdot \frac{\mathbf{a}^T e^{iN\theta_\Gamma\sigma_3/2} \mathbf{B}(x)^{-1} \mathbf{A}_\Gamma^{\nabla,R}(y) \mathbf{r}_\Gamma^{\nabla,R}(y)}{x - y}.$$
Here the notation on the right-hand side is the same as in Propositions 7.7 and 7.8.

7.3 ASYMPTOTIC FORMULAE FOR $K_{N,k}(x,y)$ AND UNIVERSALITY

Lemma 7.10. *Fix a closed interval $F \subset [a,b]$ that contains none of the band endpoints $\alpha_0, \ldots, \alpha_G$ and $\beta_0, \ldots, \beta_G$. Without loss of generality, fix the contour parameter $\epsilon > 0$ sufficiently small that F lies outside all discs $D_\Gamma^{*,*}$ and that Proposition 5.15 controls $\mathbf{E}(z) - \mathbb{I}$. Then there is a constant $C_F > 0$ such that for all N sufficiently large,*

$$\sup_{x \in F} \|\mathbf{B}'(x)\| \leq C_F \quad \text{and} \quad \sup_{x,y \in F} \frac{\|\mathbf{B}(x)^{-1}\mathbf{B}(y) - \mathbb{I}\|}{|x-y|} \leq C_F,$$

where $\|\cdot\|$ denotes a matrix norm and $\mathbf{B}(x)$ is defined by (7.9) for arbitrary $x \in [a,b]$ (note that $\mathbf{B}(x)$ depends on ϵ via $\mathbf{E}(x)$).

Proof. The matrix $\mathbf{W}(z) := \dot{\mathbf{X}}(z)e^{(\kappa g(z) + \gamma/2 - \eta(z)/2)\sigma_3}$ can be analytically continued through the interval F from the upper half-plane by a jump relation of the form (see Riemann-Hilbert Problem 5.1) $\mathbf{W}_+(z) = \mathbf{W}_-(z)\mathbf{v}$, where $\mathbf{v}$ is a constant matrix (with respect to z) whose elements are uniformly bounded as $N \to \infty$ (for z in a void or saturated region Γ_i we have $\mathbf{v} = e^{iN\theta_{\Gamma_i}\sigma_3}$, and for z in a band I we have $\mathbf{v} = -i\sigma_1$). Since $\mathbf{W}(z)$ is uniformly bounded for $z \in \mathbb{C} \setminus \Sigma_{\text{model}}$ bounded away from the band endpoints (from Proposition 5.2 and (1.18), as well as the assumption that κ remains bounded as $N \to \infty$), it follows that the analytic continuation of $\mathbf{W}_+(z)$ from F is uniformly bounded in a fixed complex neighborhood G of F as $N \to \infty$. Cauchy's Theorem applied on a closed contour in G encircling F then shows that $\mathbf{W}_+(z)$ and all its derivatives remain uniformly bounded in F as $N \to \infty$.

The same is true of the matrix $\mathbf{E}(z)$. Indeed, if F is a subinterval of a band I, then $\mathbf{E}(z)$ is already analytic in a complex neighborhood G of F and is uniformly bounded in G as $N \to \infty$ according to Proposition 5.15. The uniform boundedness of all derivatives of $\mathbf{E}_+(z) = \mathbf{E}(z)$ for $z \in F$ then follows from Cauchy's Theorem. On the other hand, if F is a subinterval of a void or saturated region, then the analytic continuation of $\mathbf{E}_+(z)$ to the neighborhood G is accomplished by the formula $\mathbf{E}_+(z) = \mathbf{F}(z)$, where $\mathbf{F}(z)$ is the solution of Riemann-Hilbert Problem 5.12. Since $\mathbf{F}(z)$ is uniformly bounded in G, again Cauchy's Theorem implies that all derivatives of $\mathbf{E}_+(z)$ are uniformly bounded for $z \in F$.

Combining these results using $\mathbf{B}(x) = \mathbf{E}_+(x)\mathbf{W}_+(x)$ establishes that $\mathbf{B}'(x)$ remains uniformly bounded in F as $N \to \infty$. The boundedness of the difference quotient follows from this result and the uniform boundedness of $\mathbf{B}(x)$ itself since $\det(\mathbf{B}(x)) = 1$. $\square$

Lemma 7.11. *Fix a value of the contour parameter $\epsilon > 0$ sufficiently small that Proposition 5.15 controls $\mathbf{E}(z) - \mathbb{I}$ on appropriate closed sets. Then for each disc $D_\Gamma^{\nabla,L}$ there is a constant $C_\Gamma^{\nabla,L} > 0$, and for each disc $D_\Gamma^{\nabla,R}$ there is a constant $C_\Gamma^{\nabla,R} > 0$ such that for all N sufficiently large,*

$$\sup_{x \in D_\Gamma^{\nabla,L} \cap \mathbb{R}} \left\| \frac{d\mathbf{A}_\Gamma^{\nabla,L}}{dx}(x) \right\| \leq C_\Gamma^{\nabla,L} \quad \text{and} \quad \sup_{x,y \in D_\Gamma^{\nabla,L} \cap \mathbb{R}} \frac{\|\mathbf{A}_\Gamma^{\nabla,L}(x)^{-1}\mathbf{A}_\Gamma^{\nabla,L}(y) - \mathbb{I}\|}{|x-y|} \leq C_\Gamma^{\nabla,L}$$

and

$$\sup_{x \in D_\Gamma^{\nabla,R} \cap \mathbb{R}} \left\| \frac{d\mathbf{A}_\Gamma^{\nabla,R}}{dx}(x) \right\| \leq C_\Gamma^{\nabla,R} \quad \text{and} \quad \sup_{x,y \in D_\Gamma^{\nabla,R} \cap \mathbb{R}} \frac{\|\mathbf{A}_\Gamma^{\nabla,R}(x)^{-1}\mathbf{A}_\Gamma^{\nabla,R}(y) - \mathbb{I}\|}{|x-y|} \leq C_\Gamma^{\nabla,R},$$

where $\|\cdot\|$ denotes a matrix norm, $\mathbf{A}_\Gamma^{\nabla,L}(x)$ is defined by (7.18), and $\mathbf{A}_\Gamma^{\nabla,R}(x)$ is defined by (7.21). Also, for the same constants and for sufficiently large N,

$$\sup_{x \in D_\Gamma^{\nabla,L} \cap \mathbb{R}} \|\mathbf{q}_\Gamma^{\nabla,L}(x)\| \leq C_\Gamma^{\nabla,L} \quad \text{and} \quad \sup_{x \in D_\Gamma^{\nabla,L} \cap \mathbb{R}} \|\mathbf{r}_\Gamma^{\nabla,L}(x)\| \leq C_\Gamma^{\nabla,L}$$

and

$$\sup_{x \in D_\Gamma^{\nabla,R} \cap \mathbb{R}} \|\mathbf{q}_\Gamma^{\nabla,R}(x)\| \leq C_\Gamma^{\nabla,R} \quad \text{and} \quad \sup_{x \in D_\Gamma^{\nabla,R} \cap \mathbb{R}} \|\mathbf{r}_\Gamma^{\nabla,R}(x)\| \leq C_\Gamma^{\nabla,R}.$$

Finally, there is a constant $K > 0$ such that for sufficiently large N,

$$\sup_{\substack{x \in D_\Gamma^{\nabla,L} \cap \mathbb{R} \\ x < \alpha}} \|\mathbf{q}_\Gamma^{\nabla,L}(x)\| \leq \frac{C_\Gamma^{\nabla,L} e^{-NK(\alpha-x)^{3/2}}}{N^{1/6}} \quad \text{and} \quad \sup_{\substack{x \in D_\Gamma^{\nabla,L} \cap \mathbb{R} \\ x < \alpha}} \|\mathbf{r}_\Gamma^{\nabla,L}(x)\| \leq \frac{C_\Gamma^{\nabla,L} e^{-NK(\alpha-x)^{3/2}}}{N^{1/6}},$$

where α is the band edge point at the center of the disc $D_\Gamma^{\nabla,L}$, and

$$\sup_{\substack{x \in D_\Gamma^{\nabla,R} \cap \mathbb{R} \\ x > \beta}} \|\mathbf{q}_\Gamma^{\nabla,R}(x)\| \leq \frac{C_\Gamma^{\nabla,R} e^{-NK(x-\beta)^{3/2}}}{N^{1/6}} \quad \text{and} \quad \sup_{\substack{x \in D_\Gamma^{\nabla,R} \cap \mathbb{R} \\ x > \beta}} \|\mathbf{r}_\Gamma^{\nabla,R}(x)\| \leq \frac{C_\Gamma^{\nabla,R} e^{-NK(x-\beta)^{3/2}}}{N^{1/6}},$$

where β is the band edge point at the center of the disc $D_\Gamma^{\nabla,R}$.

Proof. The statements concerning the matrices $\mathbf{A}_\Gamma^{\nabla,L}(x)$ and $\mathbf{A}_\Gamma^{\nabla,R}(x)$ are elementary consequences of two facts. First, from Proposition 5.15, we have that $\mathbf{E}(z)$ is analytic and remains uniformly bounded as $N \to \infty$ in each disc $D_\Gamma^{\nabla,L}$ or $D_\Gamma^{\nabla,R}$. Next (see §5.1.2), the product $\mathbf{H}_\Gamma^{\nabla,L}(z) N^{\sigma_3/6}$ is analytic in each disc $D_\Gamma^{\nabla,L}$ and remains uniformly bounded there as $N \to \infty$, while the product $\mathbf{H}_\Gamma^{\nabla,R}(z) N^{\sigma_3/6}$ is analytic in each disc $D_\Gamma^{\nabla,R}$ and remains uniformly bounded there as $N \to \infty$. It follows from Cauchy's Theorem applied on the boundary of each disc that all derivatives of $\mathbf{A}_\Gamma^{\nabla,L}(z)$ are uniformly bounded independent of N in $D_\Gamma^{\nabla,L}$, and the same holds for $\mathbf{A}_\Gamma^{\nabla,R}(z)$ in $D_\Gamma^{\nabla,R}$. The boundedness of the difference quotients then follows since $\det(\mathbf{A}_\Gamma^{\nabla,L}(x)) = 1$ in $D_\Gamma^{\nabla,L}$ and $\det(\mathbf{A}_\Gamma^{\nabla,R}(x)) = 1$ in $D_\Gamma^{\nabla,R}$.

The statements concerning the vectors $\mathbf{q}_\Gamma^{\nabla,L}(x)$, $\mathbf{r}_\Gamma^{\nabla,L}(x)$, $\mathbf{q}_\Gamma^{\nabla,R}(x)$, and $\mathbf{r}_\Gamma^{\nabla,R}(x)$ are obtained from the asymptotic formulae (5.12) and from the elementary estimates holding for all $x > 0$:

$$|\text{Ai}(x)| \leq \frac{Ce^{-2x^{3/2}/3}}{(1+x)^{1/4}} \quad \text{and} \quad |\text{Ai}'(x)| \leq C(1+x)^{1/4} e^{-2x^{3/2}/3},$$

where $C > 0$ is some appropriate constant. Then one uses the fact that in each case the argument of the Airy functions is $N^{2/3}$ times an analytic function of x that has a nonvanishing derivative and is independent of N. □

7.3.1 Asymptotic universality of statistics for particles in a band: the proofs of Theorems 3.1 and 3.2

Consider a fixed closed interval F in the interior of a band I. We can easily establish the following asymptotic formulae uniformly valid in F.

Lemma 7.12. *Let F be a fixed closed interval in the interior of a band I. Then there is a constant $C_F > 0$ such that for all sufficiently large N,*

$$\max_{x \in X_N \cap F} \left| K_{N,k}(x,x) - \frac{c}{\rho^0(x)} \frac{d\mu^c_{\min}}{dx}(x) \right| \leq \frac{C_F}{N} \tag{7.23}$$

and

$$\max_{x,y \in X_N \cap F} \left| K_{N,k}(x,y) - \frac{1}{N\pi\sqrt{\rho^0(x)\rho^0(y)}} \frac{\sin\left(\frac{N}{2}(\theta(x) - \theta(y))\right)}{x-y} \right| \leq \frac{C_F}{N}, \tag{7.24}$$

where $\theta(z)$ is defined in (4.9). Also, for some other constant $C_F' > 0$ and N sufficiently large,

$$\max_{x,y \in X_N \cap F} |(x-y) K_{N,k}(x,y)| \leq \frac{C_F'}{N}. \tag{7.25}$$

Proof. First, note that without any loss of generality we may suppose that F lies in Σ_0^∇. Indeed, if I is a transition band, this can be arranged by judicious choice of the transition points Y_∞. But even if I lies between two saturated regions, a pair of artificial transition points may be introduced in I, one on each side of F in order to "switch" F back into Σ_0^∇. Then, applying Lemma 7.10 to the exact formula (7.8) established in Proposition 7.6 yields (7.23). Similarly, applying Lemma 7.10 to (7.7) and using the identity

$$\mathbf{v}^T e^{iN(\theta(x)-\theta(y))\sigma_3/2}\mathbf{w} = 2\sin\left(\frac{N}{2}(\theta(x)-\theta(y))\right)$$

proves (7.24). Then (7.25) follows from (7.24). □

The estimate (7.25) shows that the reproducing kernel is concentrated near the diagonal, and a nonzero limit for $K_{N,k}(x,y)$ as $N \to \infty$ may be expected only for nodes x and y in F with $x-y$ of size bounded by $1/N$. To find the limit, we will now localize by considering a finite number of nodes near a certain fixed $x \in F$ (being fixed as $N \to \infty$, x is not necessarily a node). Since

$$R_1^{(N,k)}(x) \cdot N\rho^0(x) = \mathbb{E}(\text{number of particles per node near } x) \cdot \frac{\text{number of nodes}}{\text{unit length}}$$

$$= \mathbb{E}\left(\frac{\text{number of particles}}{\text{unit length}}\right),$$

from (3.3) and (7.23) we see that the asymptotic mean spacing between particles near $x \in F$ is $\delta(x)/N$, where $\delta(x)$ is defined by (3.8). For ξ_N and η_N in some bounded set D, we thus consider nodes z and w defined by

$$z := x + \xi_N \frac{\delta(x)}{N}, \qquad w := x + \eta_N \frac{\delta(x)}{N}. \tag{7.26}$$

Note that the admissible values of ξ_N and η_N are finite in number and are asymptotically equally spaced with spacing $(\rho^0(x)\delta(x))^{-1}$ (because z and w are both nodes in X_N). Since $F \subset I$ and thus neither constraint is active, we have $0 < (\rho^0(x)\delta(x))^{-1} < 1$.

Now, from Taylor's Theorem and (7.26), we have

$$\theta(z) - \theta(w) = \theta\left(x + \xi_N \frac{\delta(x)}{N}\right) - \theta\left(x + \eta_N \frac{\delta(x)}{N}\right)$$

$$= \theta\left(x + \eta_N \frac{\delta(x)}{N} + (\xi_N - \eta_N)\frac{\delta(x)}{N}\right) - \theta\left(x + \eta_N \frac{\delta(x)}{N}\right)$$

$$= \theta'\left(x + \eta_N \frac{\delta(x)}{N}\right)(\xi_N - \eta_N)\frac{\delta(x)}{N} + \frac{\theta''(\sigma)}{2}(\xi_N - \eta_N)^2 \frac{\delta(x)^2}{N^2}$$

$$= \theta'(x)(\xi_N - \eta_N)\frac{\delta(x)}{N} + \theta''(\tau)(\xi_N-\eta_N)\eta_N \frac{\delta(x)^2}{N^2} + \frac{\theta''(\sigma)}{2}(\xi_N-\eta_N)^2 \frac{\delta(x)^2}{N^2}$$

$$= \frac{2\pi}{N}(\xi_N - \eta_N) + \theta''(\tau)(\xi_N-\eta_N)\eta_N \frac{\delta(x)^2}{N^2} + \frac{\theta''(\sigma)}{2}(\xi_N-\eta_N)^2 \frac{\delta(x)^2}{N^2},$$

for some σ and τ near $x \in F$. (To arrive at the last line we use $\theta'(x) = 2\pi c d\mu_{\min}^c/dx(x)$.) Therefore

$$\frac{\sin\left(\frac{N}{2}(\theta(z)-\theta(w))\right)}{z-w} = \frac{N}{\delta(x)}\left[\frac{\sin(\pi(\xi_N - \eta_N))}{\xi_N - \eta_N} - \cos(q)\left(\theta''(\tau)\eta_N + \frac{\theta''(\sigma)}{2}(\xi_N-\eta_N)\right)\frac{\delta(x)^2}{2N}\right],$$

for some $q \in \mathbb{R}$.

Since the node density is analytic and positive, we have

$$\frac{1}{\sqrt{\rho^0(z)\rho^0(w)}} = \frac{1}{\rho^0(x)} + O\left(\frac{1}{N}\right)$$

because we are assuming that ξ_N and η_N remain bounded as $N \to \infty$. Combining these results with Lemma 7.12 proves the following.

Lemma 7.13. *Fix x in the interior of any band I and consider ξ_N and η_N to lie in a fixed bounded discrete set D such that z and w defined by (7.26) lie in the set of nodes X_N. Then there is a constant $C_D(x) > 0$ such that for all sufficiently large N,*

$$\max_{\xi_N, \eta_N \in D} \left| K_{N,k}(z,w) - \frac{c}{\rho^0(x)} \frac{d\mu_{\min}^c}{dx}(x) S(\xi_N, \eta_N) \right| \leq \frac{C_D(x)}{N}.$$

Applying Lemma 7.12 and Lemma 7.13 to the determinantal formula (3.3), we immediately obtain corresponding asymptotics for all multipoint correlation functions, which completes the proof of Theorem 3.1.

We now give the proof of Theorem 3.2. From Lemma 7.13 and the asymptotic equal spacing of ξ_N and η_N, it follows that if $x_{N,i}$ and $x_{N,j}$ are two nodes in X_N such that $x_{N,i} \to x$ and $x_{N,j} \to x$ while $i - j$ remains fixed as $N \to \infty$, and x is in the interior of a band I, then

$$K_{N,k}(x_{N,i}, x_{N,j}) = \mathcal{S}_{ij}(x) + O\left(\frac{1}{N}\right). \tag{7.27}$$

Recall the formula (3.6) for $A_m^{(N,k)}(B)$ and its interpretation (3.5) as a probability. The operator $K_{N,k}|_{B_N}$ acts on $\ell^2(B_N)$ with the kernel given by

$$K_{N,k}\left(x + (x_i - x), x + (x_j - x)\right),$$

where the x_i are the nodes in B_N. The first result is that, as $N \to \infty$,

$$\det\left(1 - tK_{N,k}|_{B_N}\right) = \det\left(1 - t\mathcal{S}(x)|_{\mathbb{B}}\right) + O\left(\frac{1}{N}\right) \tag{7.28}$$

holds uniformly for t in compact sets in $\mathbb{C}$. This behavior follows from the analytic dependence of determinants of matrices of fixed finite dimension on the matrix elements and the use of Lemma 7.13 and (7.27). The statement (3.9) then follows from the analyticity of the left-hand side of (7.28) in t.

7.3.2 Correlation functions for particles in voids: the proofs of Theorems 3.3 and 3.4

Let $F = [u,v]$ be a fixed closed interval in a void Γ such that $u \notin \{\beta_0, \ldots, \beta_G\}$ and $v \notin \{\alpha_0, \ldots, \alpha_G\}$. We admit the possibility that $u = a$ or $v = b$. Applying Lemma 7.10 to the exact formulae (7.14) and (7.15) in Proposition 7.7 and taking into account the variational inequality (2.13), we arrive at the following.

Lemma 7.14. *Let F be a fixed closed interval in a void Γ that is bounded away from all bands. Then there is a constant $C_F > 0$ such that for all N sufficiently large,*

$$\max_{x,y \in X_N \cap F} \left| K_{N,k}(x,y) e^{\frac{1}{2}N\left[\frac{\delta E_c}{\delta \mu}(x) - \ell_c\right]} e^{\frac{1}{2}N\left[\frac{\delta E_c}{\delta \mu}(x) - \ell_c\right]} \right| \leq \frac{C_F}{N}, \tag{7.29}$$

where the variational derivatives are evaluated on the equilibrium measure.

Applying this result to the formula (3.3) for the correlation functions, we complete the proof of Theorem 3.3.

Now we prove Theorem 3.4. From (7.29) we see that the reproducing kernel $K_{N,k}(x,y)$ is uniformly exponentially small for nodes in F. In particular, and unlike in the bands, the kernel is not concentrated near the diagonal. In fact, concentration of $K_{N,k}(x,y)$ for x and y in a set F bounded away from the bands requires the existence of a local minimum of $\delta E_c/\delta\mu - \ell_c$ at some point $x \in F$. Indeed, suppose first that the local minimum occurs at some point in the *interior* of F and is genuine, so that the expansion (3.11) holds with $W > 0$ and $H > 0$ as $z \to x$ (by assumption, the variational derivative is an analytic function of z in the void Γ). The proper scaling is evidently then $z - x = O(N^{-1/2})$. Thus we consider nodes z and w near x of the form

$$z = x + \frac{\xi_N}{H\sqrt{N}}, \qquad w = x + \frac{\eta_N}{H\sqrt{N}},$$

for ξ_N and η_N lying in some bounded discrete set D such that z and w are in the set of nodes X_N. We then have, from Proposition 7.7, that

$$K_{N,k}(z,w) = \frac{e^{-NW}}{N} \cdot \left[q_N(x) + O\left(\frac{1}{\sqrt{N}}\right)\right] \cdot e^{-(\xi_N^2+\eta_N^2)/2},$$

where the error is uniform for ξ_N and η_N in D and

$$q_N(x) := -\frac{T_\nabla(x)}{2\pi\rho^0(x)} \cdot \mathbf{a}^T e^{iN\theta_\Gamma\sigma_3/2} \mathbf{B}(x)^{-1}\mathbf{B}'(x)e^{-iN\theta_\Gamma\sigma_3/2}\mathbf{b}$$

is uniformly bounded as $N \to \infty$. Now the asymptotic spacing between points in the discrete set D is $H/(\sqrt{N}\rho^0(x))$, which goes to zero as $N \to \infty$. Thus for any fixed interval $[A, B] \subset \mathbb{R}$,

$$E_{\text{int}}([A,B]; x, H, N) = \sum_{\substack{\xi_N \in D \\ A \leq \xi_N \leq B}} R_1^{(N,k)}\left(x + \frac{\xi_N}{H\sqrt{N}}\right)$$

$$= \sum_{\substack{\xi_N \in D \\ A \leq \xi_N \leq B}} K_{N,k}\left(x + \frac{\xi_N}{H\sqrt{N}}, x + \frac{\xi_N}{H\sqrt{N}}\right)$$

$$= \frac{e^{-NW}}{N} \sum_{\substack{\xi_N \in D \\ A \leq \xi_N \leq B}} \left[q_N(x) + O\left(\frac{1}{\sqrt{N}}\right)\right] \cdot e^{-\xi_N^2}$$

$$= \frac{e^{-NW}\rho^0(x)}{H\sqrt{N}} \sum_{\substack{\xi_N \in D \\ A \leq \xi_N \leq B}} \left[q_N(x) + O\left(\frac{1}{\sqrt{N}}\right)\right] \cdot e^{-\xi_N^2} \frac{H}{\sqrt{N}\rho^0(x)}$$

$$= \frac{e^{-NW}\rho^0(x)}{H\sqrt{N}} \left[q_N(x) \int_A^B e^{-\xi^2} d\xi + O\left(\frac{1}{\sqrt{N}}\right)\right].$$

The statement (3.12) will be established if we can bound $q_N(x)$ away from zero as $N \to \infty$. Now, for x in a void Γ and for N sufficiently large that $\mathbf{E}(z) - \mathbb{I}$ is sufficiently small, $q_N(x)$ will be bounded away from zero if the Wronskian

$$W[\dot{X}_{11+}e^{\kappa g_+}, \dot{X}_{21+}e^{\kappa g_+}](x) := \dot{X}_{12+}(x)e^{\kappa g_+(x)}\frac{d}{dx}\dot{X}_{11+}(x)e^{\kappa g_+(x)} - \dot{X}_{11+}(x)e^{\kappa g_+(x)}\frac{d}{dx}\dot{X}_{21+}(x)e^{\kappa g_+(x)},$$

where the subscript $+$ indicates a boundary value taken from the upper half-plane, is bounded away from zero. The Wronskian is not identically zero in any subinterval of Γ for the following reasons. If $W[\dot{X}_{11+}e^{\kappa g_+}, \dot{X}_{21+}e^{\kappa g_+}](x)$ were identically zero as a function of x, then there would necessarily be an interval in which $\dot{X}_{11+}(x)$ and $\dot{X}_{21+}(x)$ are proportional by a constant multiplier. Analytically extending this proportionality toward $z = \infty$, we see from the normalization condition ($\dot{\mathbf{X}}(z) \to \mathbb{I}$ as $z \to \infty$) that we would have to have $\dot{X}_{21}(z) \equiv 0$. The jump condition for $\dot{\mathbf{X}}(z)$ in any band would then force $\dot{X}_{22}(z) \equiv 0$ in addition, contradicting the fact that $\det(\dot{\mathbf{X}}(z)) = 1$.

Since $\dot{\mathbf{X}}(z)$ takes analytic boundary values in the void Γ, it follows that $W[\dot{X}_{11+}e^{\kappa g_+}, \dot{X}_{21+}e^{\kappa g_+}](x)$ has only isolated zeros in Γ for each value of N. From the exact solution formulae given in Appendix A, the number of zeros in Γ is finite and remains uniformly bounded as $N \to \infty$, and the zeros move quasiperiodically as N varies. For a given $x \in \Gamma$, either we have $W[\dot{X}_{11+}e^{\kappa g_+}, \dot{X}_{21+}e^{\kappa g_+}](x) = 0$ for all $N \in \mathbb{Z}$, or for each sufficiently small $\varepsilon > 0$ we may extract a subsequence of N values for which $|W[\dot{X}_{11+}e^{\kappa g_+}, \dot{X}_{21+}e^{\kappa g_+}](x)| \geq \varepsilon$. The first situation may occur for only a finite number of $x \in \Gamma$. This completes the proof of Theorem 3.4.

7.3.3 Correlation functions for particles in saturated regions: the proofs of Theorems 3.5 and 3.6

Let F be a fixed closed interval in a saturated region Γ that is bounded away from the bands but which may have either a or b as an endpoint if the upper constraint is active there. We may exploit the dual ensemble

UNIVERSALITY: PROOFS OF THEOREMS STATED IN §3.3   129

(for the holes) to analyze the particle statistics in F. According to Proposition 2.6, the equilibrium measures for the particle ensemble with k particles and for the dual hole ensemble with $\bar{k} = N - k$ holes are explicitly related, and F lies in a void for the hole ensemble. Consequently, the results of our analysis for x and y in a void hold true for the dual kernel $\overline{K}_{N,\bar{k}}(x,y)$ and the corresponding hole correlation functions. To recover results for the kernel $K_{N,k}(x,y)$ and the corresponding particle correlation functions in F, we simply apply Propositions 7.2 and 7.3. This proves Theorem 3.5.

Combining the above duality arguments with the proof of Theorem 3.4 proves Theorem 3.6.

7.3.4 Asymptotic universality of statistics for particles near band/gap edges: the proofs of Theorems 3.7, 3.8, 3.9, and 3.10

Near the edge of a band, the equilibrium measure vanishes, and hence in the band the scaling $x - y = O(1/N)$ is not correct as the one-point function vanishes in the limit $N \to \infty$. Also, in the void or the saturated region near the band edge, the positive constant (3.10) or (3.13) is no longer bounded away from zero. Therefore we need to introduce a different scaling near a band edge to find the correct scaling limit.

Under the generic simplifying assumptions listed in §2.1.2, the density $d\mu_{\min}^c/dx$ of the equilibrium measure vanishes like a square root at each band edge adjacent to a void, and at a band edge adjacent to a saturated region, the "dual equilibrium measure" $\rho^0(x)/c - d\mu_{\min}^c/dx(x)$ vanishes like a square root. In this case, it turns out that the proper scaling is to consider nodes x satisfying

$$x - \alpha = O(N^{-2/3}) \qquad \text{or} \qquad x - \beta = O(N^{-2/3}),$$

depending on whether we consider a left band edge $z_0 = \alpha$ or a right band edge $z_0 = \beta$. Below, we will show that the limiting correlation function under the above scaling is given by the Airy kernel as in the edge scaling limit of the Gaussian unitary ensemble in random matrix theory, and also in the context of ensembles of more general Hermitian matrices of invariant measure (see, for example, [TraW94] and [BleI99]).

We begin with the following lemma, which is the analogue of Lemma 7.12.

Lemma 7.15. *For each disc $D_\Gamma^{\nabla,L}$, there is a constant $C_\Gamma^{\nabla,L} > 0$ such that for all sufficiently large N,*

$$K_{N,k}(x,x) = -\frac{t'(x)}{N^{1/3}\rho^0(x)} \left[\operatorname{Ai}'\left(N^{2/3}t(x)\right)^2 - \operatorname{Ai}\left(N^{2/3}t(x)\right) \operatorname{Ai}''\left(N^{2/3}t(x)\right) \right] + \varepsilon_N^{(1)}(x)$$

and

$$K_{N,k}(x,y) = -\frac{1}{N\sqrt{\rho^0(x)\rho^0(y)}} \cdot \frac{\operatorname{Ai}\left(N^{2/3}t(x)\right)\operatorname{Ai}'\left(N^{2/3}t(y)\right) - \operatorname{Ai}'\left(N^{2/3}t(x)\right)\operatorname{Ai}\left(N^{2/3}t(y)\right)}{x - y} + \varepsilon_N^{(2)}(x,y),$$

where $t(x) := -(3/4)^{2/3} N^{-2/3} \tau_\Gamma^{\nabla,L}(x)$ is a real-analytic function in $D_\Gamma^{\nabla,L}$ that is independent of N and is strictly decreasing along the real axis, and

$$\max_{x \in X_N \cap D_\Gamma^{\nabla,L}} \left|\varepsilon_N^{(1)}(x)\right| \leq \frac{C_\Gamma^{\nabla,L}}{N^{2/3}} \qquad \text{and} \qquad \max_{x,y \in X_N \cap D_\Gamma^{\nabla,L}} \left|\varepsilon_N^{(2)}(x,y)\right| \leq \frac{C_\Gamma^{\nabla,L}}{N^{2/3}}.$$

Also, for some constant $K > 0$, we have the one-sided estimates

$$\max_{\substack{x \in X_N \cap D_\Gamma^{\nabla,L} \\ x < \alpha}} \left|\varepsilon_N^{(1)}(x)\right| \leq \frac{C_\Gamma^{\nabla,L} e^{-2NK(\alpha-x)^{3/2}}}{N},$$

$$\max_{\substack{x,y \in X_N \cap D_\Gamma^{\nabla,L} \\ x,y < \alpha}} \left|\varepsilon_N^{(2)}(x,y)\right| \leq \frac{C_\Gamma^{\nabla,L} e^{-NK(\alpha-x)^{3/2}} e^{-NK(\alpha-y)^{3/2}}}{N},$$

where α is the band edge point at the center of the disc $D_\Gamma^{\nabla,L}$. Similarly, for each disc $D_\Gamma^{\nabla,R}$, there exists a constant $C_\Gamma^{\nabla,R} > 0$ such that for all sufficiently large N,

$$K_{N,k}(x,x) = \frac{t'(x)}{N^{1/3}\rho^0(x)} \left[\operatorname{Ai}'\left(N^{2/3}t(x)\right)^2 - \operatorname{Ai}\left(N^{2/3}t(x)\right) \operatorname{Ai}''\left(N^{2/3}t(x)\right) \right] + \varepsilon_N^{(1)}(x)$$

and

$$K_{N,k}(x,y) = \frac{1}{N\sqrt{\rho^0(x)\rho^0(y)}} \cdot \frac{\text{Ai}\left(N^{2/3}t(x)\right)\text{Ai}'\left(N^{2/3}t(y)\right) - \text{Ai}'\left(N^{2/3}t(x)\right)\text{Ai}\left(N^{2/3}t(y)\right)}{x-y} + \varepsilon_N^{(2)}(x,y),$$

where $t(x) := -(3/4)^{2/3}N^{-2/3}\tau_\Gamma^{\nabla,R}(x)$ is a real-analytic function in $D_\Gamma^{\nabla,R}$ that is independent of N and is strictly increasing along the real axis, and

$$\max_{x \in X_N \cap D_\Gamma^{\nabla,R}}\left|\varepsilon_N^{(1)}(x)\right| \leq \frac{C_\Gamma^{\nabla,R}}{N^{2/3}} \quad \text{and} \quad \max_{x,y \in X_N \cap D_\Gamma^{\nabla,R}}\left|\varepsilon_N^{(2)}(x,y)\right| \leq \frac{C_\Gamma^{\nabla,R}}{N^{2/3}}.$$

Also, for some constant $K > 0$, we have the one-sided estimates

$$\max_{\substack{x \in X_N \cap D_\Gamma^{\nabla,R} \\ x > \beta}}\left|\varepsilon_N^{(1)}(x)\right| \leq \frac{C_\Gamma^{\nabla,R}e^{-2NK(x-\beta)^{3/2}}}{N},$$

$$\max_{\substack{x,y \in X_N \cap D_\Gamma^{\nabla,R} \\ x,y > \beta}}\left|\varepsilon_N^{(2)}(x,y)\right| \leq \frac{C_\Gamma^{\nabla,R}e^{-NK(x-\beta)^{3/2}}e^{-NK(y-\beta)^{3/2}}}{N},$$

where β is the band edge point at the center of the disc $D_\Gamma^{\nabla,R}$.

Proof. This proof follows from Proposition 7.8 and Lemma 7.11. □

Now we localize near the diagonal by considering ξ_N and η_N to lie in a fixed bounded set such that

$$x = z_0 + \frac{\xi_N}{t'(z_0)N^{2/3}} \quad \text{and} \quad y = z_0 + \frac{\eta_N}{t'(z_0)N^{2/3}} \tag{7.30}$$

are nodes. Here $z_0 = \alpha$ or $z_0 = \beta$ is the band edge, which is independent of N. Because $t(z_0) = 0$, the Airy kernel $A(\xi_N, \eta_N)$ defined by (3.14) will appear in the asymptotics with ξ_N and η_N considered bounded. Although for each N the possible values of ξ_N and η_N are discrete, their spacing tends to zero like $N^{-1/3}$, and in this sense the Airy kernel, unlike the discrete sine kernel, may be thought of as a continuous function of two independent variables.

Now in a disc $D_\Gamma^{\nabla,L}$ centered at a left band edge $z_0 = \alpha$, a direct calculation using (2.17) shows that $t'(\alpha) = -\left(\pi c B_\alpha^L\right)^{2/3}$, where B_α^L is defined in (3.15). Similarly, in a disc $D_\Gamma^{\nabla,R}$ centered at a right band edge $z_0 = \beta$, one may use (2.18) to see that $t'(\beta) = \left(\pi c B_\beta^R\right)^{2/3}$, where B_β^R is defined in (3.16). With the help of Lemma 7.15, we may prove the following result.

Lemma 7.16. *For each fixed $M > 0$ and each left band edge α, there is a constant $C_\alpha(M) > 0$ such that for sufficiently large N,*

$$\max_{\substack{x,y \in X_N \\ \alpha - MN^{-1/2} < x,y < \alpha + MN^{-2/3}}}\left|K_{N,k}(x,y) - \frac{\left(\pi c B_\alpha^L\right)^{2/3}}{N^{1/3}\rho^0(\alpha)}A(\xi_N, \eta_N)\right| \leq \frac{C_\alpha(M)}{N^{2/3}},$$

where B_α^L is defined via a limit from the adjacent band from (3.15) and ξ_N and η_N are defined in terms of x and y using (7.30) and $t'(\alpha) = -\left(\pi c B_\alpha^L\right)^{2/3}$. Similarly, for each fixed $M > 0$ and each right band edge β, there is a constant $C_\beta(M) > 0$ such that for sufficiently large N,

$$\max_{\substack{x,y \in X_N \\ \beta - MN^{-2/3} < x,y < \beta + MN^{-1/2}}}\left|K_{N,k}(x,y) - \frac{\left(\pi c B_\beta^R\right)^{2/3}}{N^{1/3}\rho^0(\beta)}A(\xi_N, \eta_N)\right| \leq \frac{C_\beta(M)}{N^{2/3}},$$

where B_β^R is defined via a limit from the adjacent band from (3.16) and ξ_N and η_N are defined in terms of x and y using (7.30) and $t'(\beta) = \left(\pi c B_\beta^R\right)^{2/3}$.

UNIVERSALITY: PROOFS OF THEOREMS STATED IN §3.3

Proof. We show how the the computation works for x and y near a left endpoint α. The calculation near β is similar. From Lemma 7.15, we have

$$K_{N,k}(x,y) = -\frac{1}{N^{1/3}\sqrt{\rho^0(x)\rho^0(y)}} \cdot \frac{t(x)-t(y)}{x-y} \cdot A(N^{2/3}t(x), N^{2/3}t(y)) + \varepsilon_N^{(2)}(x,y).$$

With $\alpha - MN^{-1/2} < x, y < \alpha + MN^{-2/3}$, we have

$$\frac{1}{\sqrt{\rho^0(x)\rho^0(y)}} \cdot \frac{t(x)-t(y)}{x-y} = \frac{t'(\alpha)}{\rho^0(\alpha)} + O\left(\frac{1}{N^{1/2}}\right).$$

Since all partial derivatives of the Airy kernel $A(\xi_N, \eta_N)$ tend rapidly to zero as ξ_N and η_N tend to $+\infty$ (while under our assumptions about x and y, ξ_N and η_N are bounded below by a fixed constant, they may grow in the positive direction like $N^{1/6}$), we then obtain with $\alpha - MN^{-1/2} < x, y < \alpha + MN^{-2/3}$,

$$A(N^{2/3}t(x), N^{2/3}t(y)) = A(\xi_N, \eta_N) + O\left(\frac{1}{N^{1/3}}\right).$$

Combining these estimates with the uniform estimate $\varepsilon_N(x,y) = O(N^{-2/3})$ furnished by Lemma 7.15 gives the desired result. □

Applying this result to the determinantal formula (3.3) for the correlation functions completes the proof of Theorem 3.7.

Theorem 3.8 follows from Theorem 3.7 with the use of the relation (7.1) connecting the correlation functions of the particle and hole (dual) ensembles. One also uses Proposition 2.6 to change the square-root behavior of the equilibrium measure density near the upper constraint into square-root vanishing for the equilibrium measure density corresponding to the dual ensemble.

Now we turn our attention to the proof of Theorem 3.9. Here we present in detail a proof of (3.19). The proof of (3.20) is analogous and is left to the reader. The starting point is the fact that, according to (3.6), for any real s,

$$\mathbb{P}\left((x_{\min} - \alpha) \cdot (\pi N c B_\alpha^L)^{2/3} \geq -s\right)$$
$$= \mathbb{P}\left(x_{\min} \geq \alpha - \frac{s}{(\pi N c B_\alpha^L)^{2/3}}\right)$$
$$= \mathbb{P}\left(\text{there are no particles at any nodes } x_{N,j} \text{ satisfying } x_{N,j} < \alpha - \frac{s}{(\pi N c B_\alpha^L)^{2/3}}\right) \quad (7.31)$$
$$= \det(\mathbb{I} - K_{N,k}|_{L_s}),$$

where $L_s := \{y \in X_N \text{ such that } y < \alpha - s/(\pi N c B_\alpha^L)^{2/3}\}$ is the (finite, for each N) set of nodes that lie strictly to the left of $\alpha - s/(\pi N c B_\alpha^L)^{2/3}$. Since the right-hand side of (7.31) is the determinant of a finite matrix that we would like to compare with a Fredholm determinant, we will first define an integral operator $\tilde{A}_N|_{[s,\infty)}$ acting on $L^2[s,\infty)$ with a kernel $\tilde{A}_N(\xi,\eta)$ such that the Fredholm determinant $\det(1 - \tilde{A}_N|_{[s,\infty)})$ has precisely the same value for each N as the matrix determinant in (7.31). Moreover, it will be obvious from the construction that the kernel $\tilde{A}_N(\xi,\eta)$ will approximate the Airy kernel $A(\xi,\eta)$ at least pointwise.

Let $M(s)$ denote the index of the rightmost node $x_{N,M(s)}$ lying strictly to the left of $\alpha - s/(\pi N c B_\alpha^L)^{2/3}$ and let $x_{N,-1} < a$ be defined by

$$\int_{x_{N,-1}}^{a} \rho^0(x)\,dx = \frac{1}{2N}.$$

We define a kernel $\tilde{A}_N(\xi,\eta)$ on $[s,\infty) \times [s,\infty)$ by setting

$$\tilde{A}_N(\xi,\eta) := \frac{N^{1/3}}{(\pi c B_\alpha^L)^{2/3}} \sqrt{\rho^0\left(\alpha - \frac{\xi}{(\pi N c B_\alpha^L)^{2/3}}\right) \rho^0\left(\alpha - \frac{\eta}{(\pi N c B_\alpha^L)^{2/3}}\right)} K_{N,k}(x_{N,i}, x_{N,j})$$

if

$$(\pi N c B_\alpha^L)^{2/3}(\alpha - x_{N,i}) \leq \xi < (\pi N c B_\alpha^L)^{2/3}(\alpha - x_{N,i-1})$$

and
$$(\pi N c B_\alpha^L)^{2/3}(\alpha - x_{N,j}) \leq \eta < (\pi N c B_\alpha^L)^{2/3}(\alpha - x_{N,j-1}),$$
for all pairs of integers i and j satisfying $0 \leq i,j \leq M(s)$, and $\tilde{A}_N(\xi,\eta) := 0$ for all other $\xi \in [s,\infty)$ and $\eta \in [s,\infty)$.

By a direct computation, we have for each positive integer p,
$$\int_s^\infty \cdots \int_s^\infty \det(\tilde{A}_N(\xi_m,\xi_n))_{1 \leq m,n \leq p}\, d\xi_1 \ldots d\xi_p$$
$$= \sum_{i_1=0}^{M(s)} \cdots \sum_{i_p=0}^{M(s)} \det(K_{N,k}(x_{N,i_m},x_{N,i_n}))_{1 \leq m,n \leq p} \left\{ N \int_{x_{N,i_1-1}}^{x_{N,i_1}} \rho^0(x_1)\, dx_1 \cdots N \int_{x_{N,i_p-1}}^{x_{N,i_p}} \rho^0(x_p)\, dx_p \right\}$$
$$= \sum_{i_1=0}^{M(s)} \cdots \sum_{i_p=0}^{M(s)} \det(K_{N,k}(x_{N,i_m},x_{N,i_n}))_{1 \leq m,n \leq p}.$$

This calculation uses the quantization rule (1.15) that defines the positions of the nodes in terms of the function $\rho^0(x)$. The infinite series formula for the Fredholm determinant then implies that $\det(1 - \tilde{\mathcal{A}}_N|_{[s,\infty)}) = \det(\mathbb{I} - K_{N,k}|_{L_s})$.

Therefore, to prove (3.19), we need to show that as $N \to \infty$,
$$\tilde{\mathcal{A}}_N|_{[s,\infty)} \to \mathcal{A}|_{[s,\infty)} \quad \text{in trace norm.}$$

Since $\tilde{\mathcal{A}}_N|_{[s,\infty)}$ and $\mathcal{A}|_{[s,\infty)}$ are both positive trace class operators, the following two conditions [Sim79] imply convergence in trace norm:

(a) $\operatorname{tr} \tilde{\mathcal{A}}_N|_{[s,\infty)} \to \operatorname{tr} \mathcal{A}|_{[s,\infty)}$

(b) $\tilde{\mathcal{A}}_N|_{[s,\infty)} \to \mathcal{A}|_{[s,\infty)}$, in the weak-$*$ topology.

For the purpose of establishing these two conditions, the following properties of the kernels $\tilde{A}_N(\xi,\eta)$ and $A(\xi,\eta)$ are essential ingredients.

Lemma 7.17. *For each fixed ξ and η,*
$$\lim_{N \to \infty} \tilde{A}_N(\xi,\eta) = A(\xi,\eta).$$
Also, there are positive constants C and D such that the estimate
$$\left| \tilde{A}_N(\xi,\eta) \right| \leq C e^{-D(|\xi|^{3/2} + |\eta|^{3/2})}$$
holds for all $\xi > s$ and $\eta > s$ (the constants C and D depend on s but not on N). An estimate of the same form holds with $\tilde{A}_N(\xi,\eta)$ replaced by $A(\xi,\eta)$.

Proof. The pointwise convergence follows from Lemma 7.16 since ξ and η fixed corresponds to $x - \alpha$ and $y - \alpha$ of order $N^{-2/3}$. To obtain the claimed estimates, one uses Lemma 7.15 when ξ and η are of order $N^{2/3}$ such that the corresponding values of x and y are in the disc $D_\Gamma^{\nabla,L}$ surrounding the band edge α. When ξ and η are such that the corresponding x and y values are both outside the disc, one uses Lemma 7.14 to obtain an exponential estimate in terms of the variables ξ and η. Finally, when x is in the disc and y is outside the disc (or vice versa), we may use the exact representation given by Proposition 7.9 and similar calculations. □

We first prove (a). From our definition of $\tilde{\mathcal{A}}_N|_{[s,\infty)}$,
$$\operatorname{tr} \tilde{\mathcal{A}}_N|_{[s,\infty)} = \int_{[s,\infty)} \tilde{A}_N(\xi,\xi)\, d\xi$$
$$= \int_{[s,N^{1/6})} \tilde{A}_N(\xi,\xi)\, d\xi + \int_{[N^{1/6},\infty)} \tilde{A}_N(\xi,\xi)\, d\xi.$$

UNIVERSALITY: PROOFS OF THEOREMS STATED IN §3.3

Applying Lemma 7.17, we see that the second integral is exponentially small as $N \to \infty$. On the other hand, from the definition of $\tilde{A}_N$, the first integral satisfies

$$\sum_{i=M(N^{1/6})+1}^{M(s)} K_{N,k}(x_{N,i}, x_{N,i}) \leq \int_s^{N^{1/6}} \tilde{A}_N(\xi, \xi)\, d\xi \leq \sum_{i=M(N^{1/6})}^{M(s)+1} K_{N,k}(x_{N,i}, x_{N,i}).$$

Using Lemma 7.16 and the fact that each of the above sums consists of $O(N^{1/2})$ terms, we find

$$\sum_{i=M(N^{1/6})+1}^{M(s)} A(\xi_N^{(i)}, \xi_N^{(i)}) \Delta\xi + O(N^{-1/6}) \leq \int_s^{N^{1/6}} \tilde{A}_N(\xi,\xi)\, d\xi \leq \sum_{i=M(N^{1/6})}^{M(s)+1} A(\xi_N^{(i)}, \xi_N^{(i)}) \Delta\xi + O(N^{-1/6}),$$

where

$$\xi_N^{(i)} = (\pi N c B_\alpha^L)^{2/3}(\alpha - x_{N,i}) \quad \text{and} \quad \Delta\xi := \frac{(\pi c B_\alpha^L)^{2/3}}{N^{1/3}\rho^0(\alpha)}.$$

Given the asymptotic equal spacing of $\Delta\xi$ between consecutive points $\xi_N^{(i)}$ in the limit $N \to \infty$, we see that both sums above are in fact Riemann sums:

$$\lim_{N \to \infty} \sum_{i=M(N^{1/6})+1}^{M(s)} A(\xi_N^{(i)}, \xi_N^{(i)}) \Delta\xi = \lim_{N \to \infty} \sum_{i=M(N^{1/6})}^{M(s)+1} A(\xi_N^{(i)}, \xi_N^{(i)}) \Delta\xi = \int_s^\infty A(\xi, \xi)\, d\xi,$$

which proves (a).

In order to check the condition (b), we need to show that, for any $f, g \in L^2[s, \infty)$,

$$\int_s^\infty \int_s^\infty f(\xi)^* \tilde{A}_N(\xi, \eta) g(\eta)\, d\xi\, d\eta \to \int_s^\infty \int_s^\infty f(\xi)^* A(\xi, \eta) g(\eta)\, d\xi\, d\eta$$

as $N \to \infty$, where the asterisk denotes complex conjugation. But, from Lemma 7.17,

$$f(\xi)^* \tilde{A}_N(\xi, \eta) g(\eta) \to f(\xi)^* A(\xi, \eta) g(\eta)$$

as $N \to \infty$ for almost every ξ and η, and also $|f(\xi)^* \tilde{A}_N(\xi, \eta) g(\eta)| \leq C|f(\xi)| e^{-D|\xi|^{3/2}} |g(\eta)| e^{-D|\eta|^{3/2}}$, a bound that is independent of N. By Cauchy-Schwarz,

$$\int_s^\infty \int_s^\infty C|f(\xi)| e^{-D|\xi|^{3/2}} |g(\eta)| e^{-D|\eta|^{3/2}}\, d\xi\, d\eta \leq C\|f\|_2 \|g\|_2 \int_s^\infty e^{-2D|x|^{3/2}}\, dx < \infty,$$

so the desired result follows from the Lebesgue Dominated Convergence Theorem. Hence both conditions (a) and (b) hold, and this completes the proof of (3.19).

The proof of Theorem 3.10 follows from that of Theorem 3.9 by duality.

Appendix A

The Explicit Solution of Riemann-Hilbert Problem 5.1

A.1 STEPS FOR MAKING THE JUMP MATRIX PIECEWISE-CONSTANT: THE TRANSFORMATION FROM $\dot{\mathbf{X}}(z)$ TO $\mathbf{Y}^{\sharp}(z)$

The first step in solving Riemann-Hilbert Problem 5.1 is to introduce a change of variables leading to a piecewise-constant jump matrix. Suppose that $h(z)$ is a function analytic for $z \in \mathbb{C} \setminus (-\infty, \beta_G]$ and consider the change of variables

$$\mathbf{Y}(z) := \dot{\mathbf{X}}(z) e^{(\kappa g(z) - h(z))\sigma_3}. \tag{A.1}$$

Then, since by definition $-i\phi_{\Gamma_j} = \kappa g_+(z) - \kappa g_-(z)$ when z is in any gap Γ_j, for $z \in (-\infty, \beta_G)$ setting $h_\pm(z) := \lim_{\epsilon \downarrow 0} h(z \pm i\epsilon)$, we have the jump condition

$$\mathbf{Y}_+(z) = \mathbf{Y}_-(z) \begin{pmatrix} e^{iN\theta_{\Gamma_j} - h_+(z) + h_-(z)} & 0 \\ 0 & e^{-iN\theta_{\Gamma_j} + h_+(z) - h_-(z)} \end{pmatrix},$$

for $z \in \Gamma_j$ for $j = 1, \ldots, G$, and

$$\mathbf{Y}_+(z) = \mathbf{Y}_-(z) \begin{pmatrix} 0 & -ie^{\gamma - \eta(z) + h_+(z) + h_-(z)} \\ -ie^{\eta(z) - \gamma - h_+(z) - h_-(z)} & 0 \end{pmatrix},$$

for z in any band I_j for $j = 0, 1, \ldots, G$. Finally, since for all real $z < \alpha_0$ we have $g_+(z) - g_-(z) = 2\pi i$, we have introduced a new discontinuity into $\mathbf{Y}(z)$ by the change of variables (A.1):

$$\mathbf{Y}_+(z) = \mathbf{Y}_-(z) \begin{pmatrix} e^{2\pi i \kappa - h_+(z) + h_-(z)} & 0 \\ 0 & e^{-2\pi i \kappa + h_+(z) - h_-(z)} \end{pmatrix},$$

for $z \in (-\infty, \alpha_0)$.

In order to arrive at a problem with piecewise-constant jump matrices that is still normalized to the identity matrix as $z \to \infty$, we thus insist that $h(z)$ be the solution of the following scalar Riemann-Hilbert problem.

Riemann-Hilbert Problem A.1. *Find a scalar function $h(z)$ with the following properties:*

1. **Analyticity**: $h(z)$ *is an analytic function of z for $z \in \mathbb{C} \setminus (-\infty, \beta_G]$.*

2. **Normalization**: *As $z \to \infty$,*

$$h(z) = \kappa \log(z) + O\left(\frac{1}{z}\right). \tag{A.2}$$

3. **Jump Conditions**: *$h(z)$ takes piecewise-continuous boundary values on $(-\infty, \beta_G]$ with jump discontinuities allowed only at the band endpoints. For real z, let $h_\pm(z) := \lim_{\epsilon \downarrow 0} h(z \pm i\epsilon)$. For z in the gap $\Gamma_j = (\beta_{j-1}, \alpha_j)$, $j = 1, \ldots, G$, the boundary values satisfy*

$$h_+(z) - h_-(z) = ic_j, \tag{A.3}$$

where $c_1, \ldots, c_G$ are some real constants. For z in any band $I_j = (\alpha_j, \beta_j)$, $j = 0, \ldots, G$, the boundary values satisfy

$$h_+(z) + h_-(z) = \eta(z) - \gamma, \tag{A.4}$$

where γ is a real constant (the same constant for all bands). Finally, for real $z < \alpha_0$,

$$h_+(z) - h_-(z) = 2\pi i \kappa. \tag{A.5}$$

The determination of the constants $c_1, \ldots, c_G$ and the constant γ is part of the problem.

To solve Riemann-Hilbert Problem A.1 it is easiest to first solve for $h'(z)$. Evidently, the function $h'(z)$ should be analytic for $z \in \mathbb{C}\setminus\cup_j I_j$; in each band I_j the boundary values should satisfy $h'_+(z) + h'_-(z) = \eta'(z)$. As $z \to \infty$, we require the normalization condition $h'(z) = \kappa/z + O(z^{-2})$. In order to obtain a formula for $h'(z)$, recall the analytic function $R(z)$ defined for $z \in \mathbb{C} \setminus \cup_k I_k$ by (2.26) and set

$$h'(z) = \frac{k(z)}{R(z)}$$

to introduce a new unknown function $k(z)$. Evidently, $k(z)$ must be analytic in $\mathbb{C} \setminus \cup_j I_k$, and its boundary values $k_\pm(z) := \lim_{\epsilon\downarrow 0} k(z \pm i\epsilon)$ for z in a band necessarily satisfy

$$k_+(z) - k_-(z) = \eta'(z) R_+(z), \qquad \text{for } z \text{ in any band } I_j,$$

since $R_+(z) + R_-(z) = 0$ holds for z in the bands with $R_\pm(z) := \lim_{\epsilon\downarrow 0} R(z \pm i\epsilon)$. Taking into account the required asymptotic behavior of $k(z)$ for large z implied by (A.2), we solve for $k(z)$ in terms of a Cauchy integral:

$$k(z) = \frac{1}{2\pi i} \int_{\cup_j I_j} \frac{\eta'(x) R_+(x)}{x - z} dx + \kappa z^G + \sum_{p=0}^{G-1} f_p z^p.$$

This is the general solution for $k(z)$, and this establishes the formula (2.27) for $h'(z)$. The constants $f_0, \ldots, f_{G-1}$ seem at this point to be arbitrary; the next step is therefore to explain how they are determined.

Note that the inverse square-root singularities present in $h'(z)$ at the band endpoints are integrable, so $h(z)$ will indeed have piecewise-continuous boundary values as required. To complete the solution of Riemann-Hilbert Problem A.1, we must ensure that the identity $h'_+(z) + h'_-(z) = \eta'(z)$ holding in each distinct band I_j actually implies that $h_+(z) + h_-(z) = \eta(z) - \gamma$ holds with the same integration constant $-\gamma$ in each band. We thus require that the conditions (2.28) all hold. Substituting into (2.28) from (2.27), we obtain a square linear system of equations on the unknowns $f_1, \ldots, f_G$:

$$\sum_{m=0}^{G-1} f_m \int_{\Gamma_l} \frac{z^m \, dz}{R(z)} = -\int_{\Gamma_l} \left[\frac{1}{2\pi i} \int_{\cup_j I_j} \frac{\eta'(x) R_+(x)}{x - z} dx + \kappa z^G \right] \frac{dz}{R(z)}, \qquad \text{for } l = 1, \ldots, G. \tag{A.6}$$

The linear system (A.6) is invertible. The determinant of the coefficient matrix is easily seen by multilinearity to be

$$\det\left(\left\{ \int_{\Gamma_l} \frac{s^{m-1} \, ds}{R(s)} \right\}_{1 \leq l, m \leq G} \right) = \int_{\Gamma_1} \cdots \int_{\Gamma_G} D_G(s_1, \ldots, s_G) \frac{ds_G}{R(s_G)} \cdots \frac{ds_1}{R(s_1)},$$

where $D_G(s_1, \ldots, s_G)$ is the Vandermonde determinant $\det(\{s_l^{m-1}\}_{1 \leq l,m \leq G})$. Since the gaps $\Gamma_1, \ldots, \Gamma_G$ are separated from each other by the bands (i.e., $\Gamma_j = (\beta_{j-1}, \alpha_j)$ and $\alpha_j < \beta_j$ for all j), the strict inequalities $s_1 < s_2 < \cdots < s_G$ hold throughout the range of integration, which implies that $D_G(s_1, \ldots, s_G)$ is of one sign. Similarly, the product $R(s_1) R(s_2) \cdots R(s_G)$ is also of one sign. This proves that the determinant of (A.6) is nonzero. Note that the constants $f_0, \ldots, f_{G-1}$ solving (A.6) are all real because $R_+(x)$ is purely imaginary in the bands and $R(z)$ is purely real in the gaps.

With the real constants $f_0, \ldots, f_{G-1}$ determined in this way, and taking into account the normalization condition (A.2) on $h(z)$ as $z \to \infty$, we see that $h(z)$ must be given in terms of $h'(z)$ by the integral formula (2.29). Note that in (2.29), the point at infinity can be approached in any direction since $\kappa/z - h'(z) = O(z^{-2})$ as $z \to \infty$. In particular, if we consider $z < \alpha_0$ and take paths of integration with $\arg(s - \alpha_0) = \pm \pi$ to compute $h_\pm(z)$, then it is easy to see that the condition (A.5) is satisfied for $z < \alpha_0$. The formula (2.29) clearly satisfies equations of the form (A.3) for $j = 1, \ldots, G$; moreover, the constants c_j defined by (2.31), with $h(z)$ given by (2.29), are all real. Furthermore, we obtain the formula (2.30) for the integration constant γ. Clearly, γ depends only on the function $\eta(z)$ and the configuration of endpoints $\alpha_0 < \beta_0 < \alpha_1 < \beta_1 < \cdots < \alpha_G < \beta_G$.

Given a configuration of endpoints, the solution $h(z)$ of Riemann-Hilbert Problem A.1 depends additionally on the data $(\kappa, \eta(\cdot))$. A key property of the function $h(z)$, easily verified by superposition, is the following.

Proposition A.2. *Fix a configuration of endpoints. Let $h_0(z)$ be the solution of Riemann-Hilbert Problem A.1 corresponding to the data $(0, \eta(\cdot))$, with the real constants in (A.3) denoted by $c_{j,0}$ and with the integration constant in (A.4) denoted by $\gamma^{(0)}$. Let $h_1(z)$ be the solution of Riemann-Hilbert Problem A.1 corresponding to the data $(1, 0)$, with the real constants in (A.3) denoted by ω_j and with the integration constant in (A.4) denoted by $\gamma^{(1)}$. Finally, let $h(z)$ be the solution of Riemann-Hilbert Problem A.1 corresponding to general data $(\kappa, \eta(\cdot))$, with constants c_j and γ. Then,*

$$h(z) = h_0(z) + \kappa h_1(z),$$
$$c_j = c_{j,0} + \omega_j \kappa, \quad \text{for } j = 1, \ldots, G,$$
$$\gamma = \gamma^{(0)} + \gamma^{(1)} \kappa.$$

The quantities ω_j, which are interpreted as frequencies, are independent of κ and $\eta(\cdot)$, depending only on the value of the parameter $c \in (0,1)$, the functions $V(\cdot)$ and $\rho^0(\cdot)$, and the corresponding equilibrium measure. The quantities $c_{j,0}$ are similar but depend additionally on the function $\eta(\cdot)$.

The function $\mathbf{Y}(z)$ related to $\dot{\mathbf{X}}(z)$ by (A.1) is analytic for $z \in \mathbb{C} \setminus \Sigma_{\text{model}}$, and it takes boundary values on Σ_{model} that are continuous except at the band edges where inverse fourth-root singularities may exist. The boundary values $\mathbf{Y}_\pm(z) := \lim_{\epsilon \downarrow 0} \mathbf{Y}(z \pm i\epsilon)$ for $z \in \Sigma_{\text{model}}$ satisfy the jump relations

$$\mathbf{Y}_+(z) = \mathbf{Y}_-(z) \begin{pmatrix} e^{iN\theta_{\Gamma_j}} e^{-i(c_{j,0} + \omega_j \kappa)} & 0 \\ 0 & e^{-iN\theta_{\Gamma_j}} e^{i(c_{j,0} + \omega_j \kappa)} \end{pmatrix},$$

for $z \in \Gamma_j$, and

$$\mathbf{Y}_+(z) = \mathbf{Y}_-(z) \begin{pmatrix} 0 & -i \\ -i & 0 \end{pmatrix},$$

for $z \in I_j$. From (A.2) and the asymptotic relation $g(z) = \log(z) + O(1/z)$ as $z \to \infty$, which follows from (4.6), we see that $\mathbf{Y}(z) = \mathbb{I} + O(z^{-1})$ as $z \to \infty$.

To find $\mathbf{Y}(z)$ and thus to explain the solution of Riemann-Hilbert Problem 5.1, first consider the matrix $\mathbf{Y}^\sharp(z)$ related to $\mathbf{Y}(z)$ by

$$\mathbf{Y}^\sharp(z) := \begin{cases} \mathbf{Y}(z), & \Im(z) > 0, \\ \mathbf{Y}(z) \begin{pmatrix} 0 & -i \\ -i & 0 \end{pmatrix}, & \Im(z) < 0. \end{cases}$$

This matrix tends to the identity as $z \to \infty$ only when $\Im(z) > 0$. However, the advantage is that now the jump discontinuities will be characterized everywhere by piecewise-constant off-diagonal matrices. That is, if we introduce the notation Γ_0 for $\mathbb{R} \setminus \Sigma_{\text{model}} = (-\infty, \alpha_0) \cup (\beta_G, \infty)$, then $\mathbf{Y}^\sharp(z)$ is continuous and thus analytic for z in any of the bands I_j. On the other hand, letting $\mathbf{Y}^\sharp_\pm(z) := \lim_{\epsilon \downarrow 0} \mathbf{Y}^\sharp(z \pm i\epsilon)$ for real z, we see that for z in the interval Γ_j,

$$\mathbf{Y}^\sharp_+(z) = \mathbf{Y}^\sharp_-(z) \begin{pmatrix} 0 & ie^{-iN\theta_{\Gamma_j}} e^{i(c_{j,0} + \omega_j \kappa)} \\ ie^{iN\theta_{\Gamma_j}} e^{-i(c_{j,0} + \omega_j \kappa)} & 0 \end{pmatrix},$$

for $j = 1, 2, \ldots, G$, and for $z \in \Gamma_0$,

$$\mathbf{Y}^\sharp_+(z) = \mathbf{Y}^\sharp_-(z) \begin{pmatrix} 0 & i \\ i & 0 \end{pmatrix}.$$

Thus $\mathbf{Y}^\sharp(z)$ is analytic for $z \in \mathbb{C} \setminus \Sigma'_{\text{model}}$, where $\Sigma'_{\text{model}} := \Gamma_0 \cup \cdots \cup \Gamma_G$.

A.2 CONSTRUCTION OF $\mathbf{Y}^\sharp(z)$ USING HYPERELLIPTIC FUNCTION THEORY

We will first develop the solution assuming that $G > 0$. Along with the contour Σ'_{model}, we associate the hyperelliptic Riemann surface S whose model is two copies of the complex plane cut and identified along

Σ'_{model}. Such a surface comes equipped with a function $z : S \to \mathbb{C}$, $P \mapsto z(P)$ that realizes the identification of each sheet of S with the complex plane. Each point $z \in \mathbb{C}$, with the exception of the endpoints $\alpha_0, \ldots, \beta_G$, has two preimages on S. Let $y(z)$ be the function analytic for $z \in \mathbb{C} \setminus \Sigma'_{\text{model}}$ that satisfies

$$y(z)^2 = -\prod_{k=0}^{G}(z-\alpha_k)(z-\beta_k) \quad \text{and} \quad y(z) \sim iz^{G+1} \text{ as } z \to \infty \text{ with } \Im(z) > 0.$$

This function may be analytically continued to all of the Riemann surface S with the exception of the two preimages of $z = \infty$ as a function $y^S(P)$, for $P \in S$. We may distinguish the two sheets of S according to whether $y^S(P) = y(z(P))$ or $y^S(P) = -y(z(P))$, as long as $z(P) \in \mathbb{C} \setminus \Sigma'_{\text{model}}$. The polynomial relation that realizes S as an algebraic curve is then

$$y^S(P)^2 = \prod_{j=0}^{G}(z(P)-\alpha_j)(z(P)-\beta_j), \qquad P \in S.$$

We next specify for S a basis of homology cycles: closed contours a_k encircling Γ_k on the sheet where $y^S(P) = y(z(P))$, for $k = 1, \ldots, G$, in the counterclockwise direction and conjugate contours b_k oriented in the clockwise direction and chosen exactly so that b_k intersects only the closed cycle a_k, precisely once from the left of a_k. The homology basis is illustrated in Figure A.1.

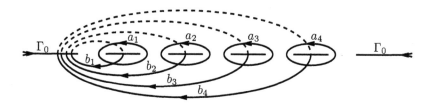

Figure A.1 *The contour Σ'_{model} and the basis $a_1, \ldots, a_G, b_1, \ldots, b_G$ of homology cycles on the associated two-sheeted Riemann surface S. The sheet on which the cycles are shown with solid curves is that on which $y^S(P) = y(z(P))$ (on the other sheet $y^S(P) = -y(z(P))$).*

Also, let a vector of holomorphic differentials $\mathbf{m}^S(P) \in \mathbb{C}^G$ be defined for $P \in S$ to have components

$$m_p^S(P) := \frac{z(P)^{p-1}}{y^S(P)} \, dz(P), \qquad \text{for } p = 1, \ldots, G.$$

A corresponding vector $\mathbf{m}(z) \in \mathbb{C}^G$ may be defined for $z \in \mathbb{C} \setminus \Sigma'_{\text{model}}$ to have components $m_p(z) := z^{p-1}/y(z)$, for $p = 1, \ldots, G$. We define a $G \times G$ constant matrix $\mathbf{A}$ of coefficients so that

$$\oint_{a_j} \mathbf{A}\mathbf{m}^S(P) = 2\pi i \mathbf{e}^{(j)}, \qquad \text{for } k = 1, \ldots, G,$$

where $\mathbf{e}^{(j)}$ are the standard unit vectors in $\mathbb{C}^G$. These equations determining the matrix $\mathbf{A}$ may be written in the equivalent form (2.33), which makes the integration concrete and also makes it clear that the elements of the matrix $\mathbf{A}$ real. We use the notation $\mathbf{a}^{(1)}, \ldots, \mathbf{a}^{(G)}$ to denote (in order) the columns of $\mathbf{A}$. The vector $\mathbf{A}\mathbf{m}^S(P)$ is the vector of *normalized holomorphic differentials* on S, the normalization being relative to the cycles $a_1, \ldots, a_G$. With $\mathbf{A}$ so determined, we construct vectors $\mathbf{b}^{(j)} \in \mathbb{C}^G$ by defining

$$\mathbf{b}^{(j)} := \oint_{b_j} \mathbf{A}\mathbf{m}^S(P), \tag{A.7}$$

and we denote by $\mathbf{B}$ the matrix whose columns are in order $\mathbf{b}^{(1)}, \ldots, \mathbf{b}^{(G)}$. The definition (A.7) may be written in the equivalent form (2.34), which makes the integration concrete. The matrix $\mathbf{B}$ is real, symmetric, and negative-definite, and thus $\mathbf{B}$ defines for $\mathbf{w} \in \mathbb{C}^G$ a Riemann theta function $\Theta(\mathbf{w})$ by the Fourier series (2.36).

Given a base point $P_0 \in S$, the Abel-Jacobi mapping $\mathbf{w}^S(P)$ is defined by
$$\mathbf{w}^S(P) := \int_{P_0}^{P} \mathbf{Am}^S,$$
and since the path of integration on S is not specified, the mapping is made well defined by taking the range to be the *Jacobian variety* $\mathrm{Jac}(S) = \mathbb{C}^G / \Lambda$, where Λ is the integer lattice with basis vectors $2\pi i \mathbf{e}^{(1)}, \ldots, 2\pi i \mathbf{e}^{(G)}$ and $\mathbf{b}^{(1)}, \ldots, \mathbf{b}^{(G)}$. The definition (2.37) of $\mathbf{w}(z)$ is a concrete version of the Abel-Jacobi mapping with base point $z = \alpha_0$. Since $\mathbf{m}(z)$ behaves like
$$\mathbf{m}(z) = -i \frac{\mathrm{sgn}(\Im(z))}{z^2} \mathbf{a}^{(G)} + O\left(\frac{1}{z^3}\right) \qquad \text{as } z \to \infty,$$
we see that $\mathbf{m}(z)$ is integrable at infinity in the two half-planes. The asymptotic behavior of $\mathbf{w}(z)$ may be easily computed:
$$\mathbf{w}(z) = \begin{cases} \mathbf{w}_+(\infty) + i\mathbf{a}^{(G)} \dfrac{1}{z} + O\left(\dfrac{1}{z^2}\right) & \text{as } z \to \infty \text{ with } \Im(z) > 0, \\ \mathbf{w}_-(\infty) - i\mathbf{a}^{(G)} \dfrac{1}{z} + O\left(\dfrac{1}{z^2}\right) & \text{as } z \to \infty \text{ with } \Im(z) < 0, \end{cases}$$
where the special values $\mathbf{w}_\pm(\infty)$ are defined by (2.38).

For $z \in \mathbb{R}$, we denote by $\mathbf{m}_\pm(z)$ and $\mathbf{w}_\pm(z)$ the boundary values taken by $\mathbf{m}(z)$ and $\mathbf{w}(z)$ on $\mathbb{R}$ from the half-planes $\mathbb{C}_\pm$. The boundary values $\mathbf{w}_\pm(z)$ are continuous functions with the following expressions:
$$\mathbf{w}_\pm(z) = \int_{\alpha_0}^{z} \mathbf{Am}_\pm(x)\, dx, \qquad \text{for } z \in \Gamma_0, \tag{A.8}$$
$$\mathbf{w}_\pm(z) = -\frac{1}{2}\mathbf{b}^{(j)} \mp \sum_{k=1}^{j-1} \pi i \mathbf{e}^{(k)} + \int_{\beta_{j-1}}^{z} \mathbf{Am}_\pm(x)\, dx, \qquad \text{for } z \in \Gamma_j,\ j = 1, \ldots, G,$$
$$\mathbf{w}_\pm(z) = \int_{\alpha_0}^{z} \mathbf{Am}(x)\, dx, \qquad \text{for } z \in I_0,$$
and
$$\mathbf{w}_\pm(z) = -\frac{1}{2}\mathbf{b}^{(j)} \mp \sum_{k=1}^{j} \pi i \mathbf{e}^{(k)} + \int_{\alpha_j}^{z} \mathbf{Am}(x)\, dx, \qquad \text{for } z \in I_j,\ j = 1, \ldots, G.$$

If z lies in the right half of Γ_0, we interpret the integral in (A.8) as always lying on Γ_0 and passing through the point at infinity. Since $\mathbf{m}(z)$ is analytic for $z \in \mathbb{C} \setminus \Sigma'_{\mathrm{model}}$ and when $z \in \Sigma'_{\mathrm{model}}$ we have $\mathbf{m}_+(z) + \mathbf{m}_-(z) = 0$, the boundary values of $\mathbf{w}(z)$ on the real axis are related as follows:
$$\mathbf{w}_+(z) = -\mathbf{w}_-(z) - \mathbf{b}^{(j)}, \qquad \text{for } z \in \Gamma_j,\ j = 1, \ldots, G, \tag{A.9}$$
and
$$\mathbf{w}_+(z) = -\mathbf{w}_-(z), \qquad \text{for } z \in \Gamma_0, \tag{A.10}$$
and
$$\mathbf{w}_+(z) = \mathbf{w}_-(z) - \sum_{k=1}^{j} 2\pi i \mathbf{e}^{(k)}, \qquad \text{for } z \in I_j,\ j = 0, \ldots, G. \tag{A.11}$$

By the $2\pi i$ periodicity of $\Theta(\mathbf{w})$ in each coordinate direction of $\mathbb{C}^G$, we see from (A.11) that for any vector $\mathbf{q} \in \mathbb{C}^G$, the function
$$f(z; \mathbf{q}) := \Theta(\mathbf{w}(z) - \mathbf{q})$$
is analytic in $\mathbb{C} \setminus \Sigma'_{\mathrm{model}}$ and in fact takes continuous boundary values on Σ'_{model}. Moreover, using the facts
$$\Theta(-\mathbf{w}) = \Theta(\mathbf{w}) \qquad \text{and} \qquad \Theta(\mathbf{w} \pm \mathbf{b}^{(j)}) = e^{-B_{jj}/2} e^{\pm w_j} \Theta(\mathbf{w})$$

holding for all $\mathbf{w} \in \mathbb{C}^G$ and easily derived directly from the Fourier series (2.36), we find from (A.10) that if for real z we define $f_\pm(z;\mathbf{q}) := \lim_{\epsilon\downarrow 0} f(z\pm i\epsilon;\mathbf{q})$, then

$$f_+(z;\mathbf{q}) = f_-(z;-\mathbf{q}), \qquad \text{for } z \in \Gamma_0, \tag{A.12}$$

and from (A.9) that for $j = 1,\ldots,G$,

$$\begin{aligned}f_+(z;\mathbf{q}) &= e^{-B_{jj}/2 - w_{j-}(z)} e^{q_j} f_-(z;-\mathbf{q}) \\ &= e^{B_{jj}/2 + w_{j+}(z)} e^{q_j} f_-(z;-\mathbf{q}),\end{aligned} \tag{A.13}$$

when $z \in \Gamma_j$. Now with the vector $\mathbf{r}$ defined componentwise by (2.32) and with the frequency vector $\mathbf{\Omega}$ having components $\omega_1,\ldots,\omega_G$, consider the quotient functions

$$g^\pm(z;\mathbf{q}) := \frac{f(z;\pm\mathbf{q}\pm i\mathbf{r}\mp i\kappa\mathbf{\Omega})}{f(z;\pm\mathbf{q})} = \frac{\Theta(\mathbf{w}(z)\mp\mathbf{q}\mp i\mathbf{r}\pm i\kappa\mathbf{\Omega})}{\Theta(\mathbf{w}(z)\mp\mathbf{q})}.$$

As long as the denominator does not vanish identically, it will have at most G zeros on $\mathbb{C}\setminus\Sigma'_\text{model}$ (more precisely, replacing $\mathbf{w}(z)$ by $\mathbf{w}^S(P)$, the resulting function of P will have exactly G zeros on the Riemann surface S, counting multiplicity, and these may occur on either of the two sheets). These quotient functions $g^\pm(z;\mathbf{q})$, when $\mathbf{q}$ is such that the denominator is not identically zero, are meromorphic functions for $z \in \mathbb{C}\setminus\Sigma'_\text{model}$. From (A.12) and (A.13), we then find that

$$g^\pm_+(z;\mathbf{q}) = e^{\pm iN\theta_{\Gamma_j}} e^{\mp i(c_{j,0}+\omega_j\kappa)} g^\mp_-(z;\mathbf{q}), \qquad \text{for } z \in \Gamma_j,\; j=1,\ldots,G,$$

and for $z \in \Gamma_0$,

$$g^\pm_+(z;\mathbf{q}) = g^\mp_-(z;\mathbf{q}).$$

The subscripts again denote boundary values taken as real z is approached from the upper and lower half-planes, just as for $f(z;\mathbf{q})$. The functions $g^\pm(z;\mathbf{q})$ also take finite values as $z \to \infty$ separately in each half-plane, and in particular we have the asymptotic formula

$$\frac{g^\pm(z;\mathbf{q})}{g^\pm_+(\infty;\mathbf{q})} = 1 + \frac{i}{z}\mathbf{a}^{(G)}\cdot\left[\frac{\nabla\Theta(\mathbf{w}_+(\infty)\mp\mathbf{q}\mp i\mathbf{r}\pm i\kappa\mathbf{\Omega})}{\Theta(\mathbf{w}_+(\infty)\mp\mathbf{q}\mp i\mathbf{r}\pm i\kappa\mathbf{\Omega})} - \frac{\nabla\Theta(\mathbf{w}_+(\infty)\mp\mathbf{q})}{\Theta(\mathbf{w}_+(\infty)\mp\mathbf{q})}\right] + O\left(\frac{1}{z^2}\right)$$

as $z \to \infty$ with $\Im(z) > 0$, where ∇ denotes the gradient vector in $\mathbb{C}^G$, and thus $\mathbf{a}^{(G)}\cdot\nabla$ is a derivative in the direction of $\mathbf{a}^{(G)}$.

Comparing the desired jump relations satisfied by $\mathbf{Y}^\sharp(z)$ in the gaps with the jump relations satisfied by the functions $g^\pm(z;\mathbf{q})$, we are led to the strategy of constructing the matrix $\mathbf{Y}^\sharp(z)$ from the quotient functions $g^\pm(z;\mathbf{q})$ by choosing $\mathbf{q}$ appropriately. The functions $g^\pm(z;\mathbf{q})$ have poles in $\mathbb{C}\setminus\Sigma'_\text{model}$ corresponding to the zeros of the denominators; however, they are also typically finite at the endpoints $\alpha_0,\ldots,\beta_G$. Since we can admit mild singularities in $\mathbf{Y}^\sharp(z)$ at the endpoints, we may introduce additional functional factors with such singularities that also have zeros that cancel any poles in $g^\pm(z;\mathbf{q})$. Thus we may seek the matrix elements of $\mathbf{Y}^\sharp(z)$ in the form of products of $g^\pm(z;\mathbf{q})$ with these functional factors and choosing the vectors $\mathbf{q}$ appropriately.

To introduce the correct functional factors, recall the function $\lambda(z)$ defined for $z \in \mathbb{C}\setminus\Sigma'_\text{model}$ by (2.39) and the corresponding functions $u(z)$ and $v(z)$ defined in the same domain by (2.40). Noting that $e^{i\pi/4}\lambda(z)$ is a real-analytic function that is positive for $z \in \mathbb{R}\setminus\Sigma'_\text{model}$, we obtain the identity

$$v(z) = -u(z^*)^*.$$

Both functions $u(z)$ and $v(z)$ are analytic throughout their domains of definition. The boundary values $u_\pm(z) := \lim_{\epsilon\downarrow 0} u(z\pm i\epsilon)$ and $v_\pm(z) := \lim_{\epsilon\downarrow 0} v(z\pm i\epsilon)$ taken on Σ'_model have mild singularities at the endpoints but are otherwise continuous and satisfy

$$u_+(z) = -v_-(z) \quad \text{and} \quad v_+(z) = u_-(z), \qquad \text{for } z \in \Sigma'_\text{model}.$$

Also, we clearly have $u(z) \to 1$ and $v(z) \to 0$ as $z \to \infty$ with $\Im(z) > 0$; more precisely,

$$u(z) = 1 + O\left(\frac{1}{z^2}\right) \quad \text{and} \quad v(z) = \frac{1}{4iz}\sum_{k=0}^G (\beta_k - \alpha_k) + \frac{1}{8iz^2}\sum_{k=0}^G (\beta_k^2 - \alpha_k^2) + O\left(\frac{1}{z^3}\right)$$

THE EXPLICIT SOLUTION OF RIEMANN-HILBERT PROBLEM 5.1

as $z \to \infty$ with $\Im(z) > 0$.

To locate the zeros of $u(z)$ and $v(z)$, note that

$$u(z)u(z^*)^* = -u(z)v(z) = \frac{i}{4\lambda(z)^2}\left[\lambda(z)^4 - 1\right], \quad (A.14)$$

which proves that any zeros of $u(z)$ must be real. But the right-hand side does not vanish for $z \in \mathbb{C}\setminus\Sigma'_{\text{model}}$ and therefore strictly speaking $u(z)$ is nonzero in its domain of definition. However, the right-hand side of (A.14) has exactly one simple zero $z = x_j$ in the interior of Γ_j, for each $j = 1,\ldots,G$, and no other zeros. These are precisely the G roots of the polynomial equation (2.41). Therefore the boundary values $u_\pm(z)$ can have zeros. Parallel arguments apply to $v(z)$. Since $\lambda_+(z) := \lim_{\epsilon\downarrow 0} \lambda(z + i\epsilon) > 0$ for $z \in \Sigma'_{\text{model}}$, we deduce finally that

$$u_-(x_j) = 0 \quad \text{and} \quad v_+(x_j) = 0, \quad \text{for } j = 1,\ldots,G,$$

where $x_1,\ldots,x_G$ are the roots of (2.41) and $\beta_{j-1} < x_j < \alpha_j$.

To build matrix elements of $\mathbf{Y}^\sharp(z)$ out of products of $g^\pm(z;\mathbf{q})$ with $u(z)$ or $v(z)$, the vector $\mathbf{q}$ should be chosen to align the poles of the functions $g^\pm(z;\mathbf{q})$ with the zeros of the boundary values of $u(z)$ or $v(z)$. An important observation at this point is that the aggregates of points $D_\pm := \{x_j \pm i0\}$ form *nonspecial divisors*, meaning that if the expression $(t_{G-1}z^{G-1} + t_{G-2}z^{G-2} + \cdots + t_0)/y(z)$ is made to vanish at all of the G points in either D_+ or D_- by an appropriate choice of the coefficients t_j, then it vanishes identically. This implies that the function $f(z;\mathbf{q})$ will have exactly the same zeros as $u(z)$, with the same multiplicity, if one takes $\mathbf{q} = \mathbf{q}_u$, with $\mathbf{q}_u$ defined by (2.42) in terms of the Abel-Jacobi mapping evaluated on the divisor D_- and the vector $\mathbf{k}$ of Riemann constants defined by (2.35). Similarly, the function $f(z;\mathbf{q}_v)$ has exactly the same zeros as $v(z)$, with the same multiplicity, when $\mathbf{q}_v$ is defined by (2.42) in terms of the Abel-Jacobi mapping evaluated on the divisor D_+ and the vector $\mathbf{k}$. Note that as a consequence of (A.9) and (2.35), $\mathbf{q}_u + \mathbf{q}_v = 0$ modulo $2\pi i \mathbb{Z}^G$. Also, $\mathbf{q}_u - \mathbf{q}_v^* = 0$ modulo $2\pi i \mathbb{Z}^G$. We therefore can see that all four functions $u(z)g^\pm(z;\pm\mathbf{q}_u)$ and $v(z)g^\pm(z;\pm\mathbf{q}_v) = v(z)g^\pm(z;\mp\mathbf{q}_u)$ are analytic for $z \in \mathbb{C}\setminus\Sigma'_{\text{model}}$ and take continuous boundary values on Σ'_{model} with the exception of the band endpoints $\alpha_0,\ldots,\beta_G$, where they all have negative one-fourth power singularities. With the help of these functions, we may now assemble the solution of Riemann-Hilbert Problem 5.1. First, we write down a formula for $\mathbf{Y}^\sharp(z)$ by setting

$$\mathbf{Y}^\sharp(z) := \begin{pmatrix} u(z)\dfrac{g^+(z;\mathbf{q}_u)}{g^+_+(\infty;\mathbf{q}_u)} & iv(z)\dfrac{g^-(z;-\mathbf{q}_v)}{g^-_-(\infty;-\mathbf{q}_v)} \\ iv(z)\dfrac{g^+(z;\mathbf{q}_v)}{g^+_-(\infty;\mathbf{q}_v)} & u(z)\dfrac{g^-(z;-\mathbf{q}_u)}{g^-_+(\infty;-\mathbf{q}_u)} \end{pmatrix}, \quad \text{for } G > 0.$$

If $G = 0$, then the Riemann theta functions are not necessary, and we have simply

$$\mathbf{Y}^\sharp(z) := \begin{pmatrix} u(z) & iv(z) \\ iv(z) & u(z) \end{pmatrix}, \quad \text{for } G = 0.$$

A.3 THE MATRIX $\dot{\mathbf{X}}(z)$ AND ITS PROPERTIES

In both cases, $G = 0$ and $G > 0$, going back to the solution $\dot{\mathbf{X}}(z)$ of Riemann-Hilbert Problem 5.1 requires multiplying $\mathbf{Y}^\sharp(z)$ on the right by $i\sigma_1$ for $\Im(z) < 0$ to recover $\mathbf{Y}(z)$, followed by multiplication on the right by $e^{(h(z)-\kappa g(z))\sigma_3}$ to obtain $\dot{\mathbf{X}}(z)$. We have proved the following.

Proposition A.3. *The unique solution of Riemann-Hilbert Problem 5.1 is given by the following explicit* *formulae:*

$$\dot{\mathbf{X}}(z) := \begin{cases} \begin{pmatrix} u(z)\dfrac{g^+(z;\mathbf{q}_u)}{g^+_+(\infty;\mathbf{q}_u)}e^{h(z)} & iv(z)\dfrac{g^-(z;-\mathbf{q}_v)}{g^-_-(\infty;-\mathbf{q}_v)}e^{-h(z)} \\ iv(z)\dfrac{g^+(z;\mathbf{q}_v)}{g^+_-(\infty;\mathbf{q}_v)}e^{h(z)} & u(z)\dfrac{g^-(z;-\mathbf{q}_u)}{g^-_+(\infty;-\mathbf{q}_u)}e^{-h(z)} \end{pmatrix} e^{-\kappa g(z)\sigma_3}, & \Im(z)>0 \text{ and } G>0, \\[2em] \begin{pmatrix} -v(z)\dfrac{g^-(z;-\mathbf{q}_v)}{g^-_-(\infty;-\mathbf{q}_v)}e^{h(z)} & iu(z)\dfrac{g^+(z;\mathbf{q}_u)}{g^+_+(\infty;\mathbf{q}_u)}e^{-h(z)} \\ iu(z)\dfrac{g^-(z;-\mathbf{q}_u)}{g^-_+(\infty;-\mathbf{q}_u)}e^{h(z)} & -v(z)\dfrac{g^+(z;\mathbf{q}_v)}{g^+_-(\infty;\mathbf{q}_v)}e^{-h(z)} \end{pmatrix} e^{-\kappa g(z)\sigma_3}, & \Im(z)<0 \text{ and } G>0, \end{cases}$$

$$\dot{\mathbf{X}}(z) := \begin{cases} \begin{pmatrix} u(z)e^{h(z)} & iv(z)e^{-h(z)} \\ iv(z)e^{h(z)} & u(z)e^{-h(z)} \end{pmatrix} e^{-\kappa g(z)\sigma_3}, & \Im(z)>0 \text{ and } G=0, \\[1em] \begin{pmatrix} -v(z)e^{h(z)} & iu(z)e^{-h(z)} \\ iu(z)e^{h(z)} & -v(z)e^{-h(z)} \end{pmatrix} e^{-\kappa g(z)\sigma_3}, & \Im(z)<0 \text{ and } G=0. \end{cases}$$

In verifying the solution, it is useful to observe in addition that $g^\pm_+(\infty;\mathbf{q}) = g^\mp_-(\infty;\mathbf{q})$, for any $\mathbf{q} \in \mathbb{C}^G$.

Proposition A.4. *Let the coefficients $B^{(1)}_{jk}$ and $B^{(2)}_{jk}$ be defined in terms of the elements of the matrix $\dot{\mathbf{X}}(z)e^{\kappa(g(z)-\log(z))\sigma_3}$ as follows:*

$$\dot{\mathbf{X}}(z)e^{\kappa(g(z)-\log(z))\sigma_3} = \mathbb{I} + \frac{1}{z}\mathbf{B}^{(1)} + \frac{1}{z^2}\mathbf{B}^{(2)} + O\left(\frac{1}{z^3}\right)$$

as $z \to \infty$. Then

$$B^{(1)}_{12} = \begin{cases} \dfrac{1}{4}(\beta_0 - \alpha_0), & \text{for } G=0, \\[1em] \dfrac{\Theta(\mathbf{w}_+(\infty) - \mathbf{q}_v + i\mathbf{r} - i\kappa\mathbf{\Omega})\Theta(\mathbf{w}_-(\infty) - \mathbf{q}_v)}{\Theta(\mathbf{w}_-(\infty) - \mathbf{q}_v + i\mathbf{r} - i\kappa\mathbf{\Omega})\Theta(\mathbf{w}_+(\infty) - \mathbf{q}_v)} \dfrac{1}{4}\sum_{j=0}^{G}(\beta_j - \alpha_j), & \text{for } G>0, \end{cases}$$

$$B^{(1)}_{21} = \begin{cases} \dfrac{1}{4}(\beta_0 - \alpha_0), & \text{for } G=0, \\[1em] \dfrac{\Theta(\mathbf{w}_+(\infty) - \mathbf{q}_v - i\mathbf{r} + i\kappa\mathbf{\Omega})\Theta(\mathbf{w}_-(\infty) - \mathbf{q}_v)}{\Theta(\mathbf{w}_-(\infty) - \mathbf{q}_v - i\mathbf{r} + i\kappa\mathbf{\Omega})\Theta(\mathbf{w}_+(\infty) - \mathbf{q}_v)} \dfrac{1}{4}\sum_{j=0}^{G}(\beta_j - \alpha_j), & \text{for } G>0. \end{cases}$$

Also, if $G=0$, then

$$B^{(1)}_{11} + \frac{B^{(2)}_{12}}{B^{(1)}_{12}} = \frac{1}{2}(\beta_0 + \alpha_0),$$

and if $G>0$ then

$$B^{(1)}_{11} + \frac{B^{(2)}_{12}}{B^{(1)}_{12}} = \frac{1}{2}\frac{\displaystyle\sum_{j=0}^{G}(\beta_j^2 - \alpha_j^2)}{\displaystyle\sum_{j=0}^{G}(\beta_j - \alpha_j)}$$

$$+ \frac{i\mathbf{a}^{(G)} \cdot \nabla\Theta(\mathbf{w}_+(\infty) + \mathbf{q}_v - i\mathbf{r} + i\kappa\mathbf{\Omega})}{\Theta(\mathbf{w}_+(\infty) + \mathbf{q}_v - i\mathbf{r} + i\kappa\mathbf{\Omega})} - \frac{i\mathbf{a}^{(G)} \cdot \nabla\Theta(\mathbf{w}_+(\infty) + \mathbf{q}_v)}{\Theta(\mathbf{w}_+(\infty) + \mathbf{q}_v)}$$

$$+ \frac{i\mathbf{a}^{(G)} \cdot \nabla\Theta(\mathbf{w}_+(\infty) - \mathbf{q}_v + i\mathbf{r} - i\kappa\mathbf{\Omega})}{\Theta(\mathbf{w}_+(\infty) - \mathbf{q}_v + i\mathbf{r} - i\kappa\mathbf{\Omega})} - \frac{i\mathbf{a}^{(G)} \cdot \nabla\Theta(\mathbf{w}_+(\infty) - \mathbf{q}_v)}{\Theta(\mathbf{w}_+(\infty) - \mathbf{q}_v)}.$$

Proof. These equations follow directly from the explicit formulae for $\dot{\mathbf{X}}(z)$ and the fact that $\mathbf{q}_u + \mathbf{q}_v = 0$ modulo $2\pi i\mathbb{Z}^G$. It is also perhaps useful to point out that it is never necessary to use an explicit expression for the leading coefficient of $h(z) - \kappa\log(z)$ as $z \to \infty$; although this coefficient appears in $B^{(2)}_{12}$ and also in $B^{(1)}_{11}$, it cancels out of the particular combination $B^{(1)}_{11} + B^{(2)}_{12}/B^{(1)}_{12}$. □

A.3.1 Completion of the proof of Proposition 5.2

The uniform boundedness of $\dot{\mathbf{X}}(z)$ for z bounded away from any band endpoints $\alpha_0, \ldots, \beta_G$ follows from the corresponding property of the functions $u(z)$ and $v(z)$ and from the manner in which the large parameter N enters into the argument of the Riemann theta functions as a real phase $\mathbf{r} - \kappa \mathbf{\Omega}$ that is independent of z (recall that $\Theta(\mathbf{w})$ is periodic with period 2π in each imaginary coordinate direction in $\mathbb{C}^G$). To see the independence of the combination $\dot{\mathbf{X}}(z)e^{\kappa g(z)\sigma_3}$ from the arbitrary locations of any transition points in Y_N, observe that the combination $\mathbf{Y}(z)e^{-h(z)\sigma_3}$ can involve the function $g(z)$ only through the endpoints of the bands, which clearly do not depend on any arbitrary choice of transition points in transition bands.

A.3.2 Completion of the proof of Proposition 5.3

Note that for real z, the product $p(z) := \dot{X}_{11}(z)\dot{X}_{12}(z)$ can equivalently be written in terms of the elements of the matrix $\mathbf{Y}^\sharp(z)$ as $p(z) = Y^\sharp_{11+}(z) Y^\sharp_{12+}(z)$. Now $\mathbf{Y}^\sharp(z)$ satisfies the symmetry

$$\mathbf{Y}^\sharp(z) = \mathbf{Y}^\sharp(z^*)^* \begin{pmatrix} 0 & -i \\ -i & 0 \end{pmatrix}. \tag{A.15}$$

Indeed, the left- and right-hand sides of (A.15) both have the same asymptotic behavior as $z \to \infty$, regardless of whether $\Im(z) > 0$ or $\Im(z) < 0$, and satisfy the same jump conditions for $z \in \Gamma_j$, $j = 0, \ldots G$. In other words, both sides of (A.15) solve the same Riemann-Hilbert problem. A uniqueness argument based on Liouville's Theorem thus proves (A.15). Using (A.15) we may also write $p(z) = -iY^\sharp_{11+}(z) Y^\sharp_{11-}(z)^*$ when z is real.

Let us now consider the zeros of the function $Y^\sharp_{11}(z)$. Recall that $u(z)$ is nonzero for $z \in \mathbb{C}\setminus\Sigma'_{\text{model}}$, and for $z \in \Sigma'_{\text{model}}$ we have that $u_+(z)$ is bounded away from zero while $u_-(z)$ vanishes only at a single point x_j in each interior gap $\Gamma_j = (\beta_{j-1}, \alpha_j)$, for $j = 1, \ldots, G$. This shows that if $G = 0$, then (since there are no interior gaps) $p(z)$ is strictly nonzero for $z \in \Sigma'_{\text{model}}$. If $G > 0$, then the zeros of $u_-(z)$ on Σ'_{model} are cancelled (by construction) by corresponding zeros of the entire function $\Theta(\mathbf{w}(z) - \mathbf{q}_u)$ in the denominator of $Y^\sharp_{11}(z)$. Thus for $G > 0$, the zeros of $u(z)$ are precisely the zeros of the numerator $\Theta(\mathbf{w}(z) - \mathbf{q}_u - i\mathbf{r} + i\kappa\mathbf{\Omega})$. Recall that $\mathbf{r}$ and $\kappa\mathbf{\Omega}$ are real. By Jacobi inversion theory, it can thus be shown that $\Theta(\mathbf{w}(z) - \mathbf{q}_u - i\mathbf{r} + i\kappa\mathbf{\Omega})$ has exactly one zero on either the upper or lower edge of each cut Γ_j, for $j = 1, \ldots, G$ (in a nongeneric situation the zero may lie at one or the other endpoint of Γ_j). As the parameter κ is varied continuously with $\mathbf{q}_u$, $\mathbf{r}$, and $\mathbf{\Omega}$ held fixed, the G zeros of $\Theta(\mathbf{w}(z) - \mathbf{q}_u - i\mathbf{r} + i\kappa\mathbf{\Omega})$ oscillate about the cuts (moving one way along the upper edge and the other way along the lower edge) in a quasiperiodic fashion. The actual behavior of the zeros is more complicated since the parameter κ can in fact be incremented only by integers (recall that the polynomial degree k is $cN + \kappa$); thus if the frequency vector $\mathbf{\Omega}$ is a rational multiple of a lattice vector in $2\pi\mathbb{Z}^G$, then the motion of the zeros will be periodic rather than quasiperiodic. In an extremely nongeneric situation the zeros of $\Theta(\mathbf{w}(z) - \mathbf{q}_u - i\mathbf{r} + i\kappa\mathbf{\Omega})$ may be located at the endpoints of the interior gaps Γ_j for all admissible κ. Therefore, if $\dot{X}_{11\pm}(z)$ is bounded away from zero for $z \in \Gamma_j$, $j = 1, \ldots, G$, then $\dot{X}_{11\mp}(z)$ necessarily has a simple zero in the interior of Γ_j. This shows that the product $p(z)$ vanishes at exactly one point $z = z_j$ in the interval $[\beta_{j-1}, \alpha_j]$, for $j = 1, \ldots, G$ and $G > 0$.

Now for real z we write

$$p(z) = -iu_+(z)u_-(z)^* \left[\frac{Y^\sharp_{11+}(z)}{u_+(z)}\right] \left[\frac{Y^\sharp_{11-}(z)}{u_-(z)}\right]^*.$$

It follows from (A.14) that for z real, $-iu_+(z)u_-(z)^*$ is a real function that satisfies

$$\begin{aligned}
-iu_+(z)u_-(z)^* &< 0, & &\text{for } z < \alpha_0, \\
-iu_+(z)u_-(z)^* &> 0, & &\text{for } z > \beta_G, \\
-iu_+(z)u_-(z)^* &\to -\infty, & &\text{as } z \uparrow \alpha_j \text{ for } j = 0, \ldots, G, \\
-iu_+(z)u_-(z)^* &\to +\infty, & &\text{as } z \downarrow \beta_j \text{ for } j = 0, \ldots, G.
\end{aligned}$$

At the same time, we have the existence of the following finite limits:
$$A_j := \lim_{z\uparrow\alpha_j} \left[\frac{Y^\sharp_{11+}(z)}{u_+(z)}\right]\left[\frac{Y^\sharp_{11-}(z)}{u_-(z)}\right]^*,$$
$$B_j := \lim_{z\downarrow\beta_j} \left[\frac{Y^\sharp_{11+}(z)}{u_+(z)}\right]\left[\frac{Y^\sharp_{11-}(z)}{u_-(z)}\right]^*.$$

We clearly have $A_j \geq 0$ and $B_j \geq 0$ for all $j = 0, \ldots, G$; the limits are strictly positive unless one of the zeros z_j occurs at an endpoint. This proves that $p(z) > 0$ for $z < z_j$ in Γ_j and that $p(z) < 0$ for $z > z_j$ in Γ_j, for $j = 1, \ldots, G$, while $p(z) < 0$ for $z < \alpha_0$ and $p(z) > 0$ for $z > \beta_G$.

Appendix B

Construction of the Hahn Equilibrium Measure: the Proof of Theorem 2.17

B.1 GENERAL STRATEGY: THE ONE-BAND ANSATZ

The main idea is to begin with an ansatz that there is only one band, a subinterval of $(0,1)$ of the form (α, β), where α and β are to be determined. Then using the ansatz we derive formulae for α and β, the "candidate" equilibrium measure, and the corresponding Lagrange multiplier. Of course, one must then check that the measure produced by this *one-band ansatz* is consistent with the variational problem.

The associated field (2.1) is

$$\varphi(x) := V^{\text{Hahn}}(x; A, B) + \int_0^1 \log|x-y|\rho^0(y)\,dy$$
$$= -(A+x)\log(A+x) - (B+1-x)\log(B+1-x) + x\log(x) + (1-x)\log(1-x)$$
$$+ A\log(A) + (B+1)\log(B+1) - 1,$$

and hence

$$\varphi'(x) = -\log(A+x) + \log(1+B-x) + \log(x) - \log(1-x).$$

In the presumed band, the candidate equilibrium measure (we will refer to its density as $\psi(x)$) satisfies the equilibrium condition (2.14). Differentiating this equation with respect to x, one finds that

$$\text{P. V.} \int_0^1 \frac{\psi(y)}{x-y}\,dy = \frac{1}{2c}\varphi'(x) \tag{B.1}$$

holds identically for x in the (as yet unknown) band $\alpha < x < \beta$. Introducing the Cauchy transform of $\psi(x)$,

$$F(z) := \int_0^1 \frac{\psi(y)}{z-y}\,dy, \qquad \text{for } z \in \mathbb{C} \setminus [0,1],$$

elementary properties of Cauchy integrals imply that if $F_+(x)$ denotes the boundary value taken on $[0,1]$ from above, and $F_-(x)$ denotes the corresponding boundary value taken from below, then

$$F_+(x) - F_-(x) = -2\pi i \psi(x),$$
$$\frac{1}{2}(F_+(x) + F_-(x)) = \text{P. V.} \int_0^1 \frac{\psi(y)}{x-y}\,dy. \tag{B.2}$$

Also, as $z \to \infty$, the condition that $\psi(x)$ should be the density of a probability measure implies that

$$F(z) = \frac{1}{z} + O\left(\frac{1}{z^2}\right) \qquad \text{as } z \to \infty.$$

According to the one-band ansatz, the remaining intervals $(0, \alpha)$ and $(\beta, 1)$ are either voids or saturated regions. So at this point, the one-band ansatz bifurcates into four distinct cases that must be investigated: these are the four configurations of void-band-void, saturated-band-void, saturated-band-saturated, and void-band-saturated. We will work out many of the details in the void-band-void case and then show how the analysis changes in the other three configurations.

B.2 THE VOID-BAND-VOID CONFIGURATION

In both voids $(0, \alpha)$ and $(\beta, 1)$, we have the lower constraint in force: $\psi(x) \equiv 0$. Hence from (B.1) and (B.2), F necessarily solves the following scalar Riemann-Hilbert problem: $F(z)$ is analytic in $\mathbb{C} \setminus [\alpha, \beta]$ and satisfies the jump condition

$$F_+(x) + F_-(x) = \frac{1}{c}\varphi'(x), \quad \text{for } \alpha < x < \beta,$$

and as $z \to \infty$,

$$F(z) = \frac{1}{z} + O\left(\frac{1}{z^2}\right). \tag{B.3}$$

The solution to this Riemann-Hilbert problem is given by the explicit formula

$$F(z) = \frac{R(z)}{2\pi i c} \int_\alpha^\beta \frac{\varphi'(y)}{R_+(y)(y-z)} \, dy, \quad \text{for } z \in \mathbb{C} \setminus [\alpha, \beta], \tag{B.4}$$

where the subscript $+$ indicates a boundary value taken from the upper half-plane, $R(z)^2 = (z-\alpha)(z-\beta)$, and the square root $R(z)$ is defined to be analytic in $\mathbb{C} \setminus [\alpha, \beta]$ with the condition that $R(z) \sim z$ as $z \to \infty$. The asymptotic condition (B.3) on F now implies, by explicit asymptotic expansion of the formula (B.4), the following two conditions on the endpoints α and β:

$$-\frac{1}{2\pi i}\int_\alpha^\beta \frac{\varphi'(y)}{R_+(y)} \, dy = 0,$$

$$-\frac{1}{2\pi i}\int_\alpha^\beta \frac{y\varphi'(y)}{R_+(y)} \, dy = c.$$

In a standard application of formulae involving Cauchy integrals, one can evaluate the above integrals and find that the endpoint equations are equivalent to

$$\cosh^{-1}\left(\frac{A+s}{d}\right) - \cosh^{-1}\left(\frac{B+1-s}{d}\right) - \cosh^{-1}\left(\frac{s}{d}\right) + \cosh^{-1}\left(\frac{1-s}{d}\right) = 0 \tag{B.5}$$

and

$$\sqrt{(A+s)^2 - d^2} - \sqrt{s^2 - d^2} + \sqrt{(1+B-s)^2 - d^2} - \sqrt{(1-s)^2 - d^2} = 2c + A + B, \tag{B.6}$$

where s and d are defined by

$$s := \frac{\beta + \alpha}{2}, \quad d := \frac{\beta - \alpha}{2}.$$

Now using the addition formula $\cosh^{-1}(a) \pm \cosh^{-1}(b) = \cosh^{-1}(ab \pm \sqrt{(a^2-1)(b^2-1)})$ twice, the equation (B.5) becomes

$$(2 + A + B)s - \sqrt{(A+s)^2 - d^2}\sqrt{s^2 - d^2} = B + 1 - \sqrt{(B+1-s)^2 - d^2}\sqrt{(1-s)^2 - d^2}. \tag{B.7}$$

Thus α and β are necessarily solutions of the system of equations (B.6) and (B.7). Now we take the square of both sides of (B.6), add two times (B.7), and find

$$\sqrt{(A+s)^2 - d^2} - \sqrt{s^2 - d^2} = \frac{(2c + A + B)^2 + A^2 - B^2}{2(2c + A + B)}.$$

To simplify upcoming formulae, we set

$$W = \sqrt{(A+s)^2 - d^2}, \tag{B.8}$$

$$X = \sqrt{s^2 - d^2}, \tag{B.9}$$

$$Y = \sqrt{(B+1-s)^2 - d^2}, \tag{B.10}$$

$$Z = \sqrt{(1-s)^2 - d^2}. \tag{B.11}$$

CONSTRUCTION OF THE HAHN EQUILIBRIUM MEASURE: THE PROOF OF THEOREM 2.17

With this notation, the equations (B.7) and (B.6) for s and d (and hence for α and β) become

$$(2 + A + B)s - WX = B + 1 - YZ, \tag{B.12}$$
$$Y - Z = 2c + A + B - (W - X). \tag{B.13}$$

By taking the square of both sides of (B.12) and adding two times (B.13), we find

$$W - X = \frac{(2c + A + B)^2 + A^2 - B^2}{2(2c + A + B)} =: K. \tag{B.14}$$

On the other hand, by the definition of W and X, $(W - X)(W + X) = A^2 + 2As$, and thus (B.14) implies that $W + Y = (A^2 + 2As)/K$. Hence

$$2W = K + \frac{A^2 + 2As}{K} \tag{B.15}$$
$$2X = -K + \frac{A^2 + 2As}{K}. \tag{B.16}$$

Also, (B.13) implies that $Y - Z = 2c + A + B - K$, and from the definitions of Y and Z, we have $(Y - Z)(Y + Z) = B^2 + 2B(1 - s)$. Therefore we find

$$2Y = 2c + A + B - K + \frac{B^2 + 2B(1 - s)}{2c + A + B - K}, \tag{B.17}$$
$$2Z = -(2c + A + B - K) + \frac{B^2 + 2B(1 - s)}{2c + A + B - K}. \tag{B.18}$$

Substituting (B.15)–(B.18) into (B.12), we find a quadratic equation in s:

$$(2 + A + B)s - \frac{1}{4}\left\{-K^2 + \left(\frac{A^2 + 2As}{K}\right)^2\right\} = B + 1 + \frac{1}{4}\left\{(2c + A + B - K)^2 - \left(\frac{B^2 + 2B(1 - s)}{2c + A + B - K}\right)^2\right\}, \tag{B.19}$$

where K is defined in (B.14). The solutions to this quadratic equation are

$$s = s_1 := \frac{A(A + B) + (A + B)(B - A + 2)c + (B - A + 2)c^2}{(A + B + 2c)^2} \tag{B.20}$$

and

$$s = s_2 := \frac{A(A + B)(1 + B) + (A + B)(A + B + 2)c + (A + B + 2)c^2}{A^2 - B^2}.$$

Since $0 \leq \alpha + \beta \leq 2$, we need to check which of these two roots actually lies in $[0, 1]$. For s_2, one sees that if $A < B$, then $s_2 < 0$, and if $A > B$, then by looking at the terms not involving c,

$$s_2 - 1 \geq \frac{A(A + B)(1 + B)}{A^2 - B^2} - 1 = \frac{A^2 B + AB^2 + AB + B^2}{A^2 - B^2} > 0.$$

On the other hand, the numerator of s_1 can be written as

$$A(A + B)(1 - c) + c\{(A + B)(B + 2) - Ac\} + (B + 2)c^2.$$

Since $0 < c < 1$, each term is positive, and thus $s_1 > 0$. Analogously, the numerator of $1 - s_1$ can be written as

$$(A + B)B(1 - c) + c\{A(A + 2) + B(A + 2 - c)\} + (A + 2)c^2,$$

and each term is positive as $c \in (0, 1)$, which implies that $s_1 < 1$. Thus $s_1 \in (0, 1)$, and we have found the root we need.

Substituting (B.20) into (B.16) and (B.18), we find

$$X = X_0, \quad Z = Z_0,$$

where
$$X_0 := \frac{-c^2 - (A+B)c + A}{(A+B+2c)^2},$$
$$Z_0 := \frac{-c^2 - (A+B)c + B}{(A+B+2c)^2}.$$

Note that
$$\begin{aligned} X_0 &> 0, & \text{for } 0 < c < c_A, \\ X_0 &< 0, & \text{for } c > c_A, \\ Z_0 &> 0, & \text{for } 0 < c < c_B, \\ Z_0 &< 0, & \text{for } c > c_B. \end{aligned}$$

Thus the conditions $X, Z > 0$ yield conditions on c which are
$$0 < c < \min(c_A, c_B).$$

Only for c satisfying these inequalities can α and β be found from the equations $\beta + \alpha = 2s$ and $\alpha\beta = s^2 - d^2 = Y$. The quadratic equation for α and β is exactly (2.74), and the explicit solutions are given by (2.75) and (2.76).

With the endpoints determined, the candidate density for the equilibrium measure in the interesting region $\alpha < x < \beta$ can be obtained by evaluating $F(z)$ and using
$$\psi(x) = -\frac{1}{2\pi i}\left(F_+(x) - F_-(x)\right).$$

First, we evaluate $F(z)$. For $z \in \mathbb{C} \setminus [\alpha, \beta]$, we have
$$F(z) = \frac{R(z)}{2\pi i c} \int_\alpha^\beta \frac{\varphi'(y)}{R_+(y)(y-z)}\,dy$$
$$= \frac{R(z)}{4\pi i c} \int_{\Gamma_0} \frac{\varphi'(y)}{R(y)(y-z)}\,dy$$

where the closed contour Γ_0 encloses the interval $[\alpha, \beta]$ once in the clockwise direction, and the inside of Γ_0 does not include any points y in the set $\{z\} \cup (-\infty, 0] \cup (1, \infty]$. Noting that $\varphi'(z)$ is analytic in $\mathbb{C} \setminus ([-A, 0] \cup [1, 1 + B])$, we deform the contour of integration so that the integral over Γ_0 becomes the integral over the union of the intervals $[-A, 0]$ and $[1, 1 + B]$. Being careful with branches of the various multivalued functions involved, we find that
$$F(z) = \frac{1}{2c}\varphi'(z) + \frac{R(z)}{2c}\left(-\int_{-A}^0 \frac{dy}{R(y)(y-z)} - \int_1^{1+B} \frac{dy}{R(y)(y-z)}\right).$$

Here, when z is in either of the intervals $(-A, 0)$ or $(1, 1 + B)$ where $F(z)$ is supposed to be analytic, the integral is interpreted as the principal value. This integral is equal to
$$F(z) = \frac{1}{2c}\varphi'(z) - \frac{R(z)}{2c}\left(\int_0^A \frac{ds}{\sqrt{(s+\alpha)(s+\beta)}(s+z)} + \int_1^{1+B} \frac{ds}{\sqrt{(s-\alpha)(s-\beta)}(s-z)}\right).$$

Now we use the following formula (see, for example, [AbrS65]),
$$\int \frac{ds}{\sqrt{(s+a)(s+b)}(s+z)} = \frac{2}{\sqrt{(z-a)(z-b)}}\left[\log\left(\sqrt{\frac{s+b}{z-b}} + \sqrt{\frac{s+a}{z-a}}\right) - \frac{1}{2}\log(s+z)\right]$$

and evaluate the two integrals exactly. The result of this calculation is that for $z \in \mathbb{C} \setminus [\alpha, \beta]$,
$$F(z) = \frac{1}{c}\log\left(\sqrt{\frac{1+B-\beta}{z-\beta}} + \sqrt{\frac{1+B-\alpha}{z-\alpha}}\right) - \frac{1}{c}\log\left(\sqrt{\frac{1-\beta}{z-\beta}} + \sqrt{\frac{1-\alpha}{z-\alpha}}\right)$$
$$- \frac{1}{c}\log\left(\sqrt{\frac{A+\beta}{z-\beta}} + \sqrt{\frac{A+\alpha}{z-\alpha}}\right) + \frac{1}{c}\log\left(\sqrt{\frac{\beta}{z-\beta}} + \sqrt{\frac{\alpha}{z-\alpha}}\right),$$

CONSTRUCTION OF THE HAHN EQUILIBRIUM MEASURE: THE PROOF OF THEOREM 2.17

where all the square root functions $\sqrt{w}$ are defined to be analytic in $w \in \mathbb{C} \setminus (-\infty, 0]$, with the condition that $\sqrt{w} > 0$ for $w > 0$, and the logarithm $\log(w)$ is defined to be analytic in $w \in \mathbb{C} \setminus (-\infty, 0]$, with the condition that $\log(w) > 0$ for $w > 1$. Now using $\log(a + ib) - \log(a - ib) = 2i \arctan(b/a)$, we obtain, for $x \in (\alpha, \beta)$, the formula cited in Theorem 2.17 for the equilibrium measure (as mentioned above, the candidate $\psi(x)$ we have just constructed turns out to be the actual density of the equilibrium measure).

The Lagrange multiplier ℓ_c can then be obtained from the variational condition (2.14), which we may evaluate for any $x \in [\alpha, \beta]$. Therefore, since $\psi(x)$ is supported only in $[\alpha, \beta]$ in the void-band-void case under consideration,

$$\ell_c = -2c \int_\alpha^\beta \log(\beta - s) \psi(s) ds + \varphi(\beta).$$

Here we have arbitrarily picked $x = \beta$. Now using the exact formula for $\psi(x)$ and the identity

$$\frac{1}{\beta - \alpha} \int_\alpha^\beta \log(\beta - s) \arctan\left(k \sqrt{\frac{\beta - s}{s - \alpha}}\right) ds = \frac{\pi k}{2(1 + k)} \left(\log(\beta - \alpha) - 1\right) + \frac{\pi k}{1 - k^2} \left[\frac{1}{k} \log(1 + k) - \log 2\right],$$

we obtain the corresponding formula for the multiplier.

B.3 THE SATURATED-BAND-VOID CONFIGURATION

Since $\psi(x) \equiv 1/c$ in the saturated region supposed to be the interval $(0, \alpha)$, from (B.1) and (B.2), the Cauchy transform of the candidate density $\psi(x)$, $F(z)$, is necessarily the solution of the following scalar Riemann-Hilbert problem: $F(z)$ is analytic in $\mathbb{C} \setminus [0, \beta]$ and satisfies the jump conditions

$$F_+(x) - F_-(x) = -\frac{2\pi i}{c},$$

for $0 < x < \alpha$, and

$$F_+(x) + F_-(x) = \frac{1}{c} \varphi(x),$$

for $\alpha < x < \beta$, and as $z \to \infty$,

$$F(z) = \frac{1}{z} + O\left(\frac{1}{z^2}\right).$$

The solution is

$$F(z) = \frac{R(z)}{2\pi i c} \left(-\int_0^\alpha \frac{2\pi i}{R(y)(y - z)} dy + \int_\alpha^\beta \frac{\varphi'(y)}{R_+(y)(y - z)} dy\right),$$

where $R(z)$ denotes the same square-root function as before. The equations for the endpoints α and β now include additional terms:

$$\int_0^\alpha \frac{1}{R(y)} dy - \frac{1}{2\pi i} \int_\alpha^\beta \frac{\varphi'(y)}{R_+(y)} dy = 0,$$

$$\int_0^\alpha \frac{y}{R(y)} dy - \frac{1}{2\pi i} \int_\alpha^\beta \frac{y \varphi'(y)}{R_+(y)} dy = c.$$

Evaluating these integrals, we see that these equations are equivalent to (compare (B.12) and (B.13))

$$(A + B + 2)s + WX = B + 1 - YZ, \tag{B.21}$$

$$Y - Z = 2c + A + B - (W + X), \tag{B.22}$$

where W, X, Y, Z are exactly as defined in (B.8)–(B.11). Similar reasoning then yields

$$2W = K + \frac{A^2 + 2As}{K},$$

$$2X = K - \frac{A^2 + 2As}{K},$$

$$2Y = 2c + A + B - K + \frac{B^2 + 2B(1-s)}{2c + A + B - K},$$

$$2Z = -(2c + A + B - K) + \frac{B^2 + 2B(1-s)}{2c + A + B - K}.$$

Substituting these formulae into (B.21)–(B.22), we obtain *exactly* the same equation as in the previous case, namely, (B.19), and we also find that we must take the root $s = s_1$. From this, we have

$$X = -X_0, \qquad Z = Z_0.$$

Thus $X, Z > 0$ implies the following conditions on c:

$$c_A < c < c_B.$$

Hence it is necessary in the saturated-band-void case that $A < B$ (see (2.73)). Under these conditions, one finds that the solutions α and β are again given by exactly the same formulae as in the previous case — only the conditions on c, A, and B are different. The candidate equilibrium measure in the band (α, β) and the corresponding Lagrange multiplier may now be found as before by evaluating the integrals in the explicit formula for $F(z)$ and taking boundary values on (α, β).

B.4 THE VOID-BAND-SATURATED CONFIGURATION

The appropriate scalar Riemann-Hilbert problem for the Cauchy transform of the candidate density is the following: $F(z)$ is analytic in $\mathbb{C} \setminus [\alpha, 1]$ and satisfies the jump conditions

$$F_+(x) - F_-(x) = -\frac{2\pi i}{c},$$

for $\beta < x < 1$, and

$$F_+(x) + F_-(x) = \frac{1}{c}\varphi(x),$$

for $\alpha < x < \beta$, and as $z \to \infty$,

$$F(z) = \frac{1}{z} + O\left(\frac{1}{z^2}\right).$$

The solution is

$$F(z) = \frac{R(z)}{2\pi i c}\left(\int_\alpha^\beta \frac{\varphi'(y)}{R_+(y)(y-z)}\,dy - \int_\beta^1 \frac{2\pi i}{R(y)(y-z)}\,dy\right).$$

By taking moments of $F(z)$ for large z and analyzing the resulting equations, we find again the same quadratic equation for s, but this time we get

$$X = X_0, \qquad Z = -Z_0.$$

These equations imply the following conditions on c:

$$c_B < c < c_A,$$

and therefore this configuration is possible only if $B < A$. Again one then finds that α and β are given by the same formulae as before, and by evaluating $F(z)$, one can calculate the candidate density for $\alpha < x < \beta$ and the Lagrange multiplier ℓ_c.

B.5 THE SATURATED-BAND-SATURATED CONFIGURATION

The scalar Riemann-Hilbert problem for the Cauchy transform of $\psi(x)$ is the following: $F(z)$ is analytic in $\mathbb{C} \setminus [0, 1]$ and satisfies the jump conditions

$$F_+(x) - F_-(x) = -\frac{2\pi i}{c},$$

for $0 < x < \alpha$ and $\beta < x < 1$, and

$$F_+(x) + F_-(x) = \frac{1}{c}\varphi(x),$$

for $\alpha < x < \beta$, and as $z \to \infty$,

$$F(z) = \frac{1}{z} + O\left(\frac{1}{z^2}\right).$$

The solution is

$$F(z) = \frac{R(z)}{2\pi i c}\left(-\int_0^\alpha \frac{2\pi i}{R(y)(y-z)}\,dy + \int_\alpha^\beta \frac{\varphi'(y)}{R_+(y)(y-z)}\,dy - \int_\beta^1 \frac{2\pi i}{R(y)(y-z)}\,dy\right).$$

By similar analysis, we arrive again at the same quadratic equation for s and find

$$X = -X_0, \qquad Z = -Z_0.$$

These equations imply the following conditions on c:

$$\max(c_A, c_B) < c < 1,$$

and again the endpoints α and β have the same expressions as before. Evaluating the integrals in $F(z)$ and taking boundary values then gives the candidate density in (α, β), and the Lagrange multiplier may then be found by direct integration.

Appendix C

List of Important Symbols

Symbol	Meaning	Page
$p^{(N,k)}(x_1,\ldots,x_k)$	Joint probability distribution of k particles	1
$\mathbb{P}(\text{event})$	Probability of an event	1
$Z_{N,k}$	Normalization constant for $p^{(N,k)}(x_1,\ldots,x_k)$	1
$\mathfrak{a}$, $\mathfrak{b}$, and $\mathfrak{c}$	Dimensions of the $\mathfrak{abc}$-hexagon	3
$P_1, P_2, P_3, P_4, P_5,$ and P_6	Vertices of the $\mathfrak{abc}$-hexagon	3
$\mathcal{L}$	Hexagonal lattice within the $\mathfrak{abc}$-hexagon	3
$\mathfrak{A}$, $\mathfrak{B}$, and $\mathfrak{C}$	Rescaled $\mathfrak{a}$, $\mathfrak{b}$, and $\mathfrak{c}$	4
$a_{N,k}(t)$ and $b_{N,k}(t)$	Solution of the finite Toda lattice	5
$A(c,T)$ and $B(c,T)$	Solution of the continuum limit of the Toda lattice	5
$\hat{a}_{N,k}(t)$ and $\hat{b}_{N,k}(t)$	Solution of the linearized Toda lattice	7
N	Number of nodes	8
X_N	Set of nodes	8
$x_{N,n}$	Node	8
$w_{N,n}$	Weight at the node $x_{N,n}$	8
$w(x)$	Weight function defined for $x \in X_N$	8
$p_{N,k}(z)$	Discrete orthonormal polynomial	8
$c_{N,k}^{(m)}$	Coefficient of z^m in $p_{N,k}(z)$	8
$\gamma_{N,k}$	Leading coefficient of $p_{N,k}(z)$	8
$\pi_{N,k}(z)$	Monic discrete orthogonal polynomial	8

Symbol	Meaning	Page
$a_{N,k}$	Diagonal recurrence coefficients	9
$b_{N,k}$	Off-diagonal recurrence coefficients	9
$\mathbf{M}(z;N,k,t)$	Jost matrix for Toda flow	15
$\mathbf{L}(z;N,k,t)$ and $\mathbf{B}(z;N,k,t)$	Matrices in Lax pair for the Toda lattice	15
$\mathbf{W}(z;N,k,t)$	Squared eigenfunction matrix	16
$\rho^0(x)$	Node density function	10
$[a,b]$	Interval containing nodes	10
$V_N(x)$	Exponent of weights	10
$V(x)$	Fixed component of $V_N(x)$ (potential)	10
$\eta(x)$	Correction to $NV(x)$	10
k	Degree of polynomial, number of particles	11
c	Asymptotic ratio of k/N	11
κ	Correction to Nc	11
$\mathbf{P}(z;N,k)$	Solution of Interpolation Problem 1.2	12
Δ	Subset of node indices where triangularity is reversed	17
$\mathbb{Z}_N$	$\{0,1,2,\ldots,N-1\}$	17
$\#\Delta$	Number of elements in Δ	17
$\mathbf{Q}(z;N,k)$	$\mathbf{P}(z;N,k)$ with residue triangularity modified	17
σ_3	Pauli matrix	17
∇	Complementary set to Δ in $\mathbb{Z}_N$	18
$\overline{\mathbf{P}}(z;N,\overline{k})$	Dual of $\mathbf{P}(z;N,k)$	18
$\overline{k}$	$N-k$, number of holes	18
σ_1	Pauli matrix	18

LIST OF IMPORTANT SYMBOLS

Symbol	Meaning	Page
$\overline{w}_{N,n}$	Dual weight at the node $x_{N,n}$	18
$\overline{\pi}_{N,\bar{k}}(z)$	Dual of $\pi_{N,k}(z)$	18
$\overline{\gamma}_{N,\bar{k}-1}$	Dual of $\gamma_{N,k}$	19
$\varphi(x)$	External field	25
$E_c[\mu]$	Energy functional	25
$\mu_{\min}^c$	Equilibrium measure	25
$F_c[\mu]$	Modified energy functional	26
ℓ_c	Lagrange multiplier (Robin constant)	26
$\underline{\mathcal{F}}$	Set where lower constraint holds	26
$\overline{\mathcal{F}}$	Set where upper constraint holds	26
G	Genus of S	28
$\alpha_0, \ldots, \alpha_G$	Left endpoints of bands	28
$\beta_0, \ldots, \beta_G$	Right endpoints of bands	28
$I_0, \ldots, I_G$	Bands	28
$\Gamma_1, \ldots, \Gamma_G$	Interior gaps (voids and saturated regions)	28
$\dfrac{\delta E_c}{\delta \mu}(x)$	Variational derivative of E_c	28
$L_c(z)$	Complex logarithmic potential of $\mu_{\min}^c$	28
$\overline{L}_c^{\Gamma}(z)$	Continuation from Γ of logarithmic potential of $\mu_{\min}^c$	28
$\overline{L}_c^{I}(z)$	Continuation from I of logarithmic potential of $\mu_{\min}^c$	28
$\xi_\Gamma(x)$	Analytic function defined in gap Γ	29
$\psi_I(x)$ and $\overline{\psi}_I(x)$	Analytic functions defined in band I	29
$\tau_\Gamma^{\nabla, L}(z)$	Conformal mapping near band/void edge α	29
$\tau_\Gamma^{\nabla, R}(z)$	Conformal mapping near band/void edge β	29

Symbol	Meaning	Page
$\tau_\Gamma^{\Delta,L}(z)$	Conformal mapping near band/saturated region edge α	29
$\tau_\Gamma^{\Delta,R}(z)$	Conformal mapping near band/saturated region edge β	30
$\theta_{\Gamma_1},\ldots,\theta_{\Gamma_G}$	Constants defined in interior gaps $\Gamma_1,\ldots,\Gamma_G$	30
$\theta_{(a,\alpha_0)}$ and $\theta_{(\beta_G,b)}$	Constants defined in exterior gaps (a,α_0) and (β_G,b)	30
$\overline{V}_N(x)$	Dual of $V_N(x)$	30
$\bar{\mu}_{\min}^{1-c}$	Dual of $\mu_{\min}^c$	30
$R(z)$	Branch of square root of $(z-\alpha_0)\cdots(z-\beta_G)$	31
$h(z)$	Solution of Riemann-Hilbert Problem A.1	31
γ	Correction to $N\ell_c$	31
$c_{j,0}$	κ-independent part of $-i(h_+(z)-h_-(z))$, for $z\in\Gamma_j$	31
ω_j	Coefficient of κ in $-i(h_+(z)-h_-(z))$, for $z\in\Gamma_j$	31
$\mathbf{r}$	Phase vector with components $N\theta_{\Gamma_j}-c_{j,0}$	31
$\mathbf{\Omega}$	Frequency vector with components ω_j	31
$y(z)$	Branch of square root of $(z-\alpha_0)\cdots(z-\beta_G)$	31
$m_1(z)\,dz,\ldots,m_G(z)\,dz$	Branches of holomorphic differentials $m_1^S(P),\ldots,m_G^S(P)$	31
$\mathbf{A}$	Matrix with columns $\mathbf{a}^{(1)},\ldots,\mathbf{a}^{(G)}$	31
$\mathbf{B}$	Riemann matrix with columns $\mathbf{b}^{(1)},\ldots,\mathbf{b}^{(G)}$	31
$\mathbf{k}$	Vector of Riemann constants	31
$\Theta(\mathbf{w})$	Riemann theta function	32
$\mathbf{w}(z)$	Branch of Abel-Jacobi mapping	32
$\mathbf{w}_\pm(\infty)$	Limiting values of $\mathbf{w}(z)$ as $z\to\infty$	32
$u(z)$ and $v(z)$	Factors in solution of Riemann-Hilbert Problem 5.1	32
$\mathbf{q}_u$ and $\mathbf{q}_v$	Vectors in the Jacobian giving zeros of $u(z)$ and $v(z)$	32

LIST OF IMPORTANT SYMBOLS

Symbol	Meaning	Page
$W(z)$	Alternate notation for $\dot{X}_{11}(z)e^{\kappa g(z)}$	32
$Z(z)$	Alternate notation for $\dot{X}_{12}(z)e^{-\kappa g(z)}$	32
$H_\Gamma^\pm(z)$	Factors in first row of $\mathbf{H}_\Gamma^{\nabla,L}(z)$, $\mathbf{H}_\Gamma^{\nabla,R}(z)$, $\mathbf{H}_\Gamma^{\Delta,L}(z)$, and $\mathbf{H}_\Gamma^{\Delta,R}(z)$	32
K_J^δ	Compact complex neighborhood of a closed interval J	34
$\theta^0(z)$	Phase variable related to $\rho^0(z)$	34
$w_{N,n}^{\text{Kraw}}(p,q)$	Krawtchouk weights	42
$V_N^{\text{Kraw}}(x;l)$	Exponent of Krawtchouk weights	42
$w_{N,n}(b,c,d)$	Weight degenerating to $w_{N,n}^{\text{Hahn}}(P,Q)$ and $w_{N,n}^{\text{Assoc}}(P,Q)$	43
$w_{N,n}^{\text{Hahn}}(P,Q)$	Hahn weights	44
$w_{N,n}^{\text{Assoc}}(P,Q)$	Associated Hahn weights	44
$V_N^{\text{Hahn}}(x;P,Q)$	Exponent of $w_{N,n}^{\text{Hahn}}(P,Q)$	44
$V^{\text{Hahn}}(x;A,B)$	Fixed component of $V_N^{\text{Hahn}}(x;NA+1,NB+1)$	44
$\eta^{\text{Hahn}}(x;P,Q)$	Correction to $NV^{\text{Hahn}}(x;A,B)$	44
$V_N^{\text{Assoc}}(x;P,Q)$	Exponent of $w_{N,n}^{\text{Assoc}}(P,Q)$	45
$V^{\text{Assoc}}(x;A,B)$	Fixed component of $V_N^{\text{Assoc}}(x;NA+1,NB+1)$	45
c_A and c_B	Critical values of c for the Hahn equilibrium measure	45
$R_m^{(N,k)}(x_1,\ldots,x_m)$	m-point correlation function of k-particle ensemble	49
$\mathbb{E}(X)$	Expected value of a random variable X	50
$K_{N,k}(x,y)$	Reproducing (Christoffel-Darboux) kernel	50
$A_m^{(N,k)}(B)$	Local particle occupation probability	50
$\bar{p}^{(N,\bar{k})}(y_1,\ldots,y_{\bar{k}})$	Joint probability distribution of $\bar{k}$ holes	51
$\overline{Z}_{N,\bar{k}}$	Normalization constant for $\bar{p}^{(N,\bar{k})}(y_1,\ldots,y_{\bar{k}})$	52
$S(\xi,\eta)$	Discrete sine kernel	52

Symbol	Meaning	Page
$\tilde{S}_{ij}(x)$	Node index form of $S(\xi,\eta)$	53
$E_{\text{int}}([A,B];x,H,N)$	Expected number of particles near x	53
$M_{\text{int}}([A,B];x,H,N)$	Number of nodes near x	54
$A(\xi,\eta)$	Airy kernel	55
$x_{\min}$ and $x_{\max}$	Nodes occupied by leftmost and rightmost particles	56
$\mathcal{A}\vert_{[s,\infty)}$	Operator acting with kernel $A(\xi,\eta)$ on $L^2[s,\infty)$	56
$y_{\min}$ and $y_{\max}$	Nodes occupied by leftmost and rightmost holes	56
$\mathcal{L}_m$	mth vertical sublattice of $\mathcal{L}$	57
$N(\mathfrak{a},\mathfrak{b},\mathfrak{c},m)$	Number of points in $\mathcal{L}_m$	57
$\mathfrak{a}_m$ and $\mathfrak{b}_m$	$\vert m-\mathfrak{a}\vert$ and $\vert m-\mathfrak{b}\vert$	57
Q_m	Lowest lattice point in $\mathcal{L}_m$	57
L_m	Number of holes in $\mathcal{L}_m$	57
$\tilde{P}_m(x_1,\ldots,x_{\mathfrak{c}})$	Probability of finding particles at $x_1,\ldots,x_{\mathfrak{c}}$ in $\mathcal{L}_m$	57
$P_m(\xi_1,\ldots,\xi_{L_m})$	Probability of finding holes at $\xi_1,\ldots,\xi_{L_m}$ in $\mathcal{L}_m$	57
τ	Rescaled location of $\mathcal{L}_m$ in the $\mathfrak{abc}$-hexagon	58
σ_2	Pauli matrix	63
$Y_\infty = \{y_1,\ldots,y_M\}$	Limiting transition points	67
$Y_N = \{y_{1,N},\ldots,y_{M,N}\}$	Transition points	67
Σ_0^∇ and Σ_0^Δ	Complementary systems of subintervals of (a,b)	67
d_N	$\#\Delta/N$	68
ϵ	Contour parameter	69
Σ	Contour of discontinuity of $\mathbf{R}(z)$	69
$\Omega_\pm^\nabla$ and $\Omega_\pm^\Delta$	Compact regions of $\mathbb{C}\setminus\Sigma$	69

LIST OF IMPORTANT SYMBOLS

Symbol	Meaning	Page
$\mathbf{R}(z)$	Matrix unknown obtained from $\mathbf{Q}(z; N, k)$	69
$\rho(x)$	Density for $g(z)$	70
$g(z)$	Complex logarithmic potential of $\rho(x)$	70
$\mathbf{S}(z)$	Matrix unknown obtained from $\mathbf{R}(z)$	70
$\theta(z)$	Phase variable related to $\rho(x)$	70
$\phi(z)$	Correction to $N\theta(z)$	70
$T_\nabla(z)$	Analytic function measuring discreteness in Σ_0^∇	71
$T_\Delta(z)$	Analytic function measuring discreteness in Σ_0^Δ	71
ϕ_Γ	Constant value of $\phi(z)$ in gap Γ	71
$\mathbf{L}_\pm(z)$	Lower-triangular factors in jump for $\mathbf{S}(z)$ in bands	72
$\mathbf{J}(z)$	Off-diagonal factor in jump for $\mathbf{S}(z)$ in bands	72
$\mathbf{U}_\pm(z)$	Upper-triangular factors in jump for $\mathbf{S}(z)$ in bands	72
$Y(z)$	Scalar function related to $T_\nabla(z)$ and $T_\Delta(z)$	72
$\theta_I^\nabla(z)$ and $\theta_I^\Delta(z)$	Analytic continuation of $\theta(z)$ from $I \cap \Sigma_0^\nabla$ and $I \cap \Sigma_0^\Delta$	78
$\mathbf{X}(z)$	Solution of Riemann-Hilbert Problem 4.6	79
Σ_{SD}	Contour of discontinuity of $\mathbf{X}(z)$	79
$\mathbf{D}(z)$	Matrix factor relating $\mathbf{X}(z)$ and $\mathbf{P}(z; N, k)$	79
$\Sigma_{0\pm}^\nabla$ and $\Sigma_{0\pm}^\Delta$	Vertical segments of Σ_{SD} connected to band endpoints	84
$\Sigma_{I\pm}$	Horizontal segments of Σ_{SD} parallel to a band I	84
$\Sigma_{\Gamma\pm}$	Horizontal segments of Σ_{SD} parallel to a gap Γ	84
$\dot{\mathbf{X}}(z)$	Solution of Riemann-Hilbert Problem 5.1	87
Σ_{model}	Contour of discontinuity of $\dot{\mathbf{X}}(z)$	88
$\Psi(z)$	$\delta E_c/\delta\mu(z) - (d_N - c)(g_+(z) + g_-(z))$	89

Symbol	Meaning	Page
h	Additional contour parameter (with ϵ)	90
$D_\Gamma^{\nabla,L}$	Disc centered at band/void edge $z = \alpha$	90
$D_{\Gamma,\mathrm{I}}^{\nabla,L}$, $D_{\Gamma,\mathrm{II}}^{\nabla,L}$, $D_{\Gamma,\mathrm{III}}^{\nabla,L}$, and $D_{\Gamma,\mathrm{IV}}^{\nabla,L}$	Quadrants of $D_\Gamma^{\nabla,L}$	91
$\mathbf{Z}_\Gamma^{\nabla,L}$	Matrix proportional to $\mathbf{X}(z)$ in $D_\Gamma^{\nabla,L}$	91
$\dot{\mathbf{Z}}_\Gamma^{\nabla,L}$	Matrix proportional to $\dot{\mathbf{X}}(z)$ in $D_\Gamma^{\nabla,L}$	91
$\mathbf{H}_\Gamma^{\nabla,L}(z)$	Holomorphic prefactor in $\dot{\mathbf{Z}}_\Gamma^{\nabla,L}(z)$	91
$\hat{\mathbf{Z}}^{\nabla,L}(\zeta)$	Explicit model for $\mathbf{Z}_\Gamma^{\nabla,L}(z)$	92
$\hat{\mathbf{X}}_\Gamma^{\nabla,L}(z)$	Local parametrix for $\mathbf{X}(z)$ in $D_\Gamma^{\nabla,L}$	93
$D_\Gamma^{\nabla,R}$	Disc centered at band/void edge $z = \beta$	94
$D_{\Gamma,\mathrm{I}}^{\nabla,R}$, $D_{\Gamma,\mathrm{II}}^{\nabla,R}$, $D_{\Gamma,\mathrm{III}}^{\nabla,R}$, and $D_{\Gamma,\mathrm{IV}}^{\nabla,R}$	Quadrants of $D_\Gamma^{\nabla,R}$	94
$\mathbf{Z}_\Gamma^{\nabla,R}$	Matrix proportional to $\mathbf{X}(z)$ in $D_\Gamma^{\nabla,R}$	94
$\dot{\mathbf{Z}}_\Gamma^{\nabla,R}$	Matrix proportional to $\dot{\mathbf{X}}(z)$ in $D_\Gamma^{\nabla,R}$	94
$\mathbf{H}_\Gamma^{\nabla,R}(z)$	Holomorphic prefactor in $\dot{\mathbf{Z}}_\Gamma^{\nabla,R}(z)$	94
$\hat{\mathbf{Z}}^{\nabla,R}(\zeta)$	Explicit model for $\mathbf{Z}_\Gamma^{\nabla,R}(z)$	94
$\hat{\mathbf{X}}_\Gamma^{\nabla,R}(z)$	Local parametrix for $\mathbf{X}(z)$ in $D_\Gamma^{\nabla,R}$	95
$D_\Gamma^{\Delta,L}$	Disc centered at band/saturated region edge $z = \alpha$	95
$D_{\Gamma,\mathrm{I}}^{\Delta,L}$, $D_{\Gamma,\mathrm{II}}^{\Delta,L}$, $D_{\Gamma,\mathrm{III}}^{\Delta,L}$, and $D_{\Gamma,\mathrm{IV}}^{\Delta,L}$	Quadrants of $D_\Gamma^{\Delta,L}$	95
$\mathbf{Z}_\Gamma^{\Delta,L}$	Matrix proportional to $\mathbf{X}(z)$ in $D_\Gamma^{\Delta,L}$	95
$\dot{\mathbf{Z}}_\Gamma^{\Delta,L}$	Matrix proportional to $\dot{\mathbf{X}}(z)$ in $D_\Gamma^{\Delta,L}$	96
$\mathbf{H}_\Gamma^{\Delta,L}(z)$	Holomorphic prefactor in $\dot{\mathbf{Z}}_\Gamma^{\Delta,L}(z)$	96
$\hat{\mathbf{Z}}^{\Delta,L}(\zeta)$	Explicit model for $\mathbf{Z}_\Gamma^{\Delta,L}(z)$	96
$\hat{\mathbf{X}}_\Gamma^{\Delta,L}(z)$	Local parametrix for $\mathbf{X}(z)$ in $D_\Gamma^{\Delta,L}$	96
$D_\Gamma^{\Delta,R}$	Disc centered at band/saturated region edge $z = \beta$	96

LIST OF IMPORTANT SYMBOLS

Symbol	Meaning	Page
$D_{\Gamma,\text{I}}^{\Delta,R}$, $D_{\Gamma,\text{II}}^{\Delta,R}$, $D_{\Gamma,\text{III}}^{\Delta,R}$, and $D_{\Gamma,\text{IV}}^{\Delta,R}$	Quadrants of $D_\Gamma^{\Delta,R}$	96
$\mathbf{Z}_\Gamma^{\Delta,R}$	Matrix proportional to $\mathbf{X}(z)$ in $D_\Gamma^{\Delta,R}$	96
$\dot{\mathbf{Z}}_\Gamma^{\Delta,R}$	Matrix proportional to $\dot{\mathbf{X}}(z)$ in $D_\Gamma^{\Delta,R}$	97
$\mathbf{H}_\Gamma^{\Delta,R}(z)$	Holomorphic prefactor in $\dot{\mathbf{Z}}_\Gamma^{\Delta,R}(z)$	97
$\hat{\mathbf{Z}}^{\Delta,R}(\zeta)$	Explicit model for $\mathbf{Z}_\Gamma^{\Delta,R}(z)$	97
$\hat{\mathbf{X}}_\Gamma^{\Delta,R}(z)$	Local parametrix for $\mathbf{X}(z)$ in $D_\Gamma^{\Delta,R}$	97
$\hat{\mathbf{X}}(z)$	Parametrix (global) for $\mathbf{X}(z)$	99
$\mathbf{E}(z)$	Error matrix $\mathbf{X}(z)\hat{\mathbf{X}}(z)^{-1}$	99
Σ_E	Contour of discontinuity of $\mathbf{E}(z)$	99
L_Γ^∇	Region of deformation below a void Γ	100
$\mathbf{F}(z)$	Solution of Riemann-Hilbert Problem 5.12	100
L_Γ^Δ	Region of deformation below a saturated region Γ	100
Σ_F	Contour of discontinuity of $\mathbf{F}(z)$	101
$\mathbf{v_F}(z)$	Jump matrix for $\mathbf{F}(z)$ on Σ_F	102
$\overline{R}_m^{(N,\bar{k})}(x_1,\ldots,x_m)$	m-point correlation function of $\bar{k}$-hole ensemble	115
$\overline{K}_{N,\bar{k}}(x,y)$	Dual of $K_{N,k}(x,y)$	115
$\mathbf{B}(x)$	Matrix factor in exact formula for $K_{N,k}(x,y)$	119
$\mathbf{v}$ and $\mathbf{w}$	Vector factors in exact formula for $K_{N,k}(x,y)$	119
$\mathbf{a}$ and $\mathbf{b}$	Vector factors in exact formula for $K_{N,k}(x,y)$	121
$\mathbf{A}_\Gamma^{\nabla,L}(x)$, $\mathbf{q}_\Gamma^{\nabla,L}(x)$, and $\mathbf{r}_\Gamma^{\nabla,L}(x)$	Factors in exact formula for $K_{N,k}(x,y)$	121
$\mathbf{A}_\Gamma^{\nabla,R}(x)$, $\mathbf{q}_\Gamma^{\nabla,R}(x)$, and $\mathbf{r}_\Gamma^{\nabla,R}(x)$	Factors in exact formula for $K_{N,k}(x,y)$	121
$\mathbf{Y}(z)$	Matrix constructed from $\dot{\mathbf{X}}(z)$ and $h(z)$	135
$\mathbf{Y}^\sharp(z)$	Matrix directly related to $\mathbf{Y}(z)$	137

Symbol	Meaning	Page
Γ_0	$(-\infty, \alpha_0) \cup (\beta_G, \infty)$	137
Σ'_{model}	Contour of discontinuity of $\mathbf{Y}^\sharp(z)$	137
S	Hyperelliptic Riemann surface	137
$z(P)$	Hyperelliptic sheet projection function	138
$y^S(P)$	Analytic continuation of $y(z(P))$ to S	138
$a_1, \ldots, a_G$ and $b_1, \ldots, b_G$	Homology basis on S	138
$m_1^S(P), \ldots, m_G^S(P)$	Holomorphic differentials (unnormalized) on S	138
$\mathbf{w}^S(P)$	Abel-Jacobi mapping on S	139
$f(z; \mathbf{q})$	Shifted Riemann theta function	139
$g^\pm(z; \mathbf{q})$	Ratios of shifted Riemann theta functions	140

Bibliography

[AbrS65] M. Abramowitz and I. Stegun, *Handbook of Mathematical Functions*, Dover, New York, 1965.

[AptV01] A. Aptekarev and W. Van Assche, "Asymptotics of discrete orthogonal polynomials and the continuum limit of the Toda lattice," *J. Phys. A: Math Gen.*, **34**, 10627–10637, 2001.

[Bai99] J. Baik, "Riemann-Hilbert problems and random permutations," Ph.D Thesis, Courant Institute of Mathematical Sciences, New York University, 1999.

[BaiDJ99] J. Baik, P. Deift, and K. Johansson, "On the distribution of the length of the longest increasing subsequence of random permutations," *J. Amer. Math. Soc.*, **12**, 1119–1178, 1999.

[BaiKMM03] J. Baik, T. Kriecherbauer, K. D. T.-R. McLaughlin, and P. D. Miller, "Uniform asymptotics for polynomials orthogonal with respect to a general class of discrete weights and universality results for associated ensembles: announcement of results," *Int. Math. Res. Not.*, **15**, 821–858, 2003.

[BleI99] P. Bleher and A. Its, "Semiclassical asymptotics of orthogonal polynomials, Riemann-Hilbert problem, and universality in the matrix model," *Ann. of Math.*, **150**, 185–266, 1999.

[BloGPU03] A. Bloch, F. Golse, T. Paul, and A. Uribe, "Dispersionless Toda and Toeplitz operators," *Duke Math. J.*, **117**, 157–196, 2003.

[Bor00] A. Borodin, "Riemann-Hilbert problem and the discrete Bessel kernel," *Int. Math. Res. Not.*, **9**, 467–494, 2000.

[Bor02] A. Borodin, "Duality of orthogonal polynomials on a finite set," *J. Statist. Phys.*, **109**, 1109–1120, 2002.

[Bor03] A. Borodin, "Discrete gap probabilities and discrete Painlevé equations," *Duke Math. J.*, **117**, 489–542, 2003.

[BorB03] A. Borodin and D. Boyarchenko, "Distribution of the first particle in discrete orthogonal polynomial ensembles," *Comm. Math. Phys.*, **234**, 287–338, 2003.

[BorO01] A. Borodin and G. Olshanski, "z-measures on partitions, Robinson-Schensted-Knuth correspondence, and $\beta = 2$ random matrix ensembles," in Random Matrix Models and Their Applications, *Math. Sci. Res. Inst. Publ.*, **40**, 71–94, Cambridge Univ. Press, Cambridge, 2001.

[BorO05] A. Borodin and G. Olshanski, "Harmonic analysis on the infinite-dimensional unitary group and determinantal point processes," *Ann. of Math.*, **161**, 1319–1422, 2005.

[BorOO00] A. Borodin, A. Okounkov, and G. Olshanski, "Asymptotics of Plancherel measures for symmetric groups," *J. Amer. Math. Soc.*, **13**, 481–515, 2000.

[CohLP98] H. Cohn, M. Larsen, and J. Propp, "The shape of a typical boxed plane partition," *New York J. Math.*, **4**, 137–165, 1998.

[CoiMM82] R. R. Coifman, A. McIntosh, and Y. L. Meyer, "L'intégrale de Cauchy définit un opérateur borné sur L^2 pour les courbes lipschitziennes [The Cauchy integral defines a bounded operator on L^2 for Lipschitz curves]," *Ann. of Math.*, **116**, 361–387, 1982.

[DahBA74] G. Dahlquist, A. Björck, and N. Anderson, *Numerical Methods*, Prentice-Hall, Englewood Cliffs, 1974.

[Dei99] P. Deift, *Orthogonal Polynomials and Random Matrices: A Riemann-Hilbert Approach*, Courant Lecture Notes in Mathematics, **3**, Courant Institute of Mathematical Sciences, New York, 1999.

[DeiKKZ96] P. Deift, S. Kamvissis, T. Kriecherbauer, and X. Zhou, "The Toda rarefaction problem," *Comm. Pure Appl. Math.*, **49**, 35–83, 1996.

[DeiKM98] P. Deift, T. Kriecherbauer, and K. T.-R. McLaughlin, "New results on the equilibrium measure for logarithmic potentials in the presence of an external field," *J. Approx. Theory*, **95**, 388–475, 1998.

[DeiKMVZ99a] P. Deift, T. Kriecherbauer, K. T.-R. McLaughlin, S. Venakides, and X. Zhou, "Strong asymptotics of orthogonal polynomials with respect to exponential weights," *Comm. Pure Appl. Math.*, **52**, 1491–1552, 1999.

[DeiKMVZ99b] P. Deift, T. Kriecherbauer, K. T.-R. McLaughlin, S. Venakides, and X. Zhou, "Uniform asymptotics for polynomials orthogonal with respect to varying exponential weights and applications to universality questions in random matrix theory," *Comm. Pure Appl. Math.*, **52**, 1335–1425, 1999.

[DeiM98] P. Deift and K. T.-R. McLaughlin, "A continuum limit of the Toda lattice," *Memoirs Amer. Math. Soc.*, **624**, 1998.

[DeiVZ97] P. Deift, S. Venakides, and X. Zhou, "New results in small dispersion KdV by an extension of the steepest-descent method for Riemann-Hilbert problems," *Int. Math. Res. Not.*, **6**, 286–299, 1997.

[DeiZ93] P. Deift and X. Zhou, "A steepest-descent method for oscillatory Riemann-Hilbert problems: asymptotics for the mKdV equation," *Ann. of Math.*, **137**, 295–368, 1993.

[DeiZ95] P. Deift and X. Zhou, "Asymptotics for the Painlevé II equation," *Comm. Pure Appl. Math.*, **48**, 277–337, 1995.

[DraS97] P. D. Dragnev and E. B. Saff, "Constrained energy problems with applications to orthogonal polynomials of a discrete variable," *J. d'Analyse Math.*, **72**, 223–259, 1997.

[DraS00] P. D. Dragnev and E. B. Saff, "A problem in potential theory and zero asymptotics of Krawtchouk polynomials," *J. Approx. Theory*, **102**, 120–140, 2000.

[FerS03] P. Ferrari and H. Spohn, "Step fluctuations for a faceted crystal," *J. Statist. Phys.*, **113**, 1–46, 2003.

[FokIK91] A. S. Fokas, A. R. Its, and A. V. Kitaev, "Discrete Painlevé equations and their appearance in quantum gravity," *Comm. Math. Phys.*, **142**, 313–344, 1991.

[IsmS98] M. E. H. Ismail and P. Simeonov, "Strong asymptotics for Krawtchouk polynomials," *J. Comput. Appl. Math.*, **100**, 121–144, 1998.

BIBLIOGRAPHY

[Joh00]   K. Johansson, "Shape fluctuations and random matrices," *Comm. Math. Phys.*, **209**, 437–476, 2000.

[Joh01]   K. Johansson, "Discrete orthogonal polynomial ensembles and the Plancherel measure," *Ann. of Math.*, **153**, 259–296, 2001.

[Joh02]   K. Johansson, "Non-intersecting paths, random tilings and random matrices," *Probab. Theory Related Fields*, **123**, 225–280, 2002.

[KamMM03]   S. Kamvissis, K. T.-R. McLaughlin, and P. D. Miller, *Semiclassical Soliton Ensembles for the Focusing Nonlinear Schrödinger Equation*, Annals of Mathematics Studies series, volume 154, Princeton University Press, Princeton, 2003.

[Kui00]   A. B. J. Kuijlaars, "On the finite-gap ansatz in the continuum limit of the Toda lattice," *Duke Math. J.*, **104**, 433–462, 2000.

[KuiM00]   A. B. J. Kuijlaars and K. T.-R. McLaughlin, "Generic behavior of the density of states in random matrix theory and equilibrium problems in the presence of real analytic external fields," *Comm. Pure Appl. Math.*, **53**, 736–785, 2000.

[KuiM01]   A. B. J. Kuijlaars and K. T.-R. McLaughlin, "Long time behavior of the continuum limit of the Toda lattice, and the generation of infinitely many gaps from C^∞ initial data," *Comm. Math. Phys.*, **221**, 305–333, 2001.

[KuiR98]   A. B. J. Kuijlaars and E. A. Rakhmanov, "Zero distributions for discrete orthogonal polynomials," *J. Comput. Appl. Math.*, **99**, 255–274, 1998.

[KuiV99]   A. B. J. Kuijlaars and W. Van Assche, "The asymptotic zero distribution of orthogonal polynomials with varying recurrence coefficients," *J. Approx. Theory*, **99**, 167–197, 1999.

[Mac60]   P. A. MacMahon, *Combinatory Analysis*, Chelsea, New York, 1960.

[Meh91]   M. L. Mehta, *Random Matrices*, Second edition, Academic Press, Boston, 1991.

[Mil02]   P. D. Miller, "Asymptotics of semiclassical soliton ensembles: rigorous justification of the WKB approximation," *Int. Math. Res. Not.*, **8**, 383–454, 2002.

[MilA98]   P. D. Miller and N. N. Akhmediev, "Modal expansions and completeness relations for some time-dependent Schrödinger equations," *Physica D*, **123**, 513–524, 1998.

[MilC01]   P. D. Miller and S. R. Clarke, "An exactly solvable model for the interaction of linear waves with Korteweg-de Vries solitons," *SIAM J. Math. Anal.*, **33**, 261–285, 2001.

[NikSU91]   A. F. Nikiforov, S. K. Suslov, and V. B. Uvarov, *Classical Orthogonal Polynomials of a Discrete Variable*, Springer Series in Computational Physics, Springer Verlag, Berlin, 1991.

[Oko01]   A. Okounkov, "Infinite wedge and random partitions," *Selecta Math. (N.S.)*, **7**, 57–81, 2001.

[OkoR03]   A. Okounkov and N. Reshetikhin, "Correlation function of a Schur process with application to local geometry of a random 3-dimensional Young diagram," *J. Amer. Math. Soc.*, **16**, 581–603, 2003.

[Rak96]   E. A. Rakhmanov, "Equilibrium measure and the distribution of zeros of the extremal polynomials of a discrete variable," *Mat. Sb.*, **187**, 109–124, 1996 (in Russian). English translation: *Sbornik:Mathematics*, **187**, 1213–1228, 1996.

[SafT97]   E. B. Saff and V. Totik, *Logarithmic Potentials with External Fields*, Springer Verlag, New York, 1997.

[Sim79]   B. Simon, *Trace Ideals and Their Applications*, London Mathematical Society lecture note series, volume 35, Cambridge, 1979.

[Sze91]   G. Szegő, *Orthogonal Polynomials*, Colloquium Publications volume 23, American Mathematical Society, Providence, 1991.

[TraW94] C. Tracy and H. Widom, "Level-Spacing distributions and the Airy kernel," *Comm. Math. Phys.*, **159**, 151–174, 1994.

[TraW98] C. Tracy and H. Widom, "Correlation functions, cluster functions, and spacing distributions for random matrices," *J. Statist. Phys.*, **92**, 809–835, 1998.

[Zho89]   X. Zhou, "The Riemann-Hilbert problem and inverse scattering," *SIAM J. Math. Anal.*, **20**, 966–986, 1989.

Index

abc-hexagon, 3, **3**, 4, 5, 57–60
    random tiling of, 2
Abel-Jacobi mapping, **32**, 139, 141
Abel mapping, *see* Abel-Jacobi mapping
Airy functions, vii, 39–41, 125
Arctic circle, *see* zone, boundary
Aztec diamond, 4, 60
    random tiling of, 2

bands, 28, **28**, 29, 30, 33, 34, 37–40, 42, 52–56, 59, 62, 63, 67, 71, 72, 78–80, 82, 84, 85, 87, 89, 90, 94–96, 99, 103, 110–114, 120, 121, 123–125, 127–130, 135–137
    adjacency, 28
    analytic continuation from, 29, 31, 78, 90
    birth of, 62
    complement of, 89
    endpoints of, **28**, 29, 33, 39–43, 55, 56, 59, 62, 79, 84, 85, 87–91, 93–103, 107, 110–114, 124, 125, 129, 130, 132, 135–138, 140, 141, 143, 146
    equilibrium condition within, 29
    interior of, 43
    interpretation as temperate zones, 59
    jump conditions in, 70
    multiple, 62, 63, 99, 100
    number of, **88**
    of the Hahn weight, 45, 145, 150
        endpoints of, 45, 148, 149, 151
    transition, 67, **67**, 68, 72, 78, 81, 82, 85, 110, 111, 126, 143
band edges, *see* bands, endpoints of

Cauchy's Theorem, 124, 125
Cauchy-Riemann equations, 22, 77, 78
Cauchy-Schwarz inequality, 133
Cauchy integral, 104, 106, 136, 145, 146
Cauchy transform, 145, 149–151
comparison matrix, **91**, 94, 96, 97
conformal mapping, 29, 30, 91, 94, 95, 97
constraints, 21, 25–27, 42, 48, 67, 70, 126
    average value of, 27, 42
    lower, vii, 26, 28, 71, 78, 89, 90, 94, 99–103, 146
    normalization, 25, 26, 47
    upper, vii, 26, 28, 35, 55, 67, 68, 71, 89, 90, 95, 96, 99–101, 103, 109, 118, 128, 131
continuum limit, 10, 11, 23, 25, 49, 50
    of the Toda lattice, 4–8, 11, 22, 60–64
        strong, vii, 7
        weak, 7
conveyor belt, 37
correlation functions, vii, 2, 49, **49**, 50, 53, 54, 56, 115, 129
    determinantal formula for, 50, 115, 127, 131
    for dual ensembles, 115, 131
    for holes, 60, 115, 129
    for particles, 127–129
    multipoint, 54, 59, 127
    one-point, 49, 53, 54, 59, 129

density of states, *see* correlation functions, one-point
discrete orthogonal polynomials, vii, 1, 2, 8, **8**, 11, 12, 18, 19, 22, 23, 25, 41, 50
    asymptotic analysis of, 87
    asymptotic formulae for, 10, 11, 23, 25, 33–41
    classical, 1, 9–11
        associated Hahn, 10, 43–48
        discrete Tchebychev, 44
        Gram, 44
        Hahn, 10, 43–48
        Krawtchouk, 10, 42–43
    coefficients of, 8, 19, 33, 106
    dual families of, 18, 19, 30, 42
    monic, 8, 9, 13, 18, 36, 105
    nonclassical, 43
    ratio asymptotics of, 34
    recurrence relation for, 9, 14, 60
        coefficients of, *see* recurrence coefficients
    zeros of
        asymptotic formulae for, 11
        defects in patterns of, 36
        exclusion principle for, 8, 26
        Hurwitz, **36**
        spurious, **34**, 36
discrete orthogonal polynomial ensemble, vii, **1**, 1–2, 4, 22, 23, 49–57, 87
    dual, 22, 52, 115–118, 128, 131
    for holes, 129, 131
    for particles, 129, 131
    for the abc-hexagon, 57
divisors
    nonspecial, 141
duality, 17–19, 22, 30, 45, 51–52, 115–118
    alternate notion of, 19

energy functional, 25, **25**, 26, 30
    variational derivative of, 27, 28, 119
equilibrium energy problem, **25**, 30, 62, 63
equilibrium measure, vii, 20–22, 25, **25–28**, 30, 34, 41, 54, 55, 62, 63, 67, 70, 71, 82, 88–90, 118, 121, 127, 129, 137
    assumptions about, 56
    candidate, 145, 150
        density of, 148–151
    deformed, 62
    density of, 26, 35, 131
    dual, 30, 129
    equilibrium condition for, 27, 29, 72, 83, 89, 121
    existence of, 26
    for holes, 129
    for particles, 56, 129

for the associated Hahn weight, 59
for the Hahn weight, vii, 22, 23, 41, 45–48, 59, 145–151
for the Krawtchouk weight, 41, 42, 62
Hölder continuity of, 27
hypothetical, 68, 69, 88
logarithmic potential of, 21, 28
quantities derived from, 28–30
regularity of, 26
scaled, 35
support of, 35, 53, 54
unconstrained support of, 33
use of, 70
variational inequalities for, 29
variations of, 29
external field, 1, 25, **25**, 30, 46, 61, 62
as a continuum limit, 25

Flaschka's variables, 4

g-function, 21, **70**, 71
gamma function, vii, 36, 77, 78
gaps, **28**, 84
exterior, **28**, 35, 37, 54, 56
interior, **28**, 30, 37, 63, 88, 99, 100
gap probabilities, 2
genus, 88
Gram-Schmidt process, 8, 9
grand canonical ensemble, 60

Hankel determinants, 6
hard edge, *see* random matrix theory, hard edges in
hole-particle duality, *see* duality
hole-particle transformation, 19
holes, 51, **51**
as positions of vertical rhombi, 57
correlations of, 60, 115, 129, 131
density of, 52, 58
discrete orthogonal polynomial ensemble for, 115, 129
equilibrium measure for, 129
saturated region for, 60
void for, 60, 118, 129
extreme, 56, 60
in the abc-hexagon, 57, 59
joint probability distribution of, 51, 52, 57
universal properties of, 52
location of, 51, 57, 59
one-point function for, 59
statistics of, 51, 118
hyperelliptic Riemann surface, 25, 137, 138, 140
holomorphic differentials on, 138, **138**
homology cycles on, 138
Jacobian variety of, 139
square-root function of, 31

interpolating polynomial
Lagrange, 19
interpolation identity, 69
interpolation problem, vii, 12, 19, 22, 23, 67
and discrete orthogonal polynomials, 19
asymptotic analysis of, 22
conditions of, 16
dual, 18, 116
residue matrices for, 17
Riemann-Hilbert problem equivalent to, 84
solution of, 12–15, 17, 18, 36, 50, 69, 79, 84, 116

inverse-spectral theory, 11

Jacobian variety, *see* hyperelliptic Riemann surface, Jacobian variety of
Jacobi matrix, 9, 61, 62
Jost matrix, **15**
jump, *see* jump discontinuity
jump conditions, 20–23, 70, 79, 84, 86, 87, 89, 91–97, 101, 102, 107, 124, 128, 135, 137, 140, 143, 146, 149–151
factorization of, 72
jump discontinuity, 20, 22, 35, 70, 81, 135, 137
absent, 82, 83
jump matrix, 21, 71, 87, 103, 104
exact, 98
near-identity, 104
oscillatory, 22, 78
piecewise-constant, 135
jump relations, *see* jump conditions

kernel
Airy, vii, **55**, 56, 60, 129–132
approximate, 131, 132
Cauchy, 104
Christoffel-Darboux, *see* kernel, reproducing
discrete incomplete beta, 60
discrete sine, vii, 52, **52**, 53, 59, 60, 130
reproducing, 11, 50, **50**, 115, 121, 126, 127, 129
asymptotic formula for, 11
dual, 115, 129
on the diagonal, 117

Lagrange multiplier, 26, **26**, 29, 30, 45, 145, 149–151
correction to, 31
for the Hahn weight, 46, 47
last-passage percolation, 2, 11, 23
Lax-Levermore method, 6
Lax Equivalence Theorem, 64
Lax pair, *see* Toda lattice, Lax pair for
Lebesgue Dominated Convergence Theorem, 133
linear stability analysis, 7
Liouville's Theorem, 12, 15, 86, 88, 89, 93, 143
logarithmic potential theory, *see* potential theory, logarithmic

MacMahon's formula, 4
measure
Borel, 25
discrete, 61
equilibrium, *see* equilibrium measure
extremal, 6
invariant, 129
Plancherel, 2
probability, 60, 70, 145
pure point, 1, 8
Schur, 2
spectral, 61
midpoint rule, 77, 112
minimizer, *see* equilibrium measure
modulation equations, *see* Whitham equations

nodes, vii, 1, 8, **8**, 9, 10, 12–14, 17–19, 36, 50, 52, 53, 67–69, 109, 115, 116, 121, 127
(Hurwitz) zeros corresponding to, 36, 37, 63, 110
adjacent, 8, 9, 20, 36, 49, 67, 110
assumptions about, 9, 56

INDEX

asymptotic properties of, 9
as eigenvalues, 61
as support of weights, 105
complementary sets of, 51
consecutive, *see* nodes, adjacent
continuous weight sampled at, 123
density of, 10, 49, 61
different sets of, 19
disjoint sets of, 52
distinct, 50, 115, 117, 120–122
dynamics under the Toda flow, 61
equally spaced, 10, 19, 42–44
interval of accumulation of, 43, 54, 56, 61, 105, 119
near a fixed point, 126, 127
neighboring, *see* nodes, adjacent
not occupied by particles, 51, 56, 115
number available, 55
number of, 9, 11, 51, 87
occupied by particles, 49, 53, 56, 115
quantization rule for, *see* quantization rule, for nodes
simple poles at, 116
spacing of, 42
unequally spaced, 9, 76
Vandermonde determinant of, 51
where the reproducing kernel is small, 127
where triangularity is reversed, 18, 20, 68
where upper constraint is inactive, 68, 108, 123
zeros located exactly at, 36
node density function, *see* nodes, density of
normalization condition, *see* constraints, normalization

one-band ansatz, 145
one-point function, *see* correlation functions, one-point

Painlevé II equation, 56, 60
parametrix
global, 19, **19**, 23, 87, 99, 102, 105, 119
local, **93**, 95–98
parititon function, 57
particles
as positions of horizontal rhombi, 57
concentration of, 49
confinement of, 1
correlations of, 53, 54, 115, 129, 131
density of, 49, 58
distinct, 54
equilibrium measure for, 56, 129
saturated region for, 60, 118
void for, 60
extreme, vii, 2, 56, 60
indistinguishability of, 1, 49
in the abc-hexagon, 57
joint probability distribution of, 1, 51, 52, 57
universal properties of, 52
location of, 2, 51, 57
minimum separation of, 49
mutual exclusion of, 49
nonexistent, 51
number of, 52
expected, 54, 55
random, 2
variance of, 50
one-point function for, 59
probability of finding, 49
separation of, 53

spacings between, 50, 126
statistics of, 118, 129
partitions, 2, 23
partition function, **1**
Plancherel-Rotach asymptotics, 2, 48, 105
potential
logarithmic, 25
complex, *see* g-function
of the equilibrium measure, 28
potential function, 10, 61, 63
potential theory
logarithmic, vii, 21, 25, 26

quantization rule
for nodes, 10, 132
for transition points, 67, 73, 82, 111

random growth models, 11
random matrix theory, 49
and orthogonal polynomials, 2
asymptotic density of eigenvalues in, 35
correlation functions of, 49
discrete analogue of, vii
extreme eigenvalues in, 56
fundamental calculation of, 50
Gaussian unitary ensemble of, 56, 129
hard edges in, 35, 109
joint probability distribution of eigenvalues in, 1
level repulsion in, 49
Tracy-Widom distribution in, 4, 60
universal eigenvalue statistics in, 23
Wishart ensembles of, 35
random permutations, 2
random tilings, 2–4, 11
recurrence coefficients, 13–16, 62, 105, 106
asymptotic formulae for, 11, 33
continuum limit of, 62
multiphase oscillations of, 62
residue matrices
triangularity of, 17, 18, 21, 23, 67, 68
rhombi, 3, 4, 57
horizontal, 3
position of, 4, 57
relation to particles and holes, 57
three types of, 4
vertical, 4
rhombus, *see* rhombi
Riemann-Hilbert problem, vii, 23, 69
classical solution of, 104
conversion into integral equations, 104
discrete, *see* interpolation problem
equivalent, 67, 84, 86, 87, 99, 102, 103
for the error, 19
solution of by Neumann series, 19
for the error (deformed), 102, 104, 124
limiting, 23
local model, 92–94, 96, 97
matrix, 19, 21, 23
with L^2 boundary values, 104
model, 87, 88, 92, 124, 135, 137, 141
on a contour, 22
scalar, 6, 135–137, 146, 149–151
simpler, 22
uniqueness of solutions of, 89, 143
with pole conditions, *see* interpolation problem

Riemann constant vector, 31, 141
Riemann invariants, 62
Riemann matrix, 31
Riemann sums, 77, 112, 133
Riemann theta function, 32, 62, 88, 138, 141, 143
Robin constant, *see* Lagrange multiplier
    correction to, *see* Lagrange multiplier, correction to

saturated-band-saturated configuration, **45**, 46, 145, 151
saturated-band-void configuration, **45**, 46, 145, 149, 150
saturated regions, vii, 20, **28**, 29, 30, 35–37, 39–41, 43, 47, 54–57, 59, 63, 67, 71, 80, 84, 85, 87, 88, 95, 99, 100, 102, 103, 109, 113, 114, 118, 124, 126, 128, 129
    analytic continuation from, 29
    defect motion in, 62
    hard edges of, 35, 109
    interior of, 43
    interpretation as polar zones, 59
    jump conditions in, 70
    location of defects in, 37
    motion of defects in, 37
    normal hole statistics in, 54
    of the Hahn weight, 45, 145, 149
    of the Krawtchouk weight, 42
    variational inequality within, 29
shadows of poles, 36
shock, vii, 6–8, 62, 64
squared eigenfunctions, 16
steepest-descent method
    classical, 2, 4
    for Riemann-Hilbert problems, 22, 78
Stirling's formula, 11, 44

Toda lattice, 4, 16, 43
    as a compatibility condition, 16
    continuum limit of, *see* continuum limit, of the Toda lattice
    hierarchy for, 14
    Lax pair for, 15, 16
    linearized, 7, 17
    zero-curvature representation of, 16
Tracy-Widom distribution, vii, 4, 56, **56**, 57, 60
Tracy-Widom law, *see* Tracy-Widom distribution
transition points, 23, 67, **67**, 68, 71, 78, 82–85, 88, 89, 98, 111, 126, 143
    artificial, *see* transition points, virtual
    virtual, 85, 126

universality, vii, **2**, 11, 22, 23, 50, 52, 87, 115, 124, 125, 129
    of the Airy kernel, 55, 56
    of the discrete sine kernel, 52
    of the Tracy-Widom distribution, 56

Vandermonde determinant, 49, 51, 136
void-band-saturated configuration, **45**, 145, 150
void-band-void configuration, **45**, 46, 145, 149
voids, **28**, 29, 30, 34, 39, 43, 47, 53–56, 59, 60, 67, 68, 71, 80, 84, 85, 87, 88, 90, 94, 99–103, 107, 108, 112, 118, 121–124, 127–129
    analytic continuation from, 29
    containing spurious zeros, 34
    interior of, 43, 54
    interpretation as polar zones, 59, 60
    jump conditions in, 70
    normal particle statistics in, 54

    not containing spurious zeros, 34
    of the Hahn weight, 45, 145, 146
    of the Krawtchouk weight, 42
    variational inequality within, 29

weights, 1, **8**, 12, 19, 50, 116
    analytic, 40
    and concentration of particles, 49
    and eigenvectors of the Jacobi matrix, 61
    and universality, 2
    associated Hahn, 4, **44**, 57
        equilibrium measure for, *see* equilibrium measure, for the associated Hahn weight
        formula for, 44
        potential function for, 45
    assumptions about, 11, 25, 56
    asymptotic properties of, 9, 10
    Charlier, 2, 8
    classical formulae for, 23
    continuous, vii, 9, 11, 12, 21, 23, 27, 40, 49, 69
    continuum limit of, 10, 11
    deformation of, 14, 62
        under the Toda flow, 15, 61, 63
    discrete, vii, 9, 21, 23, 27, 36, 50
        and poles, 23
    dual, 18, **18**, 30, 44, 52, 115, 116, 118
    for holes, 51, 52, 118
    for particles, 118
    functional notation for, 8, 123
    general, vii, 2, 9, 23, 49, 52, 87
    general Hahn, 43, **43**
        limiting, 44
        moments of, 43
    Hahn, vii, 2, 4, 23, 27, 44, **44**, 58
        alternate form of, 44
        equilibrium measure for, *see* equilibrium measure, for the Hahn weight
        formula for, 44
        special case of, 44
    Krawtchouk, 2, 4, 27, 42, **42**, 62
        alternate form of, 42
        classical formula for, 42
        equilibrium measure for, *see* equilibrium measure, for the Krawtchouk weight
        invariance under the Toda flow, 43, 62
    Meixner, 1, 8
    nodes as support of, 105
    Nth-root asymptotics of, 25, 61
    orthogonality with respect to, 2
    parameters of, 2, 14, 43, 62
    positivity of, 10, 13, 44
    potential function for, 30, 52
    specific, 67
    varying, 9, 23
Whitham equations, 6, 7, 62, **62**

zone
    boundary, 4, 59, 60
    frozen, *see* zone, polar
    polar, 4, **4**, 59, 60
    temperate, 4, **4**, 59

GPSR Authorized Representative: Easy Access System Europe - Mustamäe tee
50, 10621 Tallinn, Estonia, gpsr.requests@easproject.com